Lecture Notes in Mathematics

Edited by A. Dold and B. Eckmann

1260

Nicolae H. Pavel

Nonlinear Evolution Operators and Semigroups

Applications to Partial Differential Equations

Springer-Verlag

Lecture Notes in Mathematics

For information about Vols. 1–1062 please contact your bookseller or Springer-Verlag.

Vol. 1063: Orienting Polymers. Proceedings, 1983. Edited by J.L. Ericksen. VII, 166 pages. 1984.

Vol. 1064: Probability Measures on Groups VII. Proceedings, 1983. Edited by H. Heyer. X, 588 pages. 1984.

Vol. 1065: A. Cuyt, Padé Approximants for Operators: Theory and Applications. IX, 138 pages. 1984.

Vol. 1066: Numerical Analysis. Proceedings, 1983. Edited by D.F. Griffiths. XI, 275 pages. 1984.

Vol. 1067: Yasuo Okuyama, Absolute Summability of Fourier Series and Orthogonal Series. VI, 118 pages. 1984.

Vol. 1068: Number Theory, Noordwijkerhout 1983. Proceedings. Edited by H. Jager. V, 296 pages. 1984.

Vol. 1069: M. Kreck, Bordism of Diffeomorphisms and Related Topics. III, 144 pages. 1984.

Vol. 1070: Interpolation Spaces and Allied Topics in Analysis. Proceedings, 1983. Edited by M. Cwikel and J. Peetre. III, 239 pages. 1984.

Vol. 1071: Padé Approximation and its Applications, Bad Honnef 1983. Prodeedings. Edited by H. Werner and H.J. Bünger. VI, 264 pages. 1984.

Vol. 1072: F. Rothe, Global Solutions of Reaction-Diffusion Systems. V, 216 pages. 1984.

Vol. 1073: Graph Theory, Singapore 1983. Proceedings. Edited by K.M. Koh and H.P. Yap. XIII, 335 pages. 1984.

Vol. 1074: E.W. Stredulinsky, Weighted Inequalities and Degenerate Elliptic Partial Differential Equations. III, 143 pages. 1984.

Vol. 1075: H. Majima, Asymptotic Analysis for Integrable Connections with Irregular Singular Points. IX, 159 pages. 1984.

Vol. 1076: Infinite-Dimensional Systems. Proceedings, 1983. Edited by F. Kappel and W. Schappacher. VII, 278 pages. 1984.

Vol. 1077: Lie Group Representations III. Proceedings, 1982–1983. Edited by R. Herb, R. Johnson, R. Lipsman, J. Rosenberg. XI, 454 pages. 1984.

Vol. 1078: A.J.E.M. Janssen, P. van der Steen, Integration Theory. V, 224 pages. 1984.

Vol. 1079: W. Ruppert. Compact Semitopological Semigroups: An Intrinsic Theory. V, 260 pages. 1984

Vol. 1080: Probability Theory on Vector Spaces III. Proceedings, 1983. Edited by D. Szynal and A. Weron. V, 373 pages. 1984.

Vol. 1081: D. Benson, Modular Representation Theory: New Trends and Methods. XI, 231 pages. 1984.

Vol. 1082: C.-G. Schmidt, Arithmetik Abelscher Varietäten mit komplexer Multiplikation. X, 96 Seiten. 1984.

Vol. 1083: D. Bump, Automorphic Forms on GL (3,IR). XI, 184 pages. 1984.

Vol. 1084: D. Kletzing, Structure and Representations of Q-Groups. VI, 290 pages. 1984.

Vol. 1085: G.K. Immink, Asymptotics of Analytic Difference Equations. V, 134 pages. 1984.

Vol. 1086: Sensitivity of Functionals with Applications to Engineering Sciences. Proceedings, 1983. Edited by V. Komkov. V, 130 pages. 1984

Vol. 1087: W. Narkiewicz, Uniform Distribution of Sequences of Integers in Residue Classes. VIII, 125 pages. 1984.

Vol. 1088: A.V. Kakosyan, L.B. Klebanov, J.A. Melamed, Characterization of Distributions by the Method of Intensively Monotone Operators. X, 175 pages. 1984.

Vol. 1089: Measure Theory, Oberwolfach 1983. Proceedings. Edited by D. Kölzow and D. Maharam-Stone. XIII, 327 pages. 1984.

Vol. 1090: Differential Geometry of Submanifolds. Proceedings, 1984. Edited by K. Kenmotsu. VI, 132 pages. 1984.

Vol. 1091: Multifunctions and Integrands. Proceedings, 1983. Edited by G. Salinetti. V, 234 pages. 1984.

Vol. 1092: Complete Intersections. Seminar, 1983. Edited by S. Greco and R. Strano. VII, 299 pages. 1984.

Vol. 1093: A. Prestel, Lectures on Formally Real Fields. XI, 125 pages. 1984.

Vol. 1094: Analyse Complexe. Proceedings, 1983. Edité par E. Amar, R. Gay et Nguyen Thanh Van. IX, 184 pages. 1984.

Vol. 1095: Stochastic Analysis and Applications. Procee ngs, 1983. Edited by A. Truman and D. Williams. V, 199 pages. 198

Vol. 1096: Théorie du Potentiel. Proceedings, 1983. E é par G. Mokobodzki et D. Pinchon. IX, 601 pages. 1984.

Vol. 1097: R.M. Dudley, H. Kunita, F. Ledrappier, École d'Éte de Probabilités de Saint-Flour XII – 1982. Edité par P.L. Hennequin. X, 396 pages. 1984.

Vol. 1098: Groups – Korea 1983. Proceedings. Edited by A.C. Kim and B.H. Neumann. VII, 183 pages. 1984.

Vol. 1099: C.M. Ringel, Tame Algebras and Integral Qua tic Forms. XIII, 376 pages. 1984.

Vol. 1100: V. Ivrii, Precise Spectral Asymptotics for Ellipti Operators Acting in Fiberings over Manifolds with Boundary. V, 7 pages. 1984.

Vol. 1101: V. Cossart, J. Giraud, U. Orbanz, Resolution of Surface Singularities. Seminar. VII, 132 pages. 1984.

Vol. 1102: A. Verona, Stratified Mappings – Structure a Triangulability. IX, 160 pages. 1984.

Vol. 1103: Models and Sets. Proceedings, Logic Colloc m, 1983, Part I. Edited by G.H. Müller and M.M. Richter. VIII, 484 es. 1984.

Vol. 1104: Computation and Proof Theory. Procee , Logic Colloquium, 1983, Part II. Edited by M.M. Richter, E. Börg V. Oberschelp, B. Schinzel and W. Thomas. VIII, 475 pages. 19

Vol. 1105: Rational Approximation and Interpolation. Pr eedings, 1983. Edited by P.R. Graves-Morris, E.B. Saff and R Varga. XII, 528 pages. 1984.

Vol. 1106: C.T. Chong, Techniques of Admissible Recursio heory. IX, 214 pages. 1984.

Vol. 1107: Nonlinear Analysis and Optimization. Proceedin , 1982. Edited by C. Vinti. V, 224 pages. 1984.

Vol. 1108: Global Analysis – Studies and Applications I. E ited by Yu.G. Borisovich and Yu.E. Gliklikh. V, 301 pages. 1984.

Vol. 1109: Stochastic Aspects of Classical and Quantum Systems. Proceedings, 1983. Edited by S. Albeverio, P. Combe and M. irugue-Collin. IX, 227 pages. 1985.

Vol. 1110: R. Jajte, Strong Limit Theorems in Non-Con nutative Probability. VI, 152 pages. 1985.

Vol. 1111: Arbeitstagung Bonn 1984. Proceedings. Edited by ze-bruch, J. Schwermer and S. Suter. V, 481 pages. 1985.

Vol. 1112: Products of Conjugacy Classes in Groups. Ed by Z. Arad and M. Herzog. V, 244 pages. 1985.

Vol. 1113: P. Antosik, C. Swartz, Matrix Methods in Analy /, 114 pages. 1985.

Vol. 1114: Zahlentheoretische Analysis. Seminar. Herausge von E. Hlawka. V, 157 Seiten. 1985.

Vol. 1115: J. Moulin Ollagnier, Ergodic Theory and Statistical M ics. VI, 147 pages. 1985.

Vol. 1116: S. Stolz, Hochzusammenhängende Mannigfaltigk und ihre Ränder. XXIII, 134 Seiten. 1985.

continued on page 9

Lecture Notes in Mathematics

Edited by A. Dold and B. Eckmann

1260

Nicolae H. Pavel

Nonlinear Evolution Operators and Semigroups

Applications to Partial Differential Equations

Springer-Verlag
Berlin Heidelberg New York London Paris Tokyo

Author

Nicolae H. Pavel
Universitatea Iaşi, Facultatea de Matematică
6600 Iaşi, Romania
and
The Ohio State University, Department of Mathematics
231 West 18th Avenue, Columbus, OH 43210, USA

Mathematics Subject Classification (1980): Primary: 35A07, 35A35, 35B45, 35C99, 47H09
Secondary: 34G20, 39A10, 65J15

ISBN 3-540-17974-7 Springer-Verlag Berlin Heidelberg New York
ISBN 0-387-17974-7 Springer-Verlag New York Berlin Heidelberg

Printing and binding: Druckhaus Beltz, Hemsbach/Bergstr.
2146/3140-543210

Preface.

The first aim of this book is to present in a coherent way some of the fundamental results and recent research on nonlinear evolution operators and semigroups. The second aim is to show how to apply these abstract results to unify the treatment of several types of partial differential equations arising in physics (the heat equation, wave equation, Schrödinger equation, and so on).

The motivation of this theory is clearly pointed out in the following quotation from: Autumn Course on Semigroups, Theory and Applications, held at the International Centre For Theoretical Physics, Trieste (Italy), 12 November - 14 December 1984 (Brezis-Crandall- Kappel, Directors).

"The last two decades have witnessed a tremendous use of semigroups and evolution equations techniques in solving problems related to PDE and FDE. This allows the treatment of PDE and FDE as suitable ODE in infinite dimensional Banach spaces. This method has considerably simplified and clarified the the proofs, and has unified the treatment of several different classes of differential equations. It has solved many problems that had been left open by previously known methods, and has been very succesful in dealing with discontinuous data and regularity."

Chapter 1 deals with the construction and main properties of nonlinear evolution operator $U(t,s)$ associated with a class of nonlinear (possible multivalued) operators $A(t)$ with time dependent domain, satisfying Hypotheses $H(2.1)$ and $H(2.2)$ in Section 2. We also say that $U(t,s)$ is associated with the nonautonomous differential equation (inclusion) $x'(t) \in A(t)x(t)$. In the convergence of DS-approximate solutions (i.e., in the construction of $U(t,s)$) the fundamental estimate is given by (2.40), essentially due to Kobayashi, Kobayasi and Oharu. Among other general results, we mention Theorem 5.1 which gives a characterization of the compactness of evolution operators.

Note that $U(t,s)$ associated with the equation $x'(t) \in A(t)x(t)$ allows a unifying treatment of the existence, uniqueness and behaviour of the various types of solutions to the Cauchy problem for this equation.

Chapter 2 is devoted to nonlinear semigroups $S_A(t)$ which are generated by the DS-limit solutions associated with the dissipative operator A. In the case A - m-dissipative, $S_A(t)$ is given by the exponential formula of Crandall-Liggett. We say also that $S_A(t)$ is generated by A via the exponential formula. The semigroup approach is important in the study of the solutions of the autonomous differential equation $x' \in Ax$, which includes several different classes of PDE and FDE.

In order to avoid duplication and to reduce the length of this work, we have tried to make (as much as possible) the autonomous case as a special subcase of the time-dependent case (this was also a suggestion of the referee). Of course this is an economic way to present such a theory, but not the simplest one. For the sake of simplicity, the reader may start with the autonomous case.

In the theory of the generation of nonlinear semigroups, the fundamental estimate (given by (1.16)) due to Kobayashi, is derived from (2.40) in Chapter 1, i.e., from nonautonomous case.

In Chapter 3, one applies the results of Chapters 1 and 2, both to a class of multivalued evolution equations and to some partial differential equations modelling physical phenomena.

Most of the results here are presented for the first time in a book (e.g., Brezis' characterization of nonlinear compact semigroups in Chapter 2, the theory of nonlinear evolution operators in Chapter 1 and most of the material in Chapter 3. Some of the results are very recent and not yet published (e.g., the characterization of compactness of evolution operators given by the author, the characterization of compactness of a linear semigroup solely in terms of the resolvent of its infinitesimal generator due to Vrabie and so on).

The discussions (at the "Al.I.Cuza" University of Iasi - Romania) with my colleagues Prof. V. Barbu, C. Ursescu and I. I. Vrabie have contributed to the improvement of many sections in this book. I am expressing my thanks to all of them.

Part of this work has been written during my long stay at the International Centre for Theoretical Physics (ICTP) and SISSA, Trieste (Italy). I am very grateful to Professor Abdus Salam, Nobel Laureate, founder and the Director of the ICTP, for the pleasant hospitality and stimulating discussions.

I also wish to thank Professors A. Cellina, J. Eells, G. Vidossich and Dr. L. K. Shayo for stimulating my activity during my stay at the ICTP.

This book has been completed during my stay at The Ohio State University. I express my gratitude to Professors D. Burghelea, C. Corduneanu, J. Ferrar, T. Hallam, H. Moscovici, as well as to Ms. M. Howard, who have facilitated my activity after my arrival in the USA.

Finally, I wish to thank Springer-Verlag for their pleasant co-operation as well as Ms. Lidia Bogo and Terry England, for the professional typing of the manuscript.

Columbus, February 1987 — Nicolae H. Pavel

CONTENTS

Chapter 1

Nonlinear Evolution Operators

The aim of this chapter is to study the nonlinear evolution operators $U(t,s)$ associated with a class of nonlinear possible multivalued operators with time-dependent domain.

Preliminaries. Discrete Schemes (DS)

Let us consider the differential inclusion

$$u'(t) \in A(t)u(t), \quad s \leq t \leq T \tag{1.1}$$

with initial conditions

$$u(s) = x_o, \; x_o \in \overline{D(A(s))}, \tag{1.1}$$

where $A(t) : D(A(t)) \subset X \to 2^X$ is a time-dependent (possible multivalued) nonlinear operator acting in the real Banach space X with the time-dependent domain $D(A(t))$. (The equation (1.1) is said to be nonautonomous).

The conditions we shall impose on $A(t)$ (see (2.11)), allow even the closure $\overline{D(A(t))}$ of $D(A(t))$ to be time-dependent. We shall see that this is the case in some concrete situations (see Ch. 3, §2).

The key of the construction of $U(t,s)$ is the introduction of the "DS-approximate solution" u_n as the following step function

$$u_n(t) = \begin{cases} x_o^n, & \text{for } t = t_o^n = s \\ x_k^n, & \text{for } t \in \,]t_{k-1}^n, t_k^n] , \end{cases} \tag{1.2}$$

where n is a positive integer ($n \in N$), $k = 0,1,\dots,N_n$ and $t_k^n \in [s,T]$ $x_k^n \in D(A(t_k^n))$ are defined in that follows. (Note that DS is the abbreviation of "Discrete Schemes").

Let $s,T \in R$ with $s < T$ and $x_o \in \overline{D(A(s))}$. Suppose that there is a partition P_n of $[s,T]$

$$P_n = \{ s = t_0^n, t_1^n, \ldots, t_{N_n-1}^n, t_{N_n}^n \}$$

with

$$s = t_0^n < t_1^n < \ldots < t_{N_n-1}^n < T = t_{N_n}^n \tag{1.3}$$

and

$$d_n = \max_{1 \leq k \leq N_n} (t_k^n - t_{k-1}^n) \to o \text{ as } n \to \infty , \quad \omega d_n \leq \tfrac{1}{2} \tag{1.4}$$

(see (2.36))

Assume, in addition, that there are some elements $x_k^n \in D(A(t_k^n)) \subset X$ and $p_k^n \in X$ such that

$$y_k^n \equiv \frac{x_k^n - x_{k-1}^n}{t_k^n - t_{k-1}^n} - p_k^n \in A(t_k^n) x_k^n, \quad k = 1,2,\ldots,N_n \tag{1.5}$$

$$x_0^n \in D(A(s)), \quad x_0^n \to x_o \quad \text{as } n \to \infty \tag{1.6}$$

$$b_n = \sum_{k=1}^{N_n} (t_k^n - t_{k-1}^n) \, \|p_k^n\| \to o \text{ as } n \to \infty . \tag{1.7}$$

In the situation $N_n = \infty$, $t_{N_n}^n \equiv T \equiv \lim_{k \to \infty} t_k^n$, and $u_n(T) = \lim_{k \to \infty} x_k^n$. We shall give conditions that guarantee the existence of such elements with the properties (1.3)-(1.7), having the additional property that the corresponding "DS-approximate solution" u_n is convergent to a continuous function $u = u(t;s,x_o)$ (called DS-limit solution to (1.1)+(1.1)').

Moreover, we will prove that u is well-defined (i.e. every DS-approximate solution u_n has the same limit u) and that the operator $U(t,s)$: $\overline{D(A(s))} \to \overline{D(A(t))}$ defined by

$$U(t,s)x_o \equiv u(t;s,x_o) = \lim_{n\to\infty} u_n(t), x_o \in \overline{D(A(s))}, s \leq t \leq T \tag{1.8}$$

is an evolution operator (as in Section 3). Of course, we shall study the relationship of U with the "strong" solution to (1.1)+(1.1')(Section 3).

Various applications to some partial differential equations will be given in Chapter 3.

The DS-limit solution u

$$u(t;s,x_o) = \lim_{n\to\infty} u_n(t) \tag{1.9}$$

is also called "generalized solution", or "mild solution" (or still "weak solution") to (1.1)+(1.1').

The uniqueness of the mild solution is provable by means of the "Benilan uniqueness theorem" (Section 3).

Roughly speaking, the existence of "DS-approximate solutions" is guaranteed by the "Range condition"

$$R(I-hA(t+h)) \supset \overline{D(A(t))},\ o < h \leq h_o,\quad s \leq t < T \tag{1.10}$$

for some small $h_o > 0$. In this case (1.5) holds with $p_k^n = 0 \in X$.

As we shall see in Section 7, a strictly more general condition than (1.10) is the following "tangential condition"

$$\lim_{h\downarrow o} \frac{1}{h}\, d[x;R(I-hA(t+h))] = o,\ \forall\ x\in\overline{D(A(t))},\ s \leq t < T, \tag{1.11}$$

where $d[x;B]$ stands for the distance from $x\in X$ to the set $B\subset X$.

It is easy to check that

$$|d[x;B] - d[y;B]| \leq \|x-y\|,\quad x,y\in X, \tag{1.11'}$$

where $|r|$ is the absolute value of $r\in R$ and $\|x\|$ is the norm of $x\in X$.

Another important "tangential condition" with significant geometric intterpretation is the following one

$$\lim_{h\downarrow 0} \frac{1}{h}\, d[x+hA(t)x;D(A(t+h))] = 0,\ \forall\ x\in D(A(t)),\ s\leq t < T, \tag{1.12}$$

(where $A(t) : D(A(t))\subset X \to X$ is now supposed to be single-valued).

The relationship between (1.11) and (1.12) will be pointed out later (see § 7). Now, we only mention that if $D(A(t))$ is closed and if $(t,x) \to A(t)x$ is continuous, then (1.12) implies (1.11).

The convergence of the sequence of DS-approximate solutions u_n is guaranteed by a condition on the t-dependence of A(t) (which implies that for each t, A(t) is dissipative). Such a condition is given in the next section. See (H.(2.1)).

Remark 1.1. The condition (1.11) (and respectively (1.12)) is important in applications. Thus, in the case of the (PDE) (2.1) in Chapter 3, (1.10 is not satisfied, but (1.11) holds. The condition (1.12) plays also a crucial role in the theory of the flow-invariance of a set with respect to a differential equation (Cf. Pavel [15]).

§2. The convergence of DS-approximate solutions.

2.1. The time dependence of A(t).

For the sake of selfcontainment we start with the introduction of the functions

$$\langle y,x\rangle_s = \lim_{h\downarrow o} \frac{\|x+hy\|^2 - \|x\|^2}{2h}, \quad x,y\in X \tag{2.1}$$

$$\langle y,x\rangle_+ = \lim_{h\downarrow o} \frac{\|x+hy\| - \|x\|}{h}, \quad x,y\in X \tag{2.2}$$

$$\langle y,x\rangle_i = \lim_{h\uparrow o} \frac{\|x+hy\|^2 - \|x\|^2}{2h}, \quad x,y\in X \tag{2.3}$$

$$\langle y,x\rangle_- = \lim_{h\uparrow o} \frac{\|x+hy\| - \|x\|}{h}, \quad x,y\in X. \tag{2.4}$$

These functions are well-defined since both $h \to \|x+hy\|^2$ and $h \to \|x+hy\|$ are real convex functions. For each $h \neq o$ and $x,y \in X$ set

$$\langle y,x\rangle_h = \frac{\|x+hy\| - \|x\|}{h}. \tag{2.5}$$

The following properties are obvious

$$\langle y,x\rangle_s = \|x\|\langle y,x\rangle_+, \quad \langle y,x\rangle_i = \|x\|\langle y,x\rangle_- \tag{2.6}$$

$$\langle y,x\rangle_+ \leq \langle y,x\rangle_h \leq \|y\|, \text{ if } h > o; \quad \langle y,-x\rangle_p = \langle -y,x\rangle_p \tag{2.7}$$

where p = i or s,

$$\langle y,x\rangle_h \leq \langle y,x\rangle_- , \text{ if } h < o, \quad \langle y,x\rangle_i \leq \langle y,x\rangle_s \leq \|x\|\ \|y\|. \tag{2.8}$$

Recall also the definition of the duality mapping $F:X\to X^*$ of X, i.e.,

$$F(x) = \{x^* \in X^* ; x^*(x) = \|x\|^2 = \|x^*\|^2\} , x \in X, \tag{2.9}$$

where X^* is the dual of X. The norm on X^* is denoted also by $\|\cdot\|$.

The result below is well-known.

Proposition 2.1. For each $x,y\in X$, there are $x_i^*\in F(x)$, $i = 1,2$, such that:

$$\begin{aligned} \langle y,x\rangle_s &= x_1^*(y) = \sup \{x^*(y);\ x^*\in F(x)\} , \\ \langle y,x\rangle_i &= x_2^*(y) = \inf \{x^*(y);\ x^*\in F(x)\} . \end{aligned} \tag{2.10}$$

Here $x^*(y)$ denotes the value of $x^*\in X^*$ at $y\in X$. The proof of Proposition 2.1 is given in Appendix (Corollary 1.1 and Remark 1.1).

We are now prepared to introduce the basic hypothesis (H(2.1)) on the t-dependence of A(t).

(H(2.1)) - There exist $\omega \geq 0$, a continuous function $f:]a,b[\to X$, and a bounded (on bounded subsets) function $L:R_+ \to R_+$ such that:

$$\langle y_1-y_2,x_1-x_2\rangle_i \leq \omega\|x_1-x_2\|^2+\|f(t)-f(s)\|\ \|x_1-x_2\|L(\|x_2\|) \tag{2.11}$$

for all $a < s \leq t \leq T < b$, $[x_1,y_1]\in A(t)$, $[x_2,y_2] \in A(s)$, $-\infty \leq a < b \leq +\infty$.

(H(2.2)) - The domain D(A(t)) of A(t) depends on $t\in[s,T]$ in the following sense:

If $t_n\uparrow t$ in $]s,T]$, $x_n \in D(A(t_n))$ and $x_n \to x$ in X, then $x\in\overline{D(A(t))}$.

Remark 2.1 If D(A(t)) is a closed set for each $t \in [s,T]$, then (H(2.2)) means that the mapping $t \to D(A(t))$ is closed.

Example 2.1. Hypotheses (H(2.1)) and (H(2.2)) do not imply, in general, that $\overline{D(A(t))}$ is independent of t. For example, let $X=R=]-\infty ,+\infty[$ and

$$A(t)x = \sqrt{x-t} + 1, \text{ with } D(A(t)) = [t,+\infty[\ = \overline{D(A(t))},\ t\in R.$$

In this case (1.11) is equivalent with (1.12) which is satisfied because $R(I+hA(t)) \subset D(A(t+h)$ for all $h > 0$. Clearly (2.11) holds since

$$(*)\qquad (A(t)x-A(s)y(x-y) \leq \sqrt{t-s}|x-y|, \quad x \geq t,\ y \geq s.$$

Examples in partial differential equations in which $\overline{D(A(t))}$ is also time-dependent are given in Chapter 3, § 2.

However, if $A(t)$ is m-dissipative for every $t\in[s,T]$, then $\overline{D(A(t))}=\overline{D}$ is necessarily independent on t (see Remark 4.2). Take for Example $\overline{D}=\overline{D(A(o))}$ (in this case). Obviously, the inequality (*) is stronger than (2.11) and corresponds to the case $\|f(t)-f(s)\|\leq\sqrt{t-s}$ (see (2.45)).

<u>Remark 2.2.</u> The notation $[x,y]\in A(t)$ means $x\in D(A(t))$ and $y\in A(t)x$. For $t=s$ Condition (2.11) implies the ω-dissipativity of $A(t)$, i.e.,

$$<y_1-y_2,x_1-y_2>_i\ \leq\omega\|x_1-x_2\|^2,\ [x_j,y_j]\in A(t), \tag{2.12}$$

$j=1,2,\ t\in]a,b[$. Some details in this direction may be found in Appendix. Condition (H(2.1)) allows $\overline{D(A(t))}$ to be t-independent (see Example 2.1 above and Section 2 in Chapter 3).

In the theory of the convergence of DS-approximate solutions, the result below is essential.

<u>Propostion 2.2.</u> (1) the condition (2.11) is equivalent with

$$(1-\lambda\omega)\|x_1-x_2\|\leq\|x_1-x_2-\lambda(y_1-y_2)\|+\|f(t)-f(s)\|L(\|x_2\|) \tag{2.13}$$

for all $\lambda>o, a\leq s\leq t\leq T$, $[x_1,y_1]\in A(t)$, $[x_2,y_2]\in A(s)$.

(2) The inequality (2.13) implies

$$(\lambda+\mu-\lambda\mu\omega)\|x_1-x_2\|\leq\lambda\|x_2-\mu y_2-x_1\|+\mu\|x_1-\lambda y_1-x_2\|+$$
$$+\lambda\mu\|f(t)-f(s)\|L(\|x_2\|) \tag{2.14}$$

for all $\lambda,\mu>o$, $a\leq s\leq t\leq T$, $[x_1,y_1]\in A(t)$, $[x_2,y_2]\in A(s)$, and (2.14) implies

$$(1+\lambda\omega)\|x_1-u\|\leq\|x_1-\lambda y_1-u\|+\lambda|A(a)u|+\lambda\|f(t)-f(s)\|L(\|u\|) \tag{2.15}$$

for all $\lambda>o$, $a<s\leq t\leq T$, $[x_1,y_1]\in A(t)$, $u\in D(A(s))$, where

$$|A(s)u|=\inf\{\|v\|;\ v\in A(s)u\}. \tag{2.16}$$

<u>Remark 2.3.</u> Inequality (2.14) is equivalent to

$$<y_1,x_1-x_2>_i+<y_2,x_2-x_1>_i\ \leq\omega\|x_1-x_2\|^2+\|f(t)-f(s)\|\,\|x_1-x_2\|L(\|x_2\|) \tag{2.14'}$$

for all $a < s \le t \le T$, $[x_1,y_1]\in A(t)$, $[x_2,y_2]\in A(s)$ (see Appendix).

Proof of Proposition 2.2. (1) In view of Proposition 2.1 there is $x^*\in F(x_1-x_2)$, such that

$$\langle y_1-y_2,\ x_1-x_2\rangle_i = x^*(y_1-y_2). \tag{2.17}$$

It is now easy to check that (2.11) implies (2.13). Indeed, by (2.11) and (2.15) we have

$$\|x_1-x_2\|^2 = x^*(x_1-x_2) = x^*(x_1-x_2-\lambda(y_1-y_2))+\lambda x^*(y_1-y_2)$$
$$\le \|x_1-x_2\|\ \|x_1-x_2-\lambda(y_1-y_2)\| + \lambda\omega\|x_1-x_2\|^2+$$
$$\lambda+\|f(t)-f(s)\|L(\|x_2\|)\|x_1-x_2\|$$

which yields (2.13). We now prove that (2.13) implies (2.11). To this goal, observe that (2.13) can be written in the form

$$\frac{\|x_1-x_2-\lambda(y_1-y_2)\|-\|x_1-x_2\|}{-\lambda} \le \omega\|x_1-x_2\|+\|f(t)-f(s)\|L(\|x_2\|). \tag{2.18}$$

In view of (2.6) we see that (2.18) implies (2.11). Similarly, we show that (2.11) implies (2.14), namely

$$\begin{aligned}(\lambda+\mu)\|x_1-x_2\|^2 &= \lambda x^*(x_1-x_2)+\mu x^*(x_1-x_2) = \\ &= \mu x^*(x_1-x_2-\lambda y_1) - \lambda x^*(x_2-x_1-\mu y_2)+\lambda\mu x^*(y_1-y_2).\end{aligned} \tag{2.19}$$

Combining (2.11), (2.17) and (2.19), we get obviously (2.14). Finally, Triangular inequality, $x_2=u$, $y_2\in A(s)u$ and (2.14) imply clearly (subtracting $\lambda\|x_1-x_2\|$ and then dividing by μ)

$$(1-\lambda\omega)\|x_1-u\| \le \|x_1-\lambda y_1-u\|+\lambda\|y_2\| + \lambda\|f(t)-f(s)\|L(\|u\|),$$

$\forall y_2\in A(s)u$, which yields (2.15). The proof is complete.

2.2. A remarkable estimate.

We now consider the discrete scheme $\{\hat{P}_m, \hat{x}_j^m, \hat{y}_j^m\}$ = DS corresponding to $\hat{s}\in[o,T]$ and $\hat{x}_o\in D(A(\hat{s}))$ is the sense of (1.2)-(1.7). Therefore

$$\hat{P}_m = \{\hat{s} = \hat{t}_o^m,\ \hat{t}_1^m,\ \ldots,\ \hat{t}_{N_{m-1}}^m,\ T\} \text{ with}$$

$$\hat{s} = \hat{t}_o^m < \hat{t}_1^m < \ldots < \hat{t}_j^m < \ldots < \hat{t}_{N_{m-1}}^m < \hat{t}_{N_m}^m \equiv T \} , \quad m \in N$$

$$\hat{d}_m = \max_{1 \leq j \leq N_m} (\hat{t}_j^m - \hat{t}_{j-1}^m) \to o \text{ as } m \to \infty, \ \omega \hat{d}_m > \tfrac{1}{2} \tag{2.19}$$

$$\hat{y}_j^m = \frac{\hat{x}_j^m - \hat{x}_{j-1}^m}{\hat{t}_j^m - \hat{t}_{j-1}^m} - \hat{p}_j^m \in A(\hat{t}_j^m)\hat{x}_j^m , \quad j = 1,2,\ldots,\hat{N}_m \tag{2.20}$$

$$\hat{x}_j^m \in D(A(\hat{t}_j^m)), \ j = o,1,\ldots, N_m, \quad \hat{x}_o^m \to \hat{x}_o \in \overline{D(A(\hat{s}))} \tag{2.21}$$

$$\hat{b}_m = \sum_{j=1}^{N_m} (\hat{t}_j^m - \hat{t}_{j-1}^m) \| \hat{p}_j^m \| \to o \quad \text{as } m \to \infty \quad . \tag{2.22}$$

The DS-approximate solution $\hat{u}_m$ corresponding to the above discrete scheme is defined as u_n (see (1.2)), that is

$$u_m(t) = \begin{cases} \hat{x}_o^m , & t = \hat{t}_o^m = \hat{s} \\ \hat{x}_j^m , & t \in]\hat{t}_{j-1}^m, \hat{t}_j^m] . \end{cases} \tag{2.23}$$

For simplicity of writing, set

$$h_k^n = t_k^n - t_{k-1}^n, \ \hat{h}_j^m = \hat{t}_j^m - \hat{t}_{j-1}^m, \ k=1,2,\ldots, N_n, \ j = 1,2,\ldots, \hat{N}_m. \tag{2.24}$$

Then, by (1.6) and (2.20), we have

$$x_k^n - h_k^n y_k^n = x_{k-1}^n + h_k^n p_k^n, \quad \hat{x}_j^m - \hat{h}_j^m \hat{y}_j^m = \hat{x}_{j-1}^m + \hat{h}_j^m \hat{p}_j^m, \tag{2.25}$$

with

$$y_k^n \in A(t_k^n) x_k^n, \ \hat{y}_j^m \in A(\hat{t}_j^m)\hat{x}_j^m, \ k = 1,2,\ldots, N_n, \ j=1,2,\ldots,\hat{N}_m.$$

It is also convenient to denote by

$$a_{k,j} = \| x_k^n - \hat{x}_j^m \|, \ d_{k,j} = \| f(t_k^n) - f(\hat{t}_j^m) \| \ \leq \rho(|t_k^n - \hat{t}_j^m|) \ (\text{see}(2.38))$$

$$\alpha_{k,j} = \hat{h}_j^m/(h_k^n + \hat{h}_j^m), \ \beta_{k,j} = h_k^n/(h_k^n + \hat{h}_j^m), \ \gamma_{k,j} = h_k^n \hat{h}_j^m/(h_k^n + \hat{h}_j^m) \tag{2.26}$$

$$c_{k,j}(\eta) = [(t_k^n-\hat{t}_j^m-\eta)^2+d_n(t_k^n-s)+\hat{d}_m(\hat{t}_j^m-\hat{s})]^{1/2}, \quad o \leq |\eta| < T. \tag{2.27}$$

The next simple lemma will play an essential role in the proof of the main estimates.

Lemma 2.1. The following inequality

$$\alpha_{k,j}\, c_{k-1,j}(\eta)+\beta_{k,j}c_{k,j-1}(\eta) \leq c_{k,j}(\eta) \tag{2.28}$$

holds for $k = 1,2,\ldots,N_n$, $j=1,2,\ldots,\hat{N}_m$.

Proof. Since $\alpha_{k,j}+\beta_{k,j} = 1$ we have

$$\begin{aligned} I_1 &= \alpha_{k,j}c_{k-1,j}(\eta)+\beta_{k,j}c_{k,j-1}(\eta) \\ &\leq (\alpha_{k,j}c^2_{k-1,j}(\eta)+\beta_{k,j}c^2_{k,j-1}(\eta))^{1/2} \end{aligned} \tag{2.29}$$

Here we have used the elementary inequality

$$a_1b_1+a_2b_2 \leq (a_1^2+a_2^2)^{1/2}(b_1^2+b_2^2)^{1/2}$$

with $a_1 = (\alpha_{k,j})^{1/2}$, $a_2 = (\beta_{k,j})^{1/2}$, and so on.

For simplicity, and since there is no danger of confusion, in this proof we drop the indices m and n (i.e., we write $t_k^n = t_k$, $\hat{t}_j^m = \hat{t}_j$ and so on); Thus, according to the notations in (2.24) we have $t_{k-1}=t_k-h_k$, $\hat{t}_{j-1}=\hat{t}_j-\hat{h}_j$, and therefore

$$(t_{k-1}-\hat{t}_j-\eta)^2 = (t_k-\hat{t}_j-\eta)^2-2h_k(t_k-\hat{t}_j-\eta)+h_k^2 \tag{2.30}$$

$$(t_k-\hat{t}_{j-1}-\eta)^2 = (t_k-\hat{t}_j-\eta)^2+2\hat{h}_j(t_k-\hat{t}_j-\eta)+\hat{h}_j^2.$$

Consequently

$$I_1^2 \leq \frac{1}{h_k+\hat{h}_j}[\hat{h}_j(t_{k-1}-\hat{t}_k-\eta)^2+d_n(t_{k-1}-s)+d_m(\hat{t}_j-\hat{s}) +$$

$$\frac{1}{h_k+\hat{h}_j} \quad h_k(t_k-\hat{t}_{j-1}-\eta)^2+d_n(t_k-s)+\hat{d}_m(\hat{t}_{j-1}-\hat{s})] =$$

$$(t_k-\hat{t}_j-\eta)^2+d_n(t_k-s)+\hat{d}_m(t_j-\hat{s})+ \frac{h_k\hat{h}_j}{h_k+\hat{h}_j}(h_k-d_n+\hat{h}_j-\hat{d}_m)$$

$$\leq c^2_{k,j}(\eta). \tag{2.31}$$

Here we have used the notations in (1.4), (2.19) and (2.24) which give $h_k \leq d_n$ and $\hat{h}_j \leq \hat{d}_m$. In view of (2.31), the proof is complete.

Before proceeding to the main estimates, we need also the following result.

Lemma 2.2. Let $A(t)$ satisfy the condition (2.11) and let u_n be the DS-approximate solution (in the sense of (1.2)-(1.7)) to the problem (1.1)+(1.1)'. Then, for every $r \in [o,T]$ and $\tilde{u} \in D(A(r))$, there exists a constant $M = M(s,r,x_o,\tilde{u},T,f,\omega)$ (independent of $t \in [s,T]$ and $n \in N$) such that

$$\|u_n(t)\| \leq M, \quad \forall t \in [s,T] \text{ and } n \in N. \tag{2.32}$$

Proof. With the same convention as in (2.30), the inequality (2.15) yields

$$(1-\omega h_k)\|x_k-\tilde{u}\| \leq \|x_k-h_k y_k-\tilde{u}\| + h_k\|f(t_k)-f(r)\|L(\|\tilde{u}\|) + $$
$$+ h_k|A(r)\tilde{u}|, \ \tilde{u} \in D(A(r)), \ k = 1,2,\ldots,N_n. \tag{2.33}$$

Substituting $x_k-h_k y_k = x_{k-1}+h_k p_k$ into (2.33) we get

$$(1-\omega h_k)\|x_k-\tilde{u}\| \leq \|x_{k-1}-\tilde{u}\| + h_k|A(r)\tilde{u}| + $$
$$+ h_k(\|p_k\| + \|f(t_k)-f(r)\|L(\|\tilde{u}\|)). \tag{2.34}$$

Iterating (2.34) from k=1 to k=k one obtains (since $\sum_{i=1}^{k} h_i = t_k - s$)

$$\|x_k-\tilde{u}\| \leq \prod_{i=1}^{k}(1-\omega h_i)^{-1}[\|x_o^n-\tilde{u}\| + (t_k-s)|A(r)\tilde{u}| + $$

$$+ \sum_{i=1}^{k} h_i\|f(t_i)-f(r)\|L(\|\tilde{u}\|) + \sum_{i=1}^{k} h_i\|p_i\| . \tag{2.35}$$

But for each $\varepsilon > 0$,

$$(1-h)^{-1} \leq e^{(1+\varepsilon)h}, \quad \forall h \in [o, \ \varepsilon/(1+\varepsilon)[\tag{2.36}$$

Since we have (by hypothesis) $\omega h_i \leq \omega d_n \leq \frac{1}{2}$, it follows (for $\varepsilon = 1$) that

$$(1-\omega h_i)^{-1} \leq \exp(2|\omega|h_i), \tag{2.37}$$

and consequently (2.35) implies

$$\|x_k-\tilde{u}\| \leq \exp(2|\omega|(T-s)) \, [\, \|x_0^n-\tilde{u}\|+(t_k-s)|A(r)\tilde{u}|$$

$$+ \sum_{i=1}^{k} h_i\|f(t_i)-f(r)\|L(\|\tilde{u}\|) + \sum_{i=1}^{k} h_i\|p_i\|. \tag{2.37)'$$

Taking into account (1.6), (1.7), the continuity of f(or only the integrability of f on $[s,T]$!) and the equality $\sum_{i=1}^{k} h_i = T-s$ (2.32) follows. This completes the proof.

Now, a few remarks on "the modulus of continuity" of f. Set

$$\rho(r) \ = \sup\{\|f(t)-f(\tau)\| \ ; \ t,\tau\in[o,T], \ |t-\tau| \leq r\}, \tag{2.38}$$

$r\in[o,T]$. Obviously, $\rho: [o,T] \to R_+$ is bounded, nondecreasing and $\lim_{r\downarrow o} \rho(r)=0$. Moreover, ρ is upper semicontinuous on $[o,T]$ and right semicontinuous on $[o,T[$. The simple inequality below is useful for our later purposes:

$$\rho(r) \leq c^{-1} \rho(T)|r-r'| + \rho(\sigma), \ r\in[o,T] \tag{2.39}$$

where $o < c < \sigma \leq T$, $o \leq r' < \sigma - c$.

Let us check it. If $r \leq \sigma$, then $\rho(r) \leq \rho(\sigma)$, so (2.39) is trivially satisfied. If $r > \sigma$ and $r' < \sigma - c$, we have : $c < \sigma - r' < r - r'$, and hence $\rho(r) \leq \rho(T) \leq \frac{r-r'}{c}\rho(T)$, thereby completing the proof of (2.39).

We are now in the position to give the fundamental estimate (due to K. Kobayasi , Y. Kobayashi and S. Oharu [1]) in this book. Namely we have

<u>Lemma 2.3</u>. Let $s,\hat{s}\in[o,T]$, $x_0\in\overline{D(A(s))}$, $\hat{x}_0\in\overline{D(A(\hat{s}))}$, and let $\{P_n, x_k^n\}$ and $\{\hat{P}_m, \hat{x}_j^m\}$ be two discrete schemes in the sense of (1.3)-(1.7) and (2.19)'-(2.22), respectively. Let also (2.11) be satisfied and $o \leq |\eta| < \sigma < T$, $o < c < \sigma - |\eta|$. Assume that $d_n, \hat{d}_m < \sigma - |\eta| - c$. Then, for each $r\in[o,T]$ and every $[\tilde{u},\tilde{v}]\in A(r)$, the following inequality holds

$$\prod_{i=1}^{k} (1-\omega h_i^n) \prod_{q=1}^{j} (1-\omega\hat{h}_q^m)\|x_k^n-\hat{x}_j^m\| \leq \|x_0^n-\tilde{u}\| + \|\hat{x}_0^m-\tilde{u}\| +$$

$$c_{k,j}(s-\hat{s}) \, [\|\tilde{v}\|+M\sigma(T)] + \sum_{i=1}^{k} h_i^n\|p_i^n\| + \sum_{q=1}^{k} \hat{h}_q^m\|\hat{p}_q^m\| +$$

$$M(\hat{t}^m_j-\hat{s})\,[\,c^{-1}\quad \rho(T)c_{k,j}(\eta)+\rho(\sigma)\,], \tag{2.40}$$

for $o \le k \le N_n$ and $o \le j \le \hat{N}_m$, where

$$M = \max\{L(M(s,r,x_o,\tilde{u}),\ L(M(\hat{s},r,\hat{x}_o,\tilde{u}),L(\|\tilde{u}\|)\}$$

with M in (2.32), $L(M) = \sup\{L(\|y\|);\ \|y\| \le M\}$, and the convention $\prod_{i=1}^{o}(1-\omega h^n_i) = \prod_{i=1}^{o}(1-\omega\hat{h}^m_q) = 1.$

Proof. Set

$$\omega_{k,j} = \prod_{i=1}^{k}(1-\omega h^n_i)\ \prod_{q=1}^{j}(1-\omega\hat{h}^m_j),\ a_{k,j} = \|x^n_k-\hat{x}^m_j\|. \tag{2.41}$$

According to the inequality (2.35)

$$\omega_{k,o}\|x^n_k-\hat{x}^m_o\| \le \omega_{k,o}(\|x^n_k-\tilde{u}\| + \|\tilde{u}-x^m_o\|) \le \|x^n_o-\tilde{u}\| + \|x^m_o-\tilde{u}\|$$
$$+ (t^n_k-s)|A(r)\tilde{u}| + \sum_{i=1}^{k} h^n_i\|f(t^n_i)-f(r)\|L(\|\tilde{u}\|)+\sum_{i=1}^{k} h^n_i\|p^n_i\|. \tag{2.42}$$

Let us observe that

$$\prod_{i=1}^{k} h^n_i = t^n_k-s,\ \|f(t^n_i)-f(r)\| \le \rho(|t^n_i-r|) \le \rho(T),L(\|\tilde{u}\|) \le M,$$

$$L(\|\tilde{u}\|) \le M,\ |t^n_k-s| \le c_{k,o}(s-\hat{s}). \tag{2.43}$$

Consequently, (2.42) gives

$$\omega_{k,o}a_{k,o} \le \|x^n_o-\tilde{u}\|+\|x^m_o-\tilde{u}\|+c_{k,o}(s-\hat{s})(|A(r)\tilde{u}| + M\rho(T))$$
$$+ \prod_{i=1}^{k} h^n_i\|p^n_i\| \tag{2.44}$$

so (2.40) is fullfilled for $k=0,1,\dots,N_n$ and $j = 0$.

Clearly, in the same way one checks that (2.40) holds for k=0 and $j=0,1,\dots,N_m$. Thus (2.40) is true for the pairs (k-1,j) and (k,j-1) and then prove (2.30) for the pair (k,j). Of course, once this is accomplished, it follows that (2.40) holds for every pair (k,j), $k=o,1,\dots,N_n$, $j=o,1,\dots,N_n$.

To this aim, let us point out a relationship between $a_{k,j}$; $a_{k-1,j}$

and $a_{k,j-1}$, namely

$$(1-\omega\gamma_{k,j})a_{k,j} \le \alpha_{k,j}a_{k-1,j}+\beta_{k,j}a_{k,j-1} +$$

$$+ \gamma_{k,j}(\|p_k^n\| + \|\hat{p}_j^m\| + Md_{k,j}) \tag{2.45}$$

where $d_{k,j} \equiv \rho(|t_k^n-\hat{t}_j^m|) \ge \|f(t_k^n)-f(\hat{t}_j^m)\|$ (see (2.26)).

To get (2.45), we first note that (2.32) means $L(\|x_i^n\|)\le M$, $L(\|\hat{x}_j^m\|) \le M$. Thus, for $\lambda = h_i^n$, $x_1 = x_k^n$, $\mu = \hat{h}_j^m$, $x_2 = \hat{x}_j^m$, $t = t_k^n$ and $s = \hat{t}_j^m$, the inequality (2.14) in Proposition 2.2 leads to

$$(h_k^n+\hat{h}_j^m - \omega h_k^n\hat{h}_j^m)\|x_k^n-\hat{x}_j^m\| \le h_k^n \|\hat{x}_j^m-\hat{h}_j^m\hat{y}_j^m-x_k^n\| + \hat{h}_j^m\|x_k^n-h_k^n y_k^n-\hat{x}_j^m\| +$$

$$+ h_k^n\hat{h}_j^m M\| f(t_k^n)-f(\hat{t}_j^m)\|. \tag{2.46}$$

Substituting (2.25) into (2.46) and then dividing by $h_k^n+\hat{h}_j^m$ one obtains just (2.45). Inasmuch as $\omega \ge 0$, it is obvious that

$$0 < \omega_{k,j} \le \max\{1-\omega h_k^n,\ 1-\omega\hat{h}_j^m\} \le 1-\omega\gamma_{k,j}. \tag{2.47}$$

Multiplying (2.45) by $\omega_{k,j}$ (given by (2.41)), we have

$$(1-\omega\gamma_{k,j})\omega_{k,j} \le (1-h_k^n\omega)\omega_{k-1,j}a_{k-1,j}\alpha_{k,j} +$$

$$+ (1-\hat{h}_j^m\omega)\omega_{k,j-1}a_{k,j-1}\beta_{k,j} + \omega_{k,j}\gamma_{k,j}(\|p_k^n\| + \|\hat{p}_j^m\| + Md_{k,j}).$$

Dividing by $(1-\omega\gamma_{k,j})$ and taking into account (2.47) and $\gamma_{k,j}=h_k^n\alpha_{k,j}= =\hat{h}_j^m\beta_{k,j}$, one obtains

$$\omega_{k,j}a_{k,j} \le \alpha_{k,j}(\omega_{k-1,j}a_{k-1,j} + h_k^n\|p_k^n\|)$$

$$+ \beta_{k,j}(\omega_{k,j-1}a_{k,j-1} + \hat{h}_j^m\|\hat{p}_j^m\|) + M\gamma_{k,j}d_{k,j}. \tag{2.48}$$

Estimating $\omega_{k-1,j}a_{k-1,j}$ and $\omega_{k,j-1}a_{k,j-1}$ according to (2.40) and the induction hypothesis, we get

$$\omega_{k,j}a_{k,j} \le \|x_o^n-\tilde{u}\| + \|x_o^m-\tilde{u}\| + (\alpha_{k,j}c_{k-1,j}(\hat{s}-s) +$$

$$+ \beta_{k,j}c_{k,j-1}(s-\hat{s}))(\|\tilde{v}\|+M\rho(T)) + \sum_{i=1}^{k} h_i^n\|p_i^n\| + \sum_{q=1}^{j}\hat{h}_q^m\|\hat{p}_q^m +$$

$$+ (\hat{t}_j^m-\hat{s})[c^{-1}\rho(T)\alpha_{k,j}c_{k-1,j}(\eta) + \alpha_{k,j}\rho(\sigma)]M +$$

$$+ (\hat{t}^m_{j-1}-\hat{s}) [c^{-1} \rho(T) \beta_{k,j} c_{k,j-1}(\eta)+\beta_{k,j} \rho(\sigma)] M + M\gamma_{k,j} d_{k,j}. \qquad (2.49)$$

On the basis of the inequality (2.28) and $\hat{t}^m_{j-1} = \hat{t}^m_j-\hat{h}^m_j$, (2.49) implies

$$\omega_{k,j} a_{k,j} \leq \|x^n_o-\tilde{u}\| + \|\hat{x}^m_o-\tilde{u}\| + c_{k,j}(s-\hat{s}) (\|\tilde{v}\|+M\rho(T)) +$$

$$\sum_{i=1}^{k} h^n_i p^n_i + \sum_{q=1}^{j} \hat{h}^m_q \hat{p}^m_q + (\hat{t}^m_j-\hat{s}) [c^{-1} \rho(T)c_{k,j}(\eta)+\rho(\sigma)] M -$$

$$\hat{h}^m_j \beta_{k,j}[c^{-1} \rho(T)c_{k,j-1}(\eta) +\rho(\sigma)] M + M\gamma_{k,j} d_{k,j}. \qquad (2.50)$$

We now observe that the sum of the last two terms on the right hand side of (2.50) is less **than or equal to zero** . Indeed, with $r=|t^n_k-\hat{t}^m_j|$ and $r'=|\eta-\hat{h}^m_j|$ it follows

$$r' \leq |\eta| + \hat{h}^m_j \leq |\eta| + d_m < \sigma - c \qquad |r-r'| \leq |t^n_k-\hat{t}^m_j-\eta| \leq c_{k,j-1}(\eta)$$

so (2.39) yields

$$\gamma_{k,j} d_{k,j} = \hat{h}^m_j \beta_{k,j} d_{k,j} = \hat{h}^m_j \beta_{k,j} \rho(|t^n_k-\hat{t}^m_j|)$$

$$\leq \hat{h}^m_j \beta_{k,j}(c^{-1}\rho(T)c_{k,j-1}(\eta) + \rho(\sigma)). \qquad (2.51)$$

In view of (2.50) and (2.51) we conclude that if for the pairs (k-1,j) and (k,j-1) (2.40) holds true, then it also holds true for the pair (k,j). This completes the proof.

Remark 2.4. In the proof of Lemma 2.3 we have used only (2.14) (or equivalently (2.14)' which is strictly more general than (2.11). Therefore, (2.40) remains valid if (2.11) is replaced by (2.14).

§3. Convergence of DS-approximate solutions and generation of Evolution Operators. Integral Solutions.

3.1. Convergence of DS-approximate solutions.

Throughout this section we assume the existence of DS-approximate solution u_n defined by (1.2)-(1.7). Under this hypothesis the following basic result holds.

Theorem 3.1. Assume that u_n exists and conditions (H(2.1)) and (H(2.2)) in Section 2) are satisfied. Then there exists a continuous function

$u: [o,T] \to X$ such that for each $s \in [o,T]$ and $x_o \in \overline{D(A(s))}$, every Ds-approximate solution u_n given by (1.2)-(1.7) is uniformly convergent to u on $[s,T]$ and $u(t) \in \overline{D(A(t))}$ for all $t \in [s,T]$, $u(t_o)=x_o$. If $x_o \in D(A(s))$, and f is of bounded variation then u is Lipschitz continuous on $[s,T]$.

Proof. We shall use Lemma 2.3 with $s=\hat{s}$, $x_o=\hat{x}_o$, $\hat{x}^m_j=x^m_j$, $\hat{t}^m_j=t^m_j$, $\hat{h}^m_j=h^m_j$, $\hat{p}^m_j=p^m_j$ and $\hat{N}_m = N_m$. Let $t \in]s,T[$ and let $k=k_n$ and $j=j_m$ be such that $t \in]t^n_{k_n-1}, t^n_{k_n}] \cap [t^m_{j_m-1}, t^m_{j_m}]$.

By (2.27) we see that $c_{k_n,j_m}(o) \to o$ as $m, n \to +\infty$ (this is because $t^n_{k_n} \to t$ and $t^m_{j_m} \to t$ as $n,m \to +\infty$). On the other hand, $\omega_{k,j}$ given by (2.41) is bounded (according to (2.37)) and we have

$$\omega_{k,j} \le \exp(4|\omega|(T-s)) \equiv \bar{C} \tag{3.1}$$

By definition of u_n it follows that $u_n(t)=x^n_{k_n}$, $u_m(t)=x^m_{j_m}$ (hence $u_n(t) \in D(t^n_{k_n})$). Consequently, with $\eta=o$, (2.40) yields

$$\overline{\lim_{m,n\to\infty}} \|u_n(t)-u_m(t)\| \le \bar{C}(2\|x_o-\tilde{u}\| +M(T-s)\rho(\delta)) \tag{3.2}$$

for all $\tilde{u} \in D(A(s))$ and $\delta > o$. Taking into account that $\rho(\delta) \to o$ as $\delta \downarrow o$, (3.2) implies that actually $\lim_{m,n\to\infty} (u_n(t)-u_m(t))=o$ uniformly with respect to $t \in [s,T]$. Note also that $u_n(t) \in D(t^n_{k_n})$ and

$$u(t;s,x_o) = \lim_{n\to\infty} u_n(t), \; t \in [s,T] \tag{3.3}$$

jointly $t^n_{k_n} \to t$, imply $u(t;s,x_o) \in \overline{D(t)}$.

Arguing as above (in view of (2.40)) it is clear that any other DS-approximate $\hat{u}_n$ corresponding to s and $x_o \in \overline{D(A(s))}$ is also convergent to u. Indeed, if $\hat{x}^n_o \to \hat{x}_o = x_o$ and $\hat{s}=s$, (2.40) yield

$$\lim_{n\to\infty} \|u_n(t)-\hat{u}_n(t)\| \le \bar{C}(2\|x_o-\tilde{u}\| +M(T-s)\rho(\delta)) \tag{3.4}$$

for every $\tilde{u} \in D(A(s))$ and $\delta > o$, hence $\lim_{n\to\infty} u_n(t)= \lim_{n\to\infty} \hat{u}_n(t)$.

It remains to prove the continuity (and extendability) of u on $[s,T]$. To this aim, take $t,\bar{t} \in [s,T[$ and $n \in N$. Let k_n and j_n be such that $t^n_{k_n-1}$

$< t \leq t^n_{k_n}$, $t^n_{j_n-1} < \bar{t} \leq t^n_{j_n}$ hence $t^n_{k_n} \to t$ and $t^n_{j_n} \to \bar{t}$ as $n \to \infty$ and $x^n_{k_n} = u_n(t)$, $x^n_{j_n} = u_n(t)$. In this case $c_{k_n,j_n}(o) \to |t-\bar{t}|$. Consequently, with $s=\hat{s}$, $\hat{t}^m_j = t^m_j$, $\hat{x}^m_j = x^m_j$ (and so one), $m=n$ and $\eta=o$, (2.40) gives

$$\|u(t)-u(\bar{t})\| = \lim_{n\to\infty}\|u_n(t)-u_n(\bar{t})\| \leq \bar{C}(2\|x_o-\tilde{u}\| + \\ + |t-\bar{t}|(|A(s)\tilde{u}|+M\rho(T)) + \bar{C}M(\bar{t}-s)\, c^{-1}\rho(T)|t-\bar{t}| + \rho(\delta) \tag{3.5}$$

for every $\tilde{u} \in D(A(s))$, $o < \delta < T$ and $o < c < \delta$. Since $x_o \in \overline{D(A(s))}$, for $\varepsilon > o$ there is $\tilde{u} \in D(A(s))$ such that $\|x_o - u\| < \varepsilon / 8\bar{C}$. Let also $\rho(\varepsilon) > o$ be such that $\rho(\delta) < \frac{\varepsilon}{4}(\bar{C}M(T-s))^{-1}$, $o < \delta \leq (\varepsilon)$. It follows from (3.5) that if $o < c < \delta(\varepsilon)$ and

$$|t-\bar{t}| \leq \delta_1(\varepsilon) = \min\left\{ \frac{\varepsilon}{4(|A(s)\ \tilde{u}|+M\rho(T)\bar{C}}, \frac{\varepsilon c}{4M(T-s)\bar{C}\rho(T)} \right\}$$

we have

$$\|u(t)-u(\bar{t})\| < 3\varepsilon/4 + \rho(\delta(\varepsilon)) \leq \varepsilon$$

that is u is uniformly continuous on $[s,T[$.

Accordingly, u can be extended uniquely to a continuous function (denoted again by u) on $[s,T]$. If $x_o \in D(A(s))$, we can choose $u = x_o$ in (3.5). It follows

$$\|u(t)-u(\bar{t})\| \leq |t-\bar{t}|(|A(s)x_o|+M\rho(T)+Mc^{-1}\rho(T)(\bar{t}-s)) + \rho(\delta)M(\bar{t}-s) \tag{3.5'}$$

Set $u(t) = U(t,s)x_o$. Hence $u(s) = U(s,s)x_o = x_o$. For $t=s$, (3.5)' yields

$$\|U(t,s)x_o - x_o\| \leq (t-s)(|A(s)x_o|+M\rho(T)),\quad s \leq t \leq T \tag{3.5''}$$

for all $x_o \in D(A(s))$ where M is a constant independent of t (M depends on s and x_o, see (5.4)).

If f is of bounded variation (see Theorem 3.3), then (3.5)" is equivalent to Lipschitz continuity of $t \to U(t,s)x$.

The function u defined by (1.8)(respectively (3.3)) is called DS-limit solution of (1.1)+(1.1)'. In order to study this type of solution, we need the notion of integral solution to the problem (1.1)+(1.1)'.

3.2 Integral solution

In order to point out the idea of considering integral solutions, we start with:

Definition 3.1. By a strong solution to the Cauchy Problem (1.1)-(1.1)', we mean a function $u: [s,T] \to X$ with the properties:

(1) u is continuous on $[s,T]$ and absolutely continuous on compact subsets of $]s,T[$.

(2) $u(t) \in D(A(t))$, a.e. on $]s,T[$.

(3) u is strongly differentiable almost everywhere on $]s,T[$ and satisfies (1.1) a.e. on $]s,T[$.

Now, let us suppose that $A(t)$ satisfies (2.12). Then the problem (1.1)+(1.1)' admits at most one strong solution. Indeed, let u_1 and u_2 be solutions of (1.1)-(1.2) on $[s,T]$, i.e.

$$u_k'(t) \in A(t)u_k(t), \text{ a.e. on }]s,T[$$
$$u_k(s)=x_o, \; k=1,2. \tag{3.6}$$

On the bais if Definition 3.1, $t \to \|u_1(t)-u_2(t)\|$ is absolutely continuous on $]s,T[$, hence a.e. differentiable on $]s,T[$.

Set $y_k(t) = u_k'(t) \in A(t)u_k(t)$, $k=1,2$. According to Kato's lemma (see Appendix, Lemma 1.1),

$$\|u_1(t)-u_2(t)\| \frac{d}{dt} \|u_1(t)-u_2(t)\| = \langle y_1(t)-y_2(t),$$
$$u_1(t)-u_2(t) \rangle_i \text{ a.e. on }]s,T[\tag{3.7}$$

Clearly, (3.7) and (2.12) yield

$$\frac{d}{dt} \|u_1(t)-u_2(t)\| \le \omega\|u_1(t)-u_2(t)\|, \text{ a.e. on }]s,T[\tag{3.8}$$

which implies

$$\|u_1(t)-u_2(t)\| \le \omega\|u_1(s)-u_2(s)\| \exp(\omega(t-s)) \tag{3.9}$$

for all $t \in [s,T]$.

By hypothesis, $u_1(s)=u_2(s)=x_o$ and therefore $u_1=u_2$ on $[s,T]$, q.e.d.

To arrive to the notion of integral solution, suppose further that u is a strong solution to (1.1)-(1.1)' under the hypothesis (2.11). Take an arbitrary $r \in [s,T]$ and $[x,y] \in A(r)$. Finally, set $y(t)=u'(t)$. In view of Proposition 2.1 there exists $x^* \in F(u(t))-x)$ such that

$$x^*(y(t)-y) \leq \langle y(t)-y,\ u(t)-x \rangle_i \tag{3.10}$$

Arguing as in the proof of (3.7), we have

$$\|u(t)-x\| \frac{d}{dt}\| u(t)-x\| = x^*(y(t))=x^*(y(t)-y)+x^*(y)$$

$$\leq \langle y(t)-y,u(t)-x \rangle_i + \langle y,u(t)-x \rangle_s, \text{ a.e. on }]s,t[\tag{3.11}$$

Combining (3.11), (2.11) and (2.6) we get

$$\frac{d}{dt}\|u(t)-x\| \leq \omega\|u(t)-x\|+C\|f(t)-f(r)\|+ \langle y,u(t)-x \rangle_+ \tag{3.12}$$

where $C=C(x) \geqslant \max(L(C_1),\ L(\|x\|))$; $C_1 \geqslant \sup_{s \leq t \leq T} \|u(t)\|$.

Integrating (3.12) over $[t,\bar{t}]$ one obtains

$$\|u(\bar{t})-x\|-\|(t)-x\| \int_t^{\bar{t}} (\omega \|u(\bar{s})-x\| + \langle y,u(\bar{s})-x \rangle_+ +C \|f(\bar{s})- -f(r)\|)d\bar{s} \tag{3.13}$$

for all $s \leq t \leq \bar{t} \leq T$, $r \in [s,T]$ and $[x,y] \in A(r)$.

We are now prepared to give (similarly to Benilan [1]).

<u>Definition 3.2.</u> By an integral solution to the problem (1.1)+(1.1)' on $[s,T]$, we mean a continuous function u on $[s,T]$ satisfying the inequality (3.13) with C as in (3.12), $u(s)=x_0$ and $u(t) \in \overline{D(A(t))}$ for $t \in [s,T]$.

The notion of integral solution plays an important role in the study of the relationship between "DS-limit solutions" and "strong solutions".

Precisely, the following basic result holds.

<u>Theorem 3.2.</u> (1) Suppose that for each $t \in [s,T]$, $A(t)-\omega I$ is dissipative (i.e. (2.12) holds). Then the problem (1.1)+(1.1)' has at most one strong solution.

(2) Suppose further that Conditions (H(2.1)) and (H(2.2)) (in Section 2) are fulfilled. If u is a DS-limit solution (or a strong solution) to the problem (1.1)+(1.1)' then u is the unique integral solution to this problem.The problem (1.1)+(1.1)' has a unique integral solution.

(3) In the DS-limit solution u is strongly (or only weakly) right-differentiable at $t_o \in [s,T]$ and $A(t_o)-\omega I$ is maximal dissipative in $\overline{D(A(t_o))}$, then $u(t_o) \in D(A(t_o))$ and

$$(d^+/dt)u(t_o) \in A(t_o)u(t_o) \tag{3.14}$$

<u>Proof.</u> Part (1) has already been proved (see (3.9)).

(2) We first prove that the DS-limit solution u is an integral solution. To this goal let $[x,y] \in A(r)$. Since x_k^n is bounded (by (2.37)') there is a constant $C = \max(L(\sup\|x_k^n\|, L(\|x\|))$ such that

$$< \frac{x_k^n - x_{k-1}^n}{h_k^n} - p_k^n - y,\ x_k^n - x >_i \leq \omega\|x_k^n - x\|^2 + C\|f(t_k^n) - f(r)\|\|x_k^n - x\| \tag{3.15}$$

for $k=o,1,\dots,N_n$ (by (2.11)). In view of Proposition 2.1, there exists $x^* \in F(x_k^n - x))$ such that the left hand side of (3.15) equals $x^*(\frac{x_k^n - x_{k-1}^n}{h_k^n} - -p_k^n - y)$. Taking into account (2.6), (3.15) yields

$$x^*(x_k^n - x_{h-1}^n) \leq h_k^n\ [\|p_k^n\|\ \|x_k^n - x\| + < y, x_k - x >_s + \\ + \omega\|x_k - x\|^2 + C\ \|x_k^n - x\|\ \ \|f(t_k) - f(r)\|] \tag{3.16}$$

Observing that

$$\|x_k^n - x\|^2 - \|x_k^n - x\|\ \|x_{k-1}^n - x\| \leq x^*(x_k^n - x_{k-1}^n), \tag{3.16'}$$

and then dividing by $\|x_k^n - x\|$, (3.16) implies

$$\|x_k^n - x\| - \|x_{k-1}^n - x\| \leq h_k^n[\|p_k^n\| + < y, x_k - x >_+ + \omega\|x_k - x\| + \\ + C\|f(t_k) - f(r)\|] \tag{3.17}$$

Iterating (3.17) for $k=j+1,\dots,p$, $(j+1 < p)$ one obtains

$$\|x_p^n - x\| - \|x_j^n - x\| \leq b_n + \sum_{j+1}^{p} h_k(< y, x_k - x >_+ + \omega\|x_k - x\| + C\|f(t_k) - f(r)\|) \tag{3.18}$$

Let $p=p_n$ and $j=j_n$ be such that $\bar{t}\in]t^n_{p_n-1},t^n_{p_n}]$, $t\in]t^n_{j_n-1},t^n_{j_n}]$. Set also $a_n(\bar{s})=t_k$ for $\bar{s}\in]t_{k-1},t_k]$. According to the definition of u_n (see(1.2)), (3.18) becomes

$$\|u_n(\bar{t})-x\| - \|u_n(t)-x\| \le \int_{t^n_{j_n}}^{t^n_{p_n}} (\omega\|u_n(a_n(\bar{s}))-x\|$$

$$+\langle y,u_n(a_n(\bar{s}))-x\rangle_+ +b_n+C\|f(u_n(a_n(\bar{s})))-f(r)\|)d\bar{s} \qquad (3.19)$$

Clearly $a_n(\bar{s}) \to \bar{s}$ as $n\to\infty$ (uniformly with respect to $\bar{s}$), hence $u_n(a_n(\bar{s})) \to u(\bar{s})$ as $n\to\infty$, uniformly with respect to $\bar{s}\in[s,T[$. Passing to the limit for $n\to\infty$ in (3.19), one obtains (3.13). The fact that a strong solution is an integral solution has already been proved before Definition 3.2. The proof of the uniqueness of integral solution proceeds similarly as above, namely:

Let $\tilde{u}$ be an arbitrary integral solution to the problem (1.1)+(1.1)' and let u be the DS-limit solution to this problem (see (1.8) and (3.3)).

We will prove that $\tilde{u}=u$ on $[s,T]$. To this aim set $y^n_k\in\dfrac{x^n_k-x^n_{k-1}}{t^n_k-t^n_{k-1}}-p^n_k$
Consequently, $y^n_k\in A(t^n_k)x^n_k$ (see (1.5)). Replacing $x=x^n_k$, $y=y^n_k$ and $r=t^n_k$ in (3.13) and observing that $h^n_k y^n_k-x^n_k=-x^n_{k-1}-h^n_k p^n_k$, $h^n_k\langle y^n_k,\tilde{u}(\bar{s})-x^n_k\rangle_+\le\|\tilde{u}(\bar{s})-x^n_{k-1}\|-\|\tilde{u}(\bar{s})-x^n_k\|+h_k\|p^n_k\|$

one obtains

$$h^n_k(\|\tilde{u}(\bar{t})-x^n_k\|-\|\tilde{u}(t)-x^n_k)\|)\le\int_t^{\bar{t}}\omega h^n_k\|\tilde{u}(\bar{s})-x^n_k\|d\bar{s}$$

$$+\int_t^{\bar{t}}(\|\tilde{u}(s)-x^n_{k-1}\|-\|\tilde{u}(s)-x^n_k\|)d\bar{s}+C\int_t^{\bar{t}}h^n_k\|f(\bar{s})-f(t^n_k)\|d\bar{s}+$$

$$+h_k(\bar{t}-t)\|p^n_k\| \qquad (3.20)$$

with $\bar{C}\ge\max\{L(\sup_{s\le t\le T}\|u(t)\|),\sup\|x^n_k\|\}$
Iterating (3.20) for $k=j+1,\dots,p$ we get

$$\int_{t_j}^{t_p}(\|\tilde{u}(\bar{t})-u_n(\theta)\| - \|\tilde{u}(t)-u_n(\theta)\|)d\theta$$

$$\leq \int_t^{\bar t} (\|\tilde u(\bar s)-u_n(t_j^n)\| - \|\tilde u(\bar s)-u_n(t_p^n)\|)d\bar s + \int_{t_j^n}^{t_p^n} (\int_t^{\bar t} \omega\|\tilde u(\bar s)-u_n(\theta)\| d\bar s)d\theta$$

$$+ \bar C \int_{t_j^n}^{t_p^n} (\int_t^{\bar t} \|f(\bar s)-f(a_n(\theta))\| d\bar s)d\theta + (t_p^n-t_j^n)(\bar t-t)b_n \qquad (3.21)$$

Letting $t_j^n \to z$ and $t_p^n \to w$ as $n \to \infty$, it follows

$$\int_z^w (\|\tilde u(\bar t)-u(\theta)\| - \|\tilde u(t)-u(\theta)\|)d\theta \leq \int_t^{\bar t} (\|\tilde u(\bar s)-u(z)\| - \|\tilde u(\bar s)-u(w)\|)d\bar s$$

$$+ \int_z^w \int_t^{\bar t} \omega\|\tilde u(\bar s)-u(\theta)\| d\bar s\, d\theta + \bar C \int_z^w \int_t^{\bar t} \|f(\bar s)-f(\theta)\| d\bar s\, d\theta \qquad (3.22)$$

for all $z,w,t,\bar t \in [s,T]$ with $z \leq w$, $t \leq \bar t$.

Set

$$g(t,\tau) = \|\tilde u(t)-u(\tau)\| \qquad (3.23)$$

Of course, g is merely continuous, so it may not be differentiable. However, we may assume (without loss of generality) that g is even differentiable. This is because the usual regularization g_n of g satisfies also (3.22). Substituting g into (3.22), dividing by $\bar t-t$ and then letting $\bar t \downarrow t$ one obtains

$$\int_z^w \frac{\partial g}{\partial t}(t,\theta)d\theta + g(t,w)-g(t,z) \leq \omega\int_z^w g(t,\theta)d\theta + \bar C \int_z^w \|f(t)-f(\theta)\| d\theta \qquad (3.23)'$$

Dividing now by w-z and letting $w \downarrow z$ we get

$$\frac{\partial g}{\partial t}(t,z) + \frac{\partial g}{\partial z}(t,z) \leq \omega\, g(t,z) + \bar C\|f(t)-f(z)\| \qquad (3.24)$$

Finally, for t=z, (3.24) yields $\frac{d}{dt}g(t,t) \leq \omega g(t,t)$, i.e.

$$\|\tilde u(t)-u(t)\| \leq \|\tilde u(s)-u(s)\| \exp(\omega(t-s)) \qquad (3.25)$$

for $s \leq t \leq T$. Since $\tilde u(s) = u(s) = x_o$ it follows the desired equality $\tilde u = u$ on $[s,T]$. It remains to prove the part (3).

Take an arbitrary $x^* \in F(u(t_o)-x)$. We have

$$x^*(u(\bar{t})-u(t_o)) = x^*(u(\bar{t})-x)-x^*(u(t_o)-x) \leq$$
$$\|u(t_o)-x\| \; \|u(\bar{t})-x\| - \|u(t_o)-x\|^2 \tag{3.26}$$

Since the DS-limit solution u is an integral solution, it satisfies (3.13) with $t=t_o \in [s,T[$. A simple combination of (3.26) and (3.13) implies (for $\bar{t} \downarrow t_o$)

$$x^*(u'(t_o)) \leq \|u(t_o)-x\|(\omega\|u(t_o)-x\| + \langle y,u(t_o)-x\rangle_+ + C\|f(t_o)-f(r)\|) \tag{3.27}$$

for all $r\in[s,T[$ and $[x,y]\in A(r)$. In particular, for $r=t_o$, (3.27) in conjunction with (2.6)-(2.10) give

$$\langle u'(t_o)-y,u(t_o)-x\rangle_i \leq \omega\|u(t_o)-x\|^2, \quad [x,y]\in A(t_o) \tag{3.28}$$

In view of maximal dissipativity of $A(t_o) - \omega I$ in $\overline{D(A(t_o))}$,(3.28)implies (3.14). The proof is complete.

<u>Remark 3.1.</u> The proof above of the uniqueness of integral solution shows that any integral solution to (1.1)+(1.1') is a DS-limit solution. This remark and Theorem 3.2 prove that the notion of "integral solution" and "DS-limit solution" to (1.1)+(1.1)' are equivalent (under the hypotheses (H(2.1)) andH(2.2).) Consequently, any strong solution to the problem (1.1)+(1.1)' is a DS-limit solution.

<u>Remark 3.2.</u> The inequality (3.13) (which defines Integral Solutions) is equivalent to

$$\|u(\bar{t})-x\|^2 - \|u(t)-x\|^2 \leq 2 \int_t^{\bar{t}} (\omega\|u(\bar{s})-x\|^2 + \langle y,u(\bar{s})-x\rangle_s +$$
$$+ C\|u(\bar{s})-x\| \; \|f(\bar{s})-f(r)\|)d\bar{s} \tag{3.29}$$

Formally, this fact can be seen by multiplying (3.12) by $\|u(t)-x\|$ and then integrating over $[t,\bar{t}]$. Precisely, (3.16) and (3.16)' imply (3.29) with the same proof as for (3.19).

3.3 Evolution Operator

In order to study in a simple and systematic manner the main properties of various type of solutions e.g. the DS-limit solution $u=u(t;s,x_o)$ to the problem (1.1)+(1.1)', one associates with this problem an evolution operator defined by (1.8). Actually, the notion of evolution operator can be defined independently of differential equations, as follows.

Let $D=\{D(t);\ D(t)\subset X,\ o\le t\le T\}$ be a family of nonempty subsets of the Banach space X.

Definintion 3.3. A family $U = \{U(t,s);\ o\le s\le t\le T\}$ of (possible nonlinear) operators (or in short $U(t,s)$) is said to be an evolution operator if it satisfies the following conditions, with respect to D:

1°) For each pair (s,t) with $o\le s\le t\le T$, $U(t,s):D(s)\to D(t)$.
For each $s\in[o,T]$, $U(s,s)=I$ (the identity operators on $D(s)$).

2°) $U(t,s)U(s,r)= U(t,r)$, for $o\le r\le s\le t\le T$.

3°) For each $s\in[o,T]$ and $x\in D(s)$, the function $t\to U(t,s)x$ $u(t;s,x)$ is continuous on $[s,T]$.

If we want U to have additional properties we have to impose additional conditions on U. Let us define the continuity of $(t,s,x)\to U(t,s)x$.

4°) If $o\le s_n\le t_n\le T$, $x_n\in D(s_n)$ and $s_n\downarrow s$, $t_n\to t$, $x_n\to x$ and if $x\in D(s)$, then $U(t_n,s_n)x_n\to U(t,s)x$ (when $n\to\infty$).

A condition which implies 4°) is the following one:

5°) There exists a real number $\omega\in R$ and a function F (whose properties are listed below) such that

$$\|U(t+s,s)x-U(t+r,r)y\| \le e^{\omega t}\|x-y\| + \int_o^t e^{\omega(t-\tau)}F(\tau+s,\tau+r)d\tau \tag{3.30}$$

for $o\le r\le s\le T$, $o\le t\le T-s$, $x\in D(s)$, $y\in D(r)$ where $F:[o,T]\times[o,T]\to R$ is a continuous function, with $F(s,s) = o$, $\forall s\in[o,T]$ and $F(t,s) = F(s,t)$, $F(t,s)\le F(t,r)+F(s,r)$, $o\le r\le s\le t\le T$.

Note that we can relax the continuity of F as below

$$F(t,s)\le\|f_1(t-s)\| + \|f(t)-f(s)\|,\quad s,t\in[o,T] \tag{3.31}$$

with f_1, $f\in L^1(o,T;X)$ (a.s.o. see Iwamiya, Oharu and Takahashi[1]).

In this chapter, the conditions that are imposed to (1.1) lead to an evolution operator satisfying 1°), 2°), 3°) and 5°) (consequently 4°) too, since we will prove that 5°) implies 4°)).

Proposition 3.1. Let U satisfy the conditions 1°), 2°) and 5°). Then the following properties hold

(1) $$\|U(t+s,r)x-U(t+r,r)y\| \le e^{\omega t}\|U(s,r)x-y\| + \int_0^t e^{\omega(t-\tau)}F(\tau+s,\tau+r)d\tau$$

for $o \le r \le s \le T$, $o \le t \le T-s$, $x,y \in D(r)$

(2) $$\|U(\bar{t},s)x-U(\bar{t},s)y\| \le e^{\omega(\bar{t}-s)}\|x-y\|, \quad o \le s \le \bar{t} \le T, x,y \in D(s)$$

(3) $$\|U(t,s)x-U(t,r)y\| \le e^{\omega(t-s)}\|x-U(s,r)y\|,$$

$$o \le r \le s \le t \le T, \ x \in D(s), \ y \in D(r)$$

(4) $$\|U(\bar{t},s)x-U(t,s)x\| \le e^{\omega(t-s)}\|U(\bar{t}-t+s,s)x-x\| + \int_s^t e^{\omega(t-\tau)}F(\tau+\bar{t}-t,\tau)d\tau, \quad s \le t \le \bar{t} \le T, \ x \in D(s)$$

(5) If $o \le r \le t \le T$ and $2s-r \le T$, then

$$\|U(t,s)x-U(t,r)y\| \le e^{\omega(t-r)}\|x-y\| + e^{\omega(t-s)}\|U(2s-r,s)x-x\| + \int_r^s e^{\omega(t-\tau)}F(\tau+s-r,\tau)d\tau$$

Proof. If in (3.30) we replace x by $U(s,r)x$ we get (1). For $s=r$ and $t+s=\bar{t}$, (3.30) yields (2). Combining (2) and 2°) one obtains (3). To get (4) we write $\bar{t}=t-s+(\bar{t}-t+s)$, $t=(t-s)+s$ and then use (1) with $t-s$, $\bar{t}-t+s$ and s in place of t, s and r respectively. We now prove (5). According to (3), it remains to estimate $\|x-U(s,r)y\|$.

Clearly

$$\|x-U(s,r)y\| \le \|x-U(2s-r,s)x\| + \|U(s-r+s,s)x-U(s-r+r,r)y\| \quad (3.32)$$

Here, the second term in the right-hand side of (3.32) can be estimated on the basis of (3.30) with $s-r$ in place of t. In such a manner, the inequalities (3), (3.32) and (3.30) imply (5).

Corollary 3.1. Let U satisfy the conditions 1°), 2°), 3°) and 5°). Then U has the property 4°) (i.e. $(t,s,x) \to U(t,s)x$ is a continuous function).

Proof. Obviously,

$$\|U(t_n,s_n)x_n-U(t,s)x\| \le \|U(t_n,s_n)x_n-U(t_n,s)x\|+\|U(t_n,s)x-U(t,s)x\| \tag{3.33}$$

In view of Proposition 3.1

$$\|U(t_n,s_n)x_n-U(t_n,s)x\| \le e^{\omega(t_n-s_n)}\|x_n-U(s_n,s)x\| \tag{3.34}$$

Since $t \to U(t,s)x$ is continuous, (3.33) and (3.43) imply 4°). The result below gives a necessary and sufficient condition for $t \to U(t,s)x$ to be Lipschitz continuous. To this aim we introduce the subset $\check{D}(s)$ and the number $V_o^T \equiv d\text{-}V_o^T(F)$, as follows

$$\check{D}(s) = \{x \in D(s), L(s,x) \equiv \liminf_{h \downarrow o} h^{-1} \|U(h+s,s)x-x\| < \infty\} \tag{3.35}$$

for $o \le s \le T$.

We say that $(t,s) \to F(t,s)$ is "diagonally of bounded variation" on $[o,T] \times [o,T]$ if

$$V_o^T \equiv d\text{-}V_o^T(F) \equiv \sup \sum_{k=1}^{n} F(t_k,t_{k-1}) < +\infty \tag{3.36}$$

where the supremum above is taken over all partitions $o < t_o < t_1 < \dots < t_n = T$ of $[s,T]$. This supremum is said to be the diagonal variation of F on $[o,T] \times [o,T]$ and it is denoted by $d\text{-}V_o^T(F)$ or simply by V_o^T.

Theorem 3.3. Let U satisfy Conditions 1°), 2°), 3°) and 5°). If F is diagonally of bounded variation on $[o,T] \times [o,T]$, then a necessary and sufficient condition for $t \to U(t,s)x$ ($s \in [o,T[$, $x \in D(s)$) to be Lipschitz continuous on $[s,T]$ is that $x \in \check{D}(s)$.

Precisely, if $x \in \check{D}(s)$, then

$$\|U(\bar{t}+s,s)x-U(t+s,s)x\| \le \omega^{-1}(e^{\omega\bar{t}}-e^{\omega t})(L(s,x)+V_s^T) \tag{3.37}$$

for $o \le s \le t \le \bar{t} \le T$, $\omega > o$

Proof. The necessity is obvious. The proof of Sufficiency follows a standard device, namely. Let $s \in [o,T]$, $x \in \check{D}(s)$ and $o < t < \bar{t} < T-s$. Fix $h > o$ sufficiently small and denote by $n=n(t,h)$, $m=m(\bar{t},h)$ the greatest integers in t/h and $\bar{t}/h$ respectively (i.e. $nh \leq t \leq (n+1)h$, $mh \leq \bar{t} \leq (m+1)h$).

For $n \leq k \leq m-1$, (3.30) yields

$$\|U((k+1)h+s,s)x-U(kh+s,s)x\| = \|U(kh+h+s,h+s)U(h+s,s)x - U(kh+s,s)x\|$$
$$\leq e^{\omega kh}\|U(h+s,s)x-x\| + e^{\omega kh}\int_0^{kh} e^{-\tau}F(\tau+h+s,\tau+s)d\tau \tag{3.38}$$

Let us observe that

$$\int_0^{kh} e^{-\tau}F(\tau+h+s,\ \tau+s)d\tau \leq \sum_{j=1}^{k}\int_0^{h} F(jh+\tau+s,(j-1)h+\tau+s)d\tau$$
$$\leq \int_0^{h} V_s^T\, d\tau = h\, V_s^T \tag{3.39}$$

Substituting (3.39) into (3.38) and iterating for $k=n, n+1,\ldots,m-1$, we derive

$$\|U(mh+s,s)x-U(nh+s,s)x\| \leq \frac{e^{\omega mh}-e^{\omega nh}}{e^{\omega h}-1}\left(\|U(h+s,s)x-x\|+h\, V_s^T\right) \tag{3.40}$$

Since $(e^{\omega h}-1)/h \to \omega$, $mh \to \bar{t}$, $nh \to t$ as $h \downarrow o$, taking the inferior limit (lim inf) of both sides of the inequality (3.44) one obtains (3.37). The proof is complete.

Remark 3.3. Note that for the Lipschitz continuity of $t \to U(t,s)$ "at $t=s$", i.e. for

$$\|U(t+s,s)x-x\| \leq \tilde{L}t,\ o \leq s \leq t \leq T,\ x \in D(s) \tag{3.41}$$

with some $\tilde{L} > o$, the function F in (3.30) need not to be diagonally of bounded variation (as in Theorem 3.3 above!). Precisely, the result below holds

Theorem 3.4. Let U satisfy Conditions 1°), 2°), 3°) and 5°). Then the inequality (3.41) holds if and only if $x \in \check{D}(s)$.

Proof. Clearly, (3.41) implies $x \in \check{D}(s)$. Conversely, we will prove that if $x \in \check{D}(s)$ then for all $o \le t \le T-s$,

$$\|U(t+s,s)x-x\| \le \omega^{-1}(e^{\omega t}-1)L(s,x)+e^{\omega t}\int_0^t e^{-\omega\tau}F(\tau+s,s)d\tau \qquad (3.42)$$

To this goal, let $h>o$ be sufficiently small and let n be as in the previous proof, i.e. $nh \le t \le nh+h$. Set

$$a_p = \|U(ph+s,s)x-x\|.$$

We have

$$U((k+1)h+s,s)x-x=U(h+kh+s,kh+s)U(kh+s,s)x-U(h+s,s)x+U(h+s,s)x-x,$$

$$o \le k \le n-1$$

Consequently, on the basis of (3.30),

$$a_{k+1} \le e^{\omega h}\|U(kh+s,s)x-x\| + \int_0^h e^{\omega(h-\tau)}F(\tau+kh+s,\tau+s)d\tau + \|U(h+s,s)x-x\| \qquad (3.43)$$

Let us use the property of F below

$$F(\tau+kh+s,\tau+s) \le F(\tau+kh+s,s) + F(s,\tau+s)$$

By a change of variable ($\tau+kh=\theta$), we see that

$$\int_0^h e^{\omega(h-\tau)}F(\tau+kh+s,s)d\tau = e^{\omega(k+1)h}\int_{kh}^{(k+1)h} e^{-\omega\theta}F(\theta+s,s)d\theta$$

Set also

$$g(h)=\|U(h+s)x-x\|+\int_0^h e^{\omega(h-\tau)}F(\tau+s,s)d\tau$$

It is now clear that (3.43) yields

$$a_{k+1} \le e^{\omega h}a_k+e^{\omega(k+1)h}\int_{kh}^{(k+1)h} e^{-\omega\theta}F(\theta+s,s)d\theta \quad +g(h)$$

Iterating this inequality for $k=o,1,\dots,n-1$ (with $a_o=o$) we get

$$a_n = \|U(nh+s,s)x-x\| \leq \sum_{k=o}^{n-1} e^{\omega kh} g(h)+e^{\omega nh} \int_o^{nh} e^{-\omega\theta} F(\theta+s,s)d\theta$$

$$= \frac{e^{\omega nh}-1}{e^{\omega h}-1} g(h)+e^{\omega nh} \int_o^{nh} e^{-\omega\theta}F(\theta+s,s)d\theta \qquad (3.44$$

Taking into account that $\liminf_{h\downarrow o} h^{-1}g(h) = L(s,x)$ and $nh \to t$ as $h \downarrow o$, (3.44) implies (3.42), which completes the proof.

3.4. Generation of evolution operator.

The main problem now is to give various types of sufficient conditions in order for the family $A(t)$ in (1.1) to generate via (1.8) an evolution operator as in Section 3.3.

We will prove that H(2.1) and H(2.2) in § 2 as well as other standard conditions are such a type of sufficient conditions. Moreover, in Chapter 3, § 2 we will see that such conditions are verified in the cas of a class of PDE arising in the study of long waves of small amplitud

A main result of this section if given by

Theorem 3.5. Suppose that the family $\{A(t),\ o \leq t \leq T\}$ of (possible multivalued) operators satisfies tangential condition (1.1) (or Range condition (1.10)), dissipativity condition (H(2.1)) and the t-dependenc domain condition H(2.2). Then for each $T_1 \in\ ]o,T[$ each $s \in [o,T_1]$ and $x_o \in \overline{D(A(s))}$, the problem (1.1)+(1.1)' admits DS-approximate solutions o $[o,T_1]$ in the sense of (1.2)-(1.7)). Moreover, the family $U = \{U(t,s);\ o \leq s \leq t \leq T\}$ associated with $A(t)$ via (1.8) is an evolution operator in the sense of 1°)-5°) in Definition 3.3 with $D(s)=\overline{D(A(s))}$ and $F(t,s)=\|f(t)-f(s)\|$, where f is the function in H(2.1), and C is defined in (3.12)).

Proof. We first prove that $U(t,s)$ associated with $A(t)$ via (1.8) satisfies 1°)-5°) in Definition 3.3, with $D(t)=\overline{D(A(t))}$, $t \in [o,T]$. Actually, 1°) and 3°) are already known (Theorem 3.1). Moreover, the property 3°) follows from the uniqueness of integral solution to (1.1)-(1.1)' and Theorem 3.2 (2). Consequently, it remains to prove that $U(t,s)$ defined by (1.8) satisfies 5°). To this aim let $x \in \overline{D(A(s))}$, $y \in \overline{D(A(r))}$ and $s,r \in [o,T]$ with $o \leq r \leq s \leq T$. Set $u(t;s,x) = U(t,s)x$.

Clearly, we can use (3.22) with

$$\tilde{u}(\bar{t}) = U(\bar{t},s)x, \quad u(t)=U(t,r)y \tag{3.45}$$

and

$$\bar{C} = \max \{L(\sup_{s \le t \le T} \|\tilde{u}(t)\|),\ L(\sup_{s \le t \le T} \|u(t)\|\ \}$$

Consequently, (3.24) holds with

$$g(\tilde{t},\tilde{s}) = \|U(\tilde{t},s)x-U(\tilde{s},r)y\| \tag{3.46}$$

With t+s and z+r in place of t and z respectively, (3.24) becomes

$$\frac{\partial g}{\partial t}(t+s,z+r) + \frac{\partial g}{\partial z}(t+s,z+r) \le \omega g(t+s,z+r) + \bar{C}\|f(t+s)-f(z+r)\| \tag{3.47}$$

since the left-hand side of (3.47) (for t=z) is just $\frac{\partial}{\partial t}$ g(t+s,t+r), it follows

$$g(t+s,t+r) \le e^{\omega t} g(s,r)+\bar{C}\int_o^t e^{\omega(t-\tau)}\|f(\tau+s)-f(\tau+r)\|d\tau$$

That is (3.30) with $F(t,s) = \bar{C}\|f(t)-f(s)\|$.

3.5. Existence of DS-approximate solutions.

In this direction, Tangential condition (1.11) plays an essential role (although it has nothing to do with the convergence of these solutions). Let $s \in [o,T[$, $x_o \in \overline{D(A(s))}$ and n a positive integer. We will prove that there is an approximate solution u_n in the sense of (1.2)-(1.7) with $d_n \le \frac{1}{n}$ and $\|p_k^n\| \le \frac{1}{n}$, $k=1,2,\dots N_n$. Of course, we may assume $\omega > o$ (otherwise, one considers $\omega_o = \max\{o,\omega\}$) and $2\omega n^{-1} \le 1$. For each $x \in \overline{D(A(t))}$ set $\delta^n(x,t) = \sup\{h; o < h \le \frac{1}{n}, \exists [x_h,y_h] \in A(t+h)$, such that

$$\|x_h hy_h - x\| \le hn^{-1}\} \tag{3.47'}$$

On the basis of (1.11), $\delta^n(x,t) > o$. Choose $x_o^n \in D(A(s))$ such that $x_o^n \to x_o$ as $n \to \infty$. Set $t_o^n = s$ and inductively define $h_k^n \in]2^{-1}\delta^n(x_{k-1}^n, t_{k-1}^n), n^{-1}]$, $[x_k^n, y_k^n] \in A(t_{k-1}^n + h_k^n)$ with the properties

$$t_k^n = t_{k-1}^n + h_k^n,\ x_{k-1}^n \in D(A(t_{k-1}^n)),\ \|x_k^n - h_k^n y_k^n - x_{k-1}^n\| \le n^{-1} h_k^n \tag{3.48}$$

Denote

$$p_k^n = (x_k^n - x_{k-1}^n - h_k^n y_k^n)/h_k^n, \quad k=1,2,... \tag{3.49}$$

Consequently, $\|p_k^n\| \leq \frac{1}{n}$, (1.5) is satisfied and b_n defined by (1.7) is bounded from above by $(T-s)/s$ ($b_n \leq (T-s)/n$), hence $b_n \to o$ as $n \to \infty$. Obviously, two situations may occur:

a) either there exists a positive integer (say N_n) such that

$t^n_{N_n - 1} < T \leq t_{N_n}$, or

b) $t_k^n < T$, for all $k = o,1,...$

In the situation a) we have nothing to prove, while in the situation b) we have to prove that

$$\lim_{k \to \infty} t_k^n \equiv s + \sum_{k=1}^{\infty} h_k^n = T. \tag{3.50}$$

Suppose that b) holds. In order to prove (3.50) assume by contradiction that $\lim_{k\to\infty} t_k^n = t^* < T$. Since there is no danger of confusion, suppress the index n, i.e. $t_k^n \equiv t_k$, $x_k^n \equiv x_k$ a.s.o. We will prove that $\lim_{k\to\infty} x_k \equiv x^*$ exists. For this purpose set $a_{k,j} = \|x_k - x_j\|$. Replacing $\tilde{u} = x_o^n$, $r = t_o$ in (2.37)' (and $\omega = o$ for simplicity) we derive

$$a_{k,o} \leq (t_k - s)|A(t_o)x_o^n| + \sum_{i=1}^{k} h_i \|f(t_i) - f(t_o)\| L(\|x_o^n\|) +$$
$$+ \sum_{i=1}^{k} \|h_i\| \, \|p_i\| \tag{3.51}$$

First, this shows that there is a constant $K = K(x_o^n, t) > o$ such that $\|x_k\| \leq K$, $k=1,2,...$

We now claim that

$$a_{k,j} \leq (t_k - t_j)|A(t_q)x_q| + K_1 \sum_{m=q+1}^{k} h_m \|f(t_m) - f(t_q)\| +$$
$$K_1 \sum_{m=q+1}^{j} h_m \|f(t_m) - f(t_q)\| +$$
$$\sum_{m=q+1}^{k} h_m \|p_m\| + \sum_{m=q+1}^{j} h_m \|p_m\| \tag{3.52}$$

for all $0 \leq q < j < k$, with $K_1 \geq L(\|x_q\|)$, $q=0,1,\ldots$ To prove (3.52) we first need the general relationship between $a_{k,j}$, $a_{k-1,j}$ and $a_{k,j-1}$ below

$$(h_k+h_j)a_{k,j} \leq h_k a_{k,j-1} + h_j a_{k-1,j} + h_k h_j h_{k,j} \tag{3.53}$$

where $h_{k,j} = \|p_k\| + \|p_j\| + k_1\|f(t_k)-f(t_j)\|$.

Clearly, the proof of (3.53) is similar to that of (2.46). In view of (3.53), we can easily prove (3.52) by induction. Indeed, replacing (in (3.51)) $t_o=s$ and x_o^n by t_q and x_q respectively we see that $a_{k,q}$ satisfies (3.52) for any $k > q$. Suppose that $a_{k,j-1}$ and $a_{k-1,j}$ satisfy (3.52) and then, on the basis of (3.53), one can check that (3.52) holds. This elementary aspects are left to the reader (One uses $h_{k,j} \leq \|p_k\| + \|p_j\| + k_1\|f(t_k)-f(t_q)\| + k_1\|f(t_j)-f(t_q)\|$).

Since $\|p_m\| \leq \frac{1}{n}$, $m=1,2,\ldots$ (3.52) yields

$$\lim_{k,j\to\infty} \|x_k-x_j\| \leq 2K_1 \int_{t_q}^{t^*} \|f(t)-f(t_q)\|\, dt+2n^{-1}(t^*-t_q) \tag{3.54}$$

for every $q \geq 1$.

Letting $q \to \infty$ (i.e. $t_q \uparrow t^*$), (3.54) shows that $\lim_{k,j\to\infty} \|x_k-x_j\|=0$, that is $\lim_{\to\infty} x_k^n \equiv x^*$ exists. Inasmuch as $x_k \in D(A(t_k))$, by Hypothesis H(2.2), $x^* \in \overline{D(A(t^*))}$. Consequently, Tangential condition (1.11) can be used with $t=t^*$ and $x=x^*$. This provides $h^* \in]0, \frac{1}{n}]$ and $[\bar{x},\bar{y}] \in A(t^*+h^*)$ such that $t^*+h^* < T$ and

$$\|\bar{x}-h^*\,\bar{y}-x^*\| \leq \frac{h^*}{2n} \tag{3.55}$$

On the other hand, $t_k \uparrow t^*$ implies $h_k \downarrow 0$ as $k \to \infty$ and therefore $\delta(x_k,t_k) \leq 2h_k \to 0$ as $k \to \infty$. Let k_o be such that

$$\delta(x_k,t_k) < h^*,\ \forall\ k \geq k_o$$

Set $h_k^* = t^*+h^*-t_k$ (hence $h_k^* > h^*$, for all k).

Thus, we have $h_k^* \downarrow h^*$ and $h_k^* > h^* > \delta(x_k,t_k)$ for all $k \geq k_o$. Taking into account the definition of $\delta(x_k,t_k)$ (i.e. (3.47)) it follows

$$\|\tilde{x}_k-h^*_k\tilde{y}_k-x_k\| > n^{-1}h_k^*\ ,\quad [\tilde{x}_k,\tilde{y}_k] \in A(t_k+h_k^*) \tag{3.56}$$

for all $k \geq k_o$.

But $t_k+h_k^* = t^*+h^*$, hence $A(t_k+h_k^*) = A(t^*+h^*)$ so we may replace in (3.56) $\tilde{x}_k=\bar{x}$ and $\tilde{y}_k=\bar{y}$. Letting then $k \to \infty$ in (3.56) we get $\|\tilde{x}-h^* \bar{y}-x^*\| \geq n^{-1}h^*$, which contradicts (3.55).

Consequently, $\lim_{k\to\infty} t_k^n=T$ and $\lim_{k\to\infty} x_k^n \equiv x^*(n)$ exists. In this case, define u_n as indicated by (1.2) with $u_n(T) = \lim_{k\to\infty} x_k^n \equiv x^*(n)$. This completes the proof.

Remark 3.4. If we want to avoid the situation $N_n = +\infty$ in the definition of the DS-limit solution u_n (see (1.2)-(1.7)) we can proceed as follows:

Take an arbitrary $T_1 \in]s,T[$. Since $\lim_{k\to\infty} t_k^n = T$, there is a positive integer N_n such that $t^n_{N_n-1} < T_1 \leq t^n_{N_n}$. Define u_n as indicated by (1.2)-(1.7) with T_1 in place of T. In such a manner, $\lim_{n\to\infty} u_n(t)=u(t)$ is obtained for $t\in[s,T[$. On the other hand, since u is uniformly continuous on $[s,T]$ (in view of (3.5)), u can be uniquely extended to $[s,T]$ as a continuous function.

3.6. The case $A(t)$ m-dissipative

An important particular case of the range condition (1.10) is the following

$$R(I-A(t)) = X, \qquad t \in [o,T] \tag{3.57}$$

If $A(t)$ is dissipative, then (3.57) is equivalent to

$$R(I-hA(t)) = X, \ \forall t\in[o,T], \ \forall h> o \tag{3.57'}$$

(see Appendix, § 2 or author's book [15, p.23]). Recall that $A(t)$ is said to be m-dissipative it if is dissipative and (3.57)' holds.

Corollary 3.2. 1) Let $A(t)$ be m-dissipative for each $t\in[o,T]$. Then for each $s\in[o,T]$, and $x_o \in D(A(s))$, the problem (1.1)+(1.1)' admits a DS-approximate solution in the sense of (1.2)-(1.7), with $p_k^n = o$.

2) If in addition $\{A(t),\ o \leq t \leq T\}$ satisfies H(2.1) and H(2.2), then the evolutor $U(t,s)$ associated with $A(t)$ is given by

$$U(t,s)x_o = \lim_{n\to\infty} \prod_{k=1}^{n} (I - \frac{t-s}{n} A(s+k\frac{t-s}{n}))^{-1} x_o \tag{3.58}$$

Proof. 1) Fix an arbitrary $t \in]s,T]$. Let n be a positive integer. Define the following partition of $[s,T]$: $t_o^n = s$, $t_k^n = s+k\frac{t-s}{n}$ (hence $t_k^n - t_{k-1}^n = \frac{t-s}{n} = h_k^n$). For $x_o \in \overline{D(A(s))}$, choose $x_o^n \in D(A(s))$ such that $x_o^n \to x_o$

There exists a unique $x_1 \in D(A(t_1))$ such that

$$x_o^n \in x_1^n - \frac{t-s}{n} A(t_1^n)x_1^n \text{ , that is}$$

$$x_1^n = (I - \frac{t-s}{n} A(s + \frac{t-s}{n}))^{-1} x_o^n,$$

$$x_2^n = (I - \frac{t-s}{n} A(s+2\frac{t-s}{n}))^{-1} x_1^n$$

and in general

$$x_k^n = (I - \frac{t-s}{n} A(s+k\frac{t-s}{n}))^{-1} x_{k-1}^n, \quad k=1,2,\ldots,N_n \tag{3.59}$$

where $N_n = N_n(t,s)$ is the first positive integer such that $s+N_n\frac{t-s}{n} \geq T$. Clearly, x_k^n can be expressed in terms of x_o^n as below

$$x_k^n = \prod_{j=1}^{k} (I- \frac{t-s}{n} A(s+j\frac{t-s}{n}))^{-1} x_o^n, \quad k=1,\ldots,N_n \tag{3.60}$$

Obviously, (3.59) is equivalent to $x_k^n - x_{k-1}^n \in h_k^n A(t_k^n)x_k^n$, therefore (1.5) holds with $p_k^n = o$. In this case, the corresponding DS-approximate solution u_n is given by

$$u_n(\bar{t}) = \prod_{j=1}^{k} (I- \frac{t-s}{n} A(s+j\frac{t-s}{n}))^{-1} x_o^n, \quad s+(k-1)\frac{t-s}{n} < \bar{t} \leq s+k\frac{t-s}{n} \tag{3.61}$$

and $u_n(s)=x_o^n$. But $t=s+ n\frac{t-s}{n} = t_n^n$.
Consequently,

$$u_n(t) = \prod_{j=1}^{n} (I - \frac{t-s}{n} A(s+j\frac{t-s}{n}))^{-1} x_o^n \tag{3.62}$$

2) In view of Theorems 3.5 and 3.1

$$U(\bar{t},s)x_o = \lim_{n\to\infty} u_n(\bar{t}), \quad \bar{t} \in [s,T] .$$

In particular, for $\bar{t} = t$, (3.62) implies (3.58). Theorem 3.1 is involved here in the sense that $U(\bar{t},s)x_o$ is the uniform limit of any DS-approximate solution $u_n = u_n(\bar{t})$ (and in particular of u_n given by (3.61)). The proof is complete.

Remark 3.5. It is clear that Corollary 3.2 and (3.59)'-(3.62) remain valid if the condition (3.57) is weakened to (replaced by)

$$\overline{D(A(t))}=\bar{D} \text{ is independent of } t \in [o,T] \tag{3.62}$$

$$R(I-\lambda A(t)) \supset \bar{D}, \text{ for all } t \in [o.T] \text{ and } o < \lambda < \lambda_o \tag{3.62}$$

with $\lambda_o > o$ small enough (e.g. $\lambda_o\ |\omega| < 1/2$).
Actually, in § 4 (see (4.21)') one proves that m-dissipativity of A(t) and (2.11) imply that $\overline{D(A(t))}$ is necessarily independent of t.

3.7. Evolution operator and strong solutions

In order to discuss the relationship of the evolution operator U generated by $\{A(t)\}$ and the Cauchy problem (1.1)+(1.1)' we need the set

$$\check{D}(A(s)) = \{x\in\overline{D(A(s))};\ \liminf_{h\downarrow o} h^{-1}\ \|U(h+s,s)x-x\| \equiv L(s,x) < \infty\} \tag{3.63}$$

for $s \in [o,T[$ (see (3.35)).
On the basis of Theorems 3.3, 3.4 and 3.5 it follows

Corollary 3.3. Suppose that $\{A(t), o \le t \le T\}$ satisfies tangential condition (1.11) as well as COnditions H(2.1) and H(2.2) and f is of bounded variation on $[o,T]$. Then

$$\check{D}(A(s) = \{x \in \overline{D(A(s))};\ t\to U(t,s)x \text{ is Lipschitz continuous on } [s,T]\} \tag{3.64}$$

Other characterizations of $\check{D}(A(s))$ are given in Chapter 2, § 4. According to Theorem 3.1 (see (3.5)') (and of course, to (3.63) and (3.64)), we have

$$D(A(s)) \subset \check{D}(A(s)) \subset \overline{D(A(s))},\quad o \le s \le T \tag{3.65}$$

A situation in which (3.65) holds with strict inclusion is also given in Chapter 2, Example 4.1.

If X is reflexive, then for every $x_o \in \check{D}(A(s))$, the function $t \to U(t,s)x_o$ is differentiable a.e. on $]s,T[$. This is because every absolutely continous function $u:[o,T] \to X$, with X reflexive, is a.e. differentiable (in the strong sense). Consequently, by Theorem 3.2 (3), if X is reflexive, f is of bounded variation and $x_o \in \check{D}(A(s))$, then $u=u(t;s,x_o)$ is the strong solution to (1.1)+(1.1)' if and only if $u(t;s,x_o) = U(t,s)x_o$.

In other words, in this case there are no strong solutions except for those given by the evolution operator associated with $\{A(t)\}$ via DS-approximate solutions. Finally, we conclude this subsection with a theorem which lists the main results of this section (on the relationship of $U(t,s)$ and Cauchy problem (1.1)+(1.1)').

Theorem 3.6. Suppose that the family $\{A(t), o \leq t \leq T\}$ of (possible multivalued) operators acting in X, satisfies the conditions (1.11), H(2.1) and H(2.2). Then

1) For every $s \in [o,T[$ and $x_o \in \overline{D(A(s))}$ the Cauchy problem (1.1)+(1.1)' admits a DS-approximate solutions $u_n=u_n(t;s,x_o^n)$ (in the sense of (1.2)-(1.7)) which is uniformly convergent on $[s,T]$ to a continuous function $u=u(t;s,x_o) \in \overline{D(A(t))}$. Every such a DS-approximate solution $u_n(t;s,x_o^n)$ with $\tilde{x}_o^n \to x_o$, $\tilde{x}_o^n \in D(A(s)))$ is uniformly convergent to the same $u=u(t;s,x_o)$ which is the unique integral solution of (1.1)+(1.1)'.

2) The family $U = \{U(t,s);U(t,s):\overline{D(A(s))} \to \overline{D(A(t))}, o \leq s \leq t \leq T\}$ defined by $U(t,s)x_o = u(t;s,x_o)$ is an evolution operator (in the sense of 1°-5°) in Definition 3.3.

One has

$$D(A(t)) \subset \check{D}(A(t)) \subset \overline{D(A(t))}, \quad o \leq s \leq t \leq T$$

3) If $u=u(t;s,x_o)$ is the strong solution of (1.1)+(1.1)', then $u(t;s,x_o) = U(t,s)x_o$, $s \leq t \leq T$.

4) If $t \to U(t,s)x_o$ is strongly (or only weakly) right differentiable at $t_o \in [s,T[$ and $A(t_o) - \omega I$ is maximal dissipative, then $U(t_o,s)x_o \in D(A(t_o))$ and

$$(d^+/dt)U(t,s)x_o\Big|_{t=t_o} \in A(t_o)U(t_o,s)x_o$$

5) If X is reflexive, f is of bounded variation and $x_o \in D(A(s))$, then $u(t)=U(t,s)x_o$ is the unique strong solution to (1.1)+(1.1)'.

6) In particular, if $A(t)$ is m-dissipative, then $U(t,s)x_o$ is given by (3.58). If X is reflexive, f is bounded variation and $x_o \in D(A(s))$ then the function $u=u(t;s,x_o)$ is the strong solution to (1.1)+(1.1)' if and only if $u(t;s,x_o)$ is just the evolution operator $U(t,s)x_o$ given by (3.58).

§ 4. Other properties of evolution operators

4.1. The estimate of the difference of two integral solutions

Let us consider the DS-limit (integral) solutions to the problems

$$u'(t) \in A(t)u(t)+g_1(t), \quad o \le t \le T \qquad u(s) = x_o \in \overline{D(A(s))} \qquad (I.C)_1 \tag{4.1}$$

and

$$v'(t) \in A(t)v(t)+g_2(t) \qquad v(s)=y_o \in \overline{D(A(s))} \qquad (I.C)_2 \tag{4.2}$$

The result below provides an estimate for $\|u(t)-v(t)\|$.

Theorem 4.1. Assume that $\{A(t),\ o \le t \le T\}$ satisfies Hypotheses H(2.1) and H(2.2) in § 2 and that for each $t \in [o,T]$, $A(t)$ is m-dissipative. Moreover, assume that $g_i:[o,T] \to X$ (X a real Banach space) are continuous (i=1,2). Let $u=u(t;s,x_o)$ and $v=v(t;s,y_o)$ be the DS-limit solutions to the initial value problems (4.1) and (4.2) respectively. Then

$$\|u(t)-v(t)\| \le \|u(\tau)-v(\tau)\| + \int_\tau^t \|g_1(\theta)-g_2(\theta)\|d\theta \tag{4.3}$$

for all $o \le s \le \tau \le t \le T$.

Proof. Let $\{t_k^n\}$ be a partition of $[s,T]$

$$\{ s= t_o^n < t_1^n < \dots < t_n = T \}, \qquad n \ge 1 \tag{4.4}$$

with $h_k^n = t_k^n - t_{k-1}^n = \frac{T-s}{n}$ (for simplicity).

Since $A(t)$ is m-dissipative for each $t \in [s,T]$, there are sequences x_k^n and y_k^n satisfying the difference scheme (see § 3.6)

$$y_k^n = \frac{x_k^n - x_{k-1}^n}{h_k^n} - g_1(t_k^n) \in A(t_k^n)x_k^n \tag{4.5}$$

e.g. given x_o and t_1^n, there is x_1^n such that $x_1^n - h_1^n A(t_1^n)x_1^n \ni x_o + g_1(t_1^n)h_1^n$, i.e. $x_1^n = J_{h_1^n}(t_1^n)(x_o + g(t_1^n)h_1^n)$ and so on. Set,

$$B(t)x = A(t)x + g_1(t), \quad x \in D(A(t)), \quad t \in [o,T] \tag{4.6}$$

Then by (2.11) we have (with $\omega = o$, for simplicity)

$$\langle y_1 - y_2, x_1 - x_2 \rangle_i \le (\|f(t)-f(s)\| + \|g_1(t)-g_1(s)\|)L(\|x_2\|) \tag{4.7}$$

for $o \le s \le t \le T$, $[x_1, y_1] \in B(t)$, $[x_2, y_2] \in B(s)$. Therefore (in view of Theorem 3.1) the DS-approximate solution u_n given by

$$u_n(t) = x_k^n \quad \text{for } t_{k-1}^n < t \le t_k^n$$

and $u_n(t_o^n) = x_o^n$ (with $x_o^n \in D(A(s))$, $x_o^n \to x_o$) is convergent to the evolution operator $u(t) = U_B(t,s)x_o$ generated by $B(t) \equiv B$ (defined by (4.6)). Now let

$$\tilde{u}(t) = U(t,s)\tilde{x}_o, \quad \tilde{x}_o \in \overline{D(A(s))} \tag{4.8}$$

where $U(t,s)$ is the evolution operator generated by $A(t)$. Then $\tilde{u}$ satisfies (3.13) (i.e. $\tilde{u}$ is an integral solution to (1.1)). It is now clear that (3.20) holds with $g_1(t_k^n)$ in place of p_k^n. Then it is easily seen that (3.21) holds with $(\bar{t}-t)\int_{t_p^n}^{t_j^n} \|g^n(\theta)\| d$ in place of $(t_p^n - t_j^n)(\bar{t}-t)b_n$ where

$$g^n(\theta) = \begin{cases} g_1(t_o^n), & \text{if } \theta = t_o^n \equiv s \\ g_1(t_k^n), & \text{if } t_{k-1}^n < \theta \le t_k^n, \ k=1,2,\dots,n \end{cases} \tag{4.9}$$

In our case here, it follows that in the right-hand side of (3.22) we have to add $(\bar{t}-t)\int_z^w \|g_1(\theta)\| d\theta$. Consequently, in this case (3.24) becomes

$$\frac{\partial g}{\partial t}(t,z) + \frac{\partial g}{\partial z}(t,z) \le \text{Const}\|f(t)-f(z)\| + \|g_1(z)$$

which yields (for $t=z$) $\frac{d}{dt}(g(t,t)) \le \|g_1(t)\|$ for $s \le t \le T$. Integrating over $[\tau,t] \subset [s,t]$ we get

$$\|\tilde{u}(t) - u(t)\| \le \|\tilde{u}(\tau) - u(\tau) \quad + \int_\tau^t \|g_1(\theta)\| d\theta \tag{4.10}$$

for all $o \le s \le \tau \le t \le T$.

We conclude that if $A(t)$ satisfies a condition of the form (2.11) and B(t) is given by (4.6), then (4.10) holds, where $\tilde{u}(t)=U(t,s)\tilde{x}_o$ is the DS-limit solution to

$$\tilde{u}'(t) \in A(t)\tilde{u}(t), \qquad \tilde{u}(s)=\tilde{x}_o \in \overline{D(A(s))} \tag{4.11}$$

Therefore, with B(t) in place of A(t), and C(t) in place of B(t), where

$$C(t)x = B(t)x+f_2(t)-f_1(t) = A(t)x+f_2(t)$$

we have (4.10) with $g_1(t) = f_2(t)-f_1(t)$, and thus (4.3) is proved.

Remark 4.1. It is easy to see that the DS-limit solution $u(t)=U_B(t,s)x_o$ with B=B(t) given by (4.6) is the integral solution to (4.1) in the sense below (of (4.11)'). Moreover (4.3) remains valid in Conditions of Crandal-Pazy-Evans (§ 6)

$$\|u(\bar{t})-x\| - \|u(t)-x\| \leq \int_t^{\bar{t}} (< y+g_1(\tau),u(\tau)-x >_+ + C\|f(\tau)-f(r)\|)d\tau \tag{4.11)'}$$

for all $s \leq t \leq \bar{t} \leq T$, $r \in [s,T]$ and $[x,y] \in A(r)$ where $C=\max(L(\|x\|), L(\sup_{s \leq t \leq T} \|U_B(t,s)x_o\|))$.

Indeed, in view of (4.5) we have (3.15) with $p_k^n=g_1(t_k^n)$. Consequently (3.15) yields (similarly to (3.16))

$$x^*(x_k-x_{k-1}) \leq <y+g_1(t_k),x_k-x >_s + C\|x_k-x\| \, \|f(t_k)-f(r)\|$$

which implies (3.19) with $y+g_1(a_n(\tau))$ in place of y. Passing to the limit as $n \to \infty$ in (3.19) one obtains (4.11)'. In other words we have proved that the following result holds:

Theorem 4.1' Let A(t) be as in Theorem 4.1 and let $g: [o,T] \to X$ be continuous (or only $g \in L^1(o,T;X)$). Then the DS-limit solution $u: [o,T] \to X$ of the problem

$$u'(t) \in A(t)u(t)+g(t), \; u(s)=x_o \in \overline{D(A(s))},$$

is the unique integral solution of this problem, i.e. it satisfies: $u(s)=x_o$, $u(t) \in \overline{D(A(t))}$, $t \in [s,T]$,

$$\|u(\bar{t})-x\| - \|u(t)-x\| \leq \int_t^{\bar{t}} (< y+g(\tau), u(\tau)-x >_+ + C\|f(\tau)-f(r)\|)d\tau \tag{4.11)''}$$

for all $s \leq t \leq \bar{t} \leq T$, $r \in [s,T]$ and $[x,y] \in A(r)$ with $C=\max\{L(\|x\|), L(\sup_{s \leq t \leq T} \|u(t)\|)\}$.

4.2. The relationship between $U(s+h,s)x$ and $S_s(h)x$.

Let $s \in [o,T]$. Denote by $S_s(t)$ the semigroup genenrated by $A(s)$ (which is supposed to be m-dissipative), via the exponential formula of Crandall Liggett (see Ch. 2). If $U(t,s)$ is the evolution operator corresponding to $\{A(t),\ o \le t \le T\}$, then the following result holds.

Proposition 4.1. Let $\{A(t),\ o \le t \le T\}$ satisfy H(2.1) and H(2.2). Suppose in addition that for each $t \in [o,T]$, $A(t)$ is m-ddispative. Then for each $o \le s < T$,

$$\|U(s+h,s)x_o - S_s(h)x_o\| \le C \int_s^{s+h} \|f(t)-f(s)\|dt \qquad (4.12)$$

for all $x_o \in D(A(s))$ and $o < h < T-s$, where C is a constant independent of h (C depends on s and x_o).

Proof. According to Theorem 3.2, $u(t)=U(t,s)x_o$ is the unique integral solution to (4.11), that is it satisfies (3.13). Take r=s, and

$$y_k^n = \frac{x_k^n - x_{k-1}^n}{t_k^n - t_{k-1}^n} \in A(s)x_k^n, \quad x_o^n \equiv x_o \in D(A(s)) \qquad (4.13)$$

where $\{t_k^n\}$ can be chosen as in (4.4).

The existence of x_k^n is obviously guaranteed by m-dissipativity of A(s). Since $[x_k^n, y_k^n] \in A(s)$ it is clear that in this case (3.20) holds with $p_k^n = o$ and $f(t_k^n)$ replaced by $f(s)$, for all $k=1,2,\dots,n$. Consequently, (3.23)' holds with $f(s)$ in place of $f(\theta)$, which means that in (3.24) we have to replace now $f(z)$ by $f(s)$. We conclude that

$$g(t,\tau) = U(t,s)x_o - S_s(\tau-s)x_o \qquad (4.14)$$

(with $\tilde{u}(t)=U(t,s)x_o$, $u(\tau)=S_s(\tau-s)x_o$, $s \le \tau$) satisfies (3.22), (3.23)'and (3.24) with $f(s)$ in place of $f(\theta)$. Hence

$$\frac{d}{dt} g(t,t) \le C\|f(t)-f(s)\| \qquad (4.15)$$

Integrating over $[s,t]$, we get

$$g(t,t) \le g(s,s) + C \int_s^t \|f(\theta)-f(s)\|d\theta \qquad (4.16)$$

With t=s+h, (4.16) is just (4.12) and the proof is complete. From (4.12) it follows

Corollary 4.1. 1)Let Conditions of Proposition 4.1 be fulfilled. Then

$$\lim_{h\downarrow o} h^{-1}\|U(s+h,s)x_o-S_s(h)x_o\| = o \tag{4.17}$$

for each $s\in[o,T]$ and $x_o\in\overline{D(A(s))}$.

2) Consequently,

$$\limsup_{h\downarrow o} \|Us+h,s)x-x\|h^{-1}= \limsup_{h\downarrow o} \|S_s(h)x-x\|h^{-1} \tag{4.18}$$

for $s\in[o,T]$ and $x\in\overline{D(A(s))}$.

Remark 4.2. If $\{A(t),\ o\le t\le T\}$ satisfies Hypotheses H(2.1) and H(2.2), and if for each t, A(t) is m-dissipative, then $\overline{D(A(t))}=\bar{D}$ is independent of t (so we choose e.g. D = D(A(o))). Let us prove this fact. Suppose $o\le s<t<T$. Take $x_2\in D(A(s))$ and $x_1 = J_\lambda(t)x_2\in D(A(t))$. We have

$$J_\lambda(t)x_2 - \lambda A_\lambda(t)x_2 = x_2, \quad (J_\lambda(t)x_2 = (1-\lambda A(t)))^{-1}x_2)$$

With $u = x_2$ and $y_1 = A_\lambda(t)x_2\in AJ_\lambda(t)x_2$ (ω =o for simplicity) the inequality (2.15) yields

$$\|A_\lambda(t)x_2\|\le|A(s)x_2| + \|f(t)-f(s)\|\ L(\|x_2\|) \tag{4.19}$$

This shows that $A_\lambda(t)x_2$ is bounded with respect to $\lambda>o$, and therefore

$$x_2 = \lim_{\lambda\downarrow o} J_\lambda(t)x_2\in\overline{D(A(t))},\ \text{i.e.}\ D(A(s))\subset\overline{D(A(t))} \tag{4.20}$$

Clearly, we now take $x_1\in D(A(t))$ and $x_2 = J_\lambda(s)x_1\in D(A(s))$. Since we have (see (2.13))

$$\|x_2-x_1\|\le\|x_2-\lambda y_2-x_1\| + \lambda|A(t)x_1| + \lambda\|f(s)-f(t)\|L(\|x_2\|)$$

$$\forall\ \lambda>o,\ [x_2,y_2]\in A(s),\ \text{and}\ x_1\in D(A(t)) \tag{4.21}$$

it follows

$$\|A_\lambda(s)x_1\| < |A(t)x_1| + \|f(s)-f(t)\|\ L(\|J_\lambda(s)x_1\|)$$

But $\|J_\lambda(s)x_1\|$ is bounded with respect to $\lambda>o$, hence $A_\lambda(s)x_1$ is also bounded. Consequently,

$$x_1 = \lim J_\lambda(s)x_1 \in D(A(s)), \text{ i.e. } D(A(t)) \subset \overline{D(A(s))} \tag{4.21)'}$$

so $\overline{D(A(t))} = \overline{D(A(s))}$.

In Chapter 2 (see also § 6) one introduces the generalized domain $\hat{D}(A)$ of the semigroup generator $A:D(A) \subset X \to 2^X$. namely, $\hat{D}(A)$ can be defined as below

$$\hat{D}(A) = \left\{ x \in \overline{D(A)}, \ \limsup_{t \downarrow o} \frac{\|S(t)x-x\|}{t} < +\infty \right\} \tag{4.22}$$

where $S = S_A(t)$ is the semigroup generated by A. In view of (4.18) we see that

Corollary 4.2. In the conditions of Proposition 4.1, the set $\check{D}(A(s))$ defined by (3.63) us just the generalized domain $\hat{D}(A(s))$ of $A(s)$.

4.3. The quasi-autonomous case

Let us discuss some results of the subsection 4.1, in the particular case

$$A(t)x = Ax+f(t), \quad t \geq o, \ x \in D(A) \tag{4.23}$$

where $A:D(A) \subset X \to 2^X$ is an m-dissipative (possible multivalued) operator, and $f:R_+ \to X$ is continuous. Clearly, in this case $D(A(t)) = D(A)$, $\forall t \geq o$ so Hypothesis H(2.2) in § 2 Ch. 1 is trivially satisfied. Moreover, Condition (2.11) in Ch. 1 holds with $L(r) = 1, r \geq o$.

Let us consider the initial value problem

$$u'(t) \in Au(t)+f(t), \ u(s)= x_o \in \overline{D(A)}, \ o \leq s \leq t \tag{4.24}$$

In this case we can give a more precise definition of the integral solution on $[s,+\infty[$ to (4.24) (see (4.11)' with $g_1(t)=f(t)$. Namely, by an integral solution on $[s,+\infty[$ to (4.24) we mean a continuous function $u: [s,+\infty[\to \overline{D(A)}$ satisfying $u(s) = x_o$ and the integral inequality

$$\|u(\bar{t})-x\| \leq \|u(t)-x\| + \int_t^{\bar{t}} \langle y+f(\tau), u(\tau)-x \rangle_+ \, d\tau \tag{4.25}$$

for all $s \leq t \leq \bar{t}$, $x \in D(A)$ and $y \in Ax$.

A strong solution u to (4.24) on $[s,+\infty[$ is automatically an integral solution on $[s,+\infty[$ to (4.24). Indeed, let x be arbitrary in D(A) and $y \in Ax$. By standard argumetns, (4.24) yields

$$\frac{d}{dt}\|u(t)-x\| \le \langle y+f(t),\, u(t)-x\rangle_+ , \text{ a.e. on } [s,+\infty[\tag{4.26}$$

Integrating (4.26) over $[t,\bar{t}]$ one obtains (4.25). A careful check of the proof of Theorems 3.2 (see (4.11)') and 4.1 shows that we have

Theorem 4.2. Let $A:D(A)\subset X\to 2^X$ be m-dissipative and let $f: [0,+\infty[\to X$ be continuous. Then the initial value problem (4.24) admits a unique integral solution $u=u(t;s,x_o)$ in the sense of (4.25) and all the conclusions of Th. 3.1 (with $A(t)x$ given by (4.23)) hold. Moreover, if $g:[0,+\infty[\ \to X$ is continuous and $v=v(t;s,y_o)$ is the integral solution to the problem

$$v'(t) \in Av(t)+g(t),\ v(s)=y_o\in \overline{D(A)},\ o\le s\le t \tag{4.27}$$

then

$$\|u(t)-v(t)\| \le \|u(\tau)-v(\tau)\| + \int_\tau^t \|f(\theta)-g(\theta)\|d\theta \tag{4.28}$$

for all $o \le s \le \tau \le t$.

Remark 4.3. The conclusion of Theorem 4.2 remains valid if $f,g\in L^1_{loc}(o,\infty;X)$ (see Theorem 2.2 in Ch. 2). In view of (4.23) and (3.58), the integral solution to (4.24) is given by

$$u(t;s,x_o)=U(t,s)x_o = \lim_{n\to\infty} \prod_{k=1}^{n} \left(I-\frac{t-s}{n}\left(A-f\left(s+k\,\frac{t-s}{n}\right)\right)\right)^{-1} x_o \tag{4.29}$$

$o \le s \le t$, $x_o \in \overline{D(A)}$.

If $f\in L^1(o,T;X)$, there exists a sequence $f_n\in C^1(o,T;X)$ such that $f_n\to f$ in $L^1(o,T;X)$. In view of Theorem 4.2, the initial value problem

$$u_n' \in Au_n + f_n(t),\ u_n(o) = x_o\in\overline{D(A)},\ o\le t \tag{4.30}$$

admits a unique integral solution u_n on $[o, T]$ i.e.

$$\|u_n(\bar{t})-x\| \le \|u_n(t)-x\| + \int_t^{\bar{t}} \langle y+f_n(\tau),\, u_n(\tau)-x\rangle_+\, d\tau \tag{4.31}$$

for all $o\le t\le \bar{t}$, $[x,y]\in A$, $n=1,2,\ldots$

On the other hand, according to (4.28),

$$\|u_n(t)-u_m(t)\| \le \int_o^t \|f_n(\tau)-f_m(\tau)\|d\tau ,\quad o\le t\le T \tag{4.32}$$

which shows that $u_n(t)$ is uniformly convergent to a (continuous) function $u: [o,T]\to \overline{D(A)}$. Passing to the limit as $n\to\infty$ in (4.31) we get (4.25). Thus we have proved that the conclusions of Theorem 4.2 hold with

$f \in L^1_{loc}(o,\infty;X)$.

Finally, in the case $A-\omega I$ dissipative (4.25) has the form

$$\|u(\bar t)-x\| \le \|u(t)-x\| + \int_t^{\bar t} \omega\|u(\tau)-x\| + \langle y+f(\tau),u(\tau)-x\rangle_+ d\tau \tag{4.33}$$

for all $o \le s \le t \le \bar t \le T$, $x \in D(A)$ and $y \in Ax$

which is equivalent to

$$\|u(\bar t)-x\| \le \|u(t)-x\| \; e^{\omega(\bar t-t)} + \int_t^{\bar t} e^{\omega(\bar t-\tau)} \langle y+f(\tau),u(\tau)-x\rangle_+ d\tau \tag{4.34}$$

for all $o \le s \le \bar t \le T$ and $[x,y] \in A$.

Similarly, (4.28) becomes

$$\|u(t)-v(t)\| \le \|u(\tau)-v(\tau)\| e^{\omega(t-\tau)} + \int_\tau^t e^{\omega(t-\theta)} \|f(\theta)-g(\theta)\| d\theta \tag{4.35}$$

for all $o \le s \le \tau \le t \le T$.

§ 5. Compact Evolution Operators

5.1 Necessary conditions for compactness

Let $U = \{U(t,s),\ o \le s \le t \le T\}$ be an evolution operator in the sense of 1°), 2°), 3°) and 4°) in § 3.3 with respect to $\{D(t), o \le t \le T\}$.

Definition 5.1. U is said to be compact if for each $o \le s < t \le T$, the operator $U(t,s):D(s) \to D(t)$ maps bounded subsets of $D(s)$ into relatively compact (precompact) subsets of $D(t)$ (In short, $U(t,s)$ is compact for every $t \in]s,T]$). Of course, for $t = s$, $U(s,s)$ is the identity (on $D(s)$) which is not compact if X is of infinite dimension. Moreover, if $U(t_o,s)$ is compact for $t_o > s$, then $U(t,s)$ is also compact for all $t > t_o$ (this is because $U(t,s) = U(t,t_o)U(t_o,s)$).

Definition 5.2. $U(t,s)$ is said to be equicontinuous at $t_o > s$ on a subset $Y \subset D(s)$ if for every $\varepsilon > o$ there is $\delta = \delta(\varepsilon,t,t_o,s,Y)$ such that

$$\|U(t,s)y-U(t_o s)y\| < \varepsilon \;,\ \forall\ |t-t_o| < \delta \;,\ \forall\ y \in Y \tag{5.1}$$

Proposition 5.1. If U is compact, then for each $o \le s < t < T$, $U(t,s)$ is equicontinuous at $t_o > s$, on bounded subsets $Y \subset D(s)$.

Proof. Let $\varepsilon > o$ and $s < t_o$. Choose $\bar t$ such that $s < \bar t < t_o < T$. If Y is a sbounded subset of $D(s)$, then $U(\bar t,s)Y$ is a relatively compact subset of D(t). Since the function $(t,z) \to U(t,\bar{t})z$ is continuous on $[s,T] \times U(\bar{t},s)Y$ it is uniformly continuous on this subset, i.e. there is $\delta = \delta(\varepsilon,\bar{t},s,Y)$ such that

$$\|U(t,\bar{t})z-U(t_0,\bar{t})z\| < \varepsilon \quad , \forall |t-t_0| < \delta, \ \forall \ z \in U(t,s)Y \tag{5.2}$$

Since $\bar{t}$ depends on t_0, $z = U(\bar{t},s)y$ with $y \in Y$ and $U(t,\bar{t})U(\bar{t},s)y=U(t,s)y$ we see that (5.2) is just (5.1). The proof is complete.

For other results we assume that U is generated by the family $\{A(t), o \le t \le T\}$ via Theorem 3.5. For simplicity, assume that A(t) is m-dissipative. The following result holds

Proposition 5.2. Assume that the evolution operator U is generated by $\{A(t)\}$ via Theorem 3.5. Suppose in addition that U is compact and that for each t, A(t) is m-dissipative. Then for each $\lambda > 0$ and $s \in [o,T[$, $J_\lambda(s) = (I-\lambda A(s))^{-1}$ is compact.

Proof. The main point in this proof is the inequality (3.5)", i.e.

$$\|U(t,s)x-x\| \le (t-s)(\|y\|+M\,\rho(T)) \tag{5.2)'$$

for all $y \in A(s)x$ and $t \in [s,T]$. Here M is a constant depending on x (M=M(x)), but M is bounded on bounded subsets. Indeed, let $k=k_n$ be such that $t^n_{k-1} < t \le t^n_{k_n}$. Then (2.37)' becomes (with $\omega = o, r = s$ and $p_i = o$)

$$\|u_n(t)-\tilde{u}\| \le \|x^n_{k_n}-\tilde{u}\| + (t^n_{k_n}-s)\|\tilde{y}\| + \\ + L(\|\tilde{u}\|)\int_s^{t^n_{k_n}} \|f_n(\theta)-f(s)\|\,d\theta, \ \forall \ \tilde{y} \in A(s)\tilde{u} \tag{5.3}$$

where $f_n(\theta) = f(t^n_i)$ for $t^n_{i-1} < \theta \le t^n_i$, $f_n(s) = f(s)$.

Letting $n \to \infty$ we get

$$\|U(t,s)x_0-\tilde{u}\| \le \|x_0-\tilde{u}\| + (t-s)\|\tilde{y}\| + L(\|\tilde{u}\|)\int_s^t \|f(\theta)-f(s)\|\,d\theta$$

If $x_0=x=\tilde{u} \in D(A(s))$ and $\tilde{y} \in A(s)s$, then

$$\|U(t,s)x-x\| < (t-s)\|\tilde{y}\|+L(\|x\|)(t-s)\rho(T), \forall x \in D(A(s)),\ \tilde{y} \in A(s)x \tag{5.4}$$

hence (5.2)' holds with $M(x)=L(\|x\|)$ and L bounded on bounded subsets (by hypothesis). Let Y be a bounded subset of X. Then $J_\lambda(s)Y$ is bounded (for each $\lambda > 0$). With $x=J_\lambda(s)y$, and $\tilde{y}=A_\lambda(s)y$, (5.4) becomes

$$\|U(t,s)J_\lambda(s)y-J_\lambda(s)y\| \leq (t-s)(\|A_\lambda(s)y\| + \rho(T)L(\|J_\lambda(s)y\|))$$

$$\leq (t-s)C, \; s < t < T, \; y \in Y \qquad (5.5)$$

with $C = C(s,\lambda,y)$ bounded with respect to $y \in Y$.

Indeed, $A_\lambda(s)y = \dfrac{J_\lambda(s)y-y}{\lambda}$ is bounded wiht respect to y . Fix $\lambda > 0$. Then for each $\varepsilon > 0$, there is $\delta = \delta(\varepsilon, \lambda, s, Y) > 0$ such that for $t=s+\delta$ (with $\delta C \leq \varepsilon$),

$$\|U(s+\delta,s)J_\lambda(s)y- J_\lambda(s)y\| \leq \varepsilon \quad , \forall\, y \in Y \qquad (5.6)$$

Since $K_\varepsilon \equiv U(s+\delta,s)J_\lambda(s)Y$ is relatively compact, from (5.6) it follows that $J_\lambda(s)Y$ is also relatively compact (see Lemma 5.1 below). This completes the proof.

For the conclusion that $J_\lambda(s)Y$ is relatively compact, we have obviously used

<u>Lemma 5.1.</u> Let K be a nonempty subset of the Banach space X. If for every $\varepsilon > 0$, there is a relatively compact subset K_ε with the property that for each $x \in K$, there is $x_\varepsilon \in K_\varepsilon$ such that $\|x_\varepsilon - x\| < \varepsilon$, then K is relatively compact.

Proof. Let $\varepsilon > 0$. Since $K_{\varepsilon/2}$ is relatively compact (as in the hypothesis above) then it can be covered by a finite union of balls $B(x_\varepsilon^i, \frac{\varepsilon}{2})$ of radius $\frac{\varepsilon}{2}$ about x_ε^i, i.e. $K_{\varepsilon/2} \subset \bigcup_{i=1}^{n} B(x_\varepsilon^i, \frac{\varepsilon}{2})$. Then $K \subset \bigcup_{i=1}^{n} B((x_\varepsilon^i, \varepsilon)$. This is because for every $x \in K$, there is $x_{\varepsilon/2} \in K_{\varepsilon/2}$ such that $\|x-x_{\varepsilon/2}\| < \frac{\varepsilon}{2}$. On the other hand, $x_{\varepsilon/2} \in B(x_\varepsilon^{i_o}, \frac{\varepsilon}{2})$ for some $i_o \in \{1,\ldots,n_\varepsilon\}$, i.e. $\|x_{\varepsilon/2}-x_\varepsilon^{i_o}\| < \frac{\varepsilon}{2}$.

It follows $\|x-x_\varepsilon^{i_o}\| < \varepsilon$ which completes the proof.

From Lemma 5.1 we easily derive

<u>Corollary 5.1.</u> Let $\{T_n;\, T_n : D \subset X \to X,\, n=1,2,\ldots\}$ be a family of compact (possible nonlinear) operators from the (nonempty) subset D of X into X. Let $T:D \to X$ be defined by $Tx = \lim_{n\to\infty} T_n x$, $x \in D$. If $T_n x \to Tx$ as $n \to \infty$, uniformly on bounded subsets of D, then T is compact.

Obviously in the linear case (i;e. D is a closed subspace of X and all T_n are linear operators). Corollary 5.1 can be restated as

<u>Corollary 5.2.</u> The uniform limit of compact operators is a compact operator.

Of course, in linear case, the fact that $\lim_{n\to\infty} T_n x = Tx$ uniformly on bounded subsets of X is equivalent to the fact that $T_n \to T$ as $n\to\infty$ in the uniform operator topology (i.e. T is the uniform limit of T_n).

5.2. The extension of Brezis' theorem

We have seen that if the evolution operator U generated by $\{A(t), o \le t \le T\}$ is compact, then encessarily

(I) For each $s < t_o$, $U(t,s)$ is equi-continuous at t_o on bounded subsets $Y \subset \overline{D(A(s))}$ (in the sense of Definition 5.2 with $D(s) = \overline{D(A(s))}$).

(II) For each $s \in [o,T]$ and $\lambda > o$, $J_\lambda(s) = (I - \lambda A(s))^{-1}$ is compact.

The question is:

Are Conditions (I) and (II) sufficient in order for $U = \{U(t,s), o \le s < t < T\}$ to be compact? The answer is yes. In other words, we have

Theorem 5.1. Let U be the evolution operator generated by $\{A(t), o \le t \le T\}$ via Theorem 3.5. Suppose also that for each $t \in [o,T]$, $A(t)$ is m-dissipative. Then U is compact if and only if Conditions (I) and (II) hold.

The key of the proof of this theorem is the estimate below (which was suggested to the author by a similar estimate of Brezis [4], i.e. the inequality (3.12) in Chapter 2 of this book).

Lemma 5.2 Suppose that Hypotheses of Theorem 5.1 are fulfilled. Then

$$\|J_\lambda(s)x - x\| \le \frac{2}{t-s}\int_s^t \|U(\tau,s)x - x\|\, d\tau + \frac{\lambda}{t-s}\|U(t,s)x - x\| + \frac{\lambda C}{t-s}\int_s^t \|f(\tau) - f(s)\|\, d\tau \tag{5.7}$$

for all $\lambda > o$, $o \le s \le t \le T$, $x \in \overline{D(A(s))}$, where $C = C(x,s,\lambda)$ is a positive constant with the property that $x \to C(x,s,\lambda)$ is bounded on bounded subsets.

Proof. For r=s, the constant C in (3.15) is just $C = L(\|x\|)$ (see (2.11)) Consequently, (3.13) yields (with $u(t) = U(t,s)x_o$ and $\omega = o$)

$$\|U(t,s)x_o - x\| \le \|x_o - x\| + L(\|x\|)\int_s^t \|f(\tau) - f(s)\|\, d\tau + \int_s^t \langle y, U(\tau,s)x_o - x\rangle_+\, d\tau \tag{5.8}$$

for all $x \in D(A(s))$ and $y \in A(s)x$, $x_o \in \bar{D}$ with $\bar{D}=\overline{D(A(s))} \equiv \overline{D(A(o))}$ $s \in [o,T]$ see Remark 4.1). Now let us recall that

$$\langle y, U(\tau,s)x_o - x \rangle_+ \leq \frac{\|U(\tau,s)x_o - x + \lambda y\| - \|(\tau,s)x_o - x\|}{\lambda} \tag{5.9}$$

for all $\lambda > o$((2.5) and (2.7)).

Take $x=J_\lambda(s)x_o$ with $x_o \in \bar{D}$ and $y=A_\lambda(s)x_o \in AJ_\lambda(s)x_o$. Since $J_\lambda(s)x_o - -\lambda A\ (s)x_o = x_o$, (5.8) and (5.9) lead us to

$$\|U(t,s)x_o - J_\lambda(s)x_o\| \leq \|J_\lambda(s)x_o - x_o\| + L(\|J_\lambda(s)x_o\|)\int_s^t \|f(\tau)-f(s)\|d\tau +$$

$$+ \int_s^t \lambda^{-1}(\|U(\tau,s)x_o - x_o\|) - \|U(\tau,s)x_o - J_\lambda(s)x_o\|)d\tau \tag{5.10}$$

But

$$\|J_\lambda(s)x_o - x_o\| \leq \|J_\lambda(s)x_o - U(t,s)x_o\| + \|U(t,s)x_o - x_o\|$$

so (5.10) yields

$$- \|U(t,s)x_o - x_o\| \leq \frac{2}{\lambda}\int_s^t 2\|U(\tau,s)x_o - x_o\|d\tau - \frac{1}{\lambda}(t-s)\|J_\lambda(s)x_o - x_o\|$$

$$+ L(\|J_\lambda(s)x_o\|)\int_s^t f\|(\tau)-f(s)\|d\tau \tag{5.10'}$$

for all $x_o \in \bar{D}$, which is equivalent to (5.7) (with $C=L(\|J_\lambda(s)x\|)$, $x \in \bar{D}$).

We shall now return to the <u>proof of Theorem</u> 5.1.(Sufficiency part). We have only to prove that Conditions (I) and (II) imply the compactness of U(the necessity of (I) and (II) has been proved in the subsection 5.1). Therefore let $o \leq r < s < T$, and let Y be a bounded subset of $\bar{D}=\overline{D(A(r))}$. We must prove that $U(s,r)Y$ is relatively compact.

Indeed, for $x=U(s,r)y$ (5.7) becomes

$$\|J_\lambda(s)U(s,r)y - U(s,r)y\| \leq \frac{2}{t-s}\int_s^t \|U(\tau,r)y - U(s,r)y\|\,d\tau$$

$$+ \frac{\lambda}{t-s}\|U(t,r)y - U(s,r)y\| + \tag{5.11}$$

$$+ \frac{\lambda}{t-s}L(\|J_\lambda(s)(U(s,r)y)\|)\int_s^t \|f(\tau)-f(s)\|d\tau$$

for all $y \in Y$ and $t > s$. Let $\varepsilon > o$.Conditions (I) guarantees the existence of a positive $\delta = \delta(\varepsilon,s,r,Y)$ such that

$$\|U(\tau,r)y-U(s,r)y\| < \frac{\varepsilon}{8}, \tag{5.12}$$

for all $y \in Y$ and $|\tau-s| < \delta$ (equicontinuity of $\tau \to U(\tau,r)y$ at $s > r$ with respect to $y \in Y$).

In view of the nonexpansivity of the operators $x \to J_\lambda(s)x, y \to U(s,r)y$, it follows that for each $x_0 \in D(A(s))$ and $\lambda < 1$, $\|J_\lambda(s)U(s,r)y-x_0\| \le \|U(s,r)y-x_0\| + |A(s)x_0| \le K$ for all $y \in Y$, with $K = K(s,x_0,r)$ independent of $y \in Y$ and λ . Accordingly, the constant $L=L(y,\lambda,s,r)$ in (5.11) is bounded from above by a constant $\tilde{K}(s,x_0,r) \equiv \tilde{K}$ independent of $y \in Y$ and $\lambda < 1$. It follows that for $t=s+d$, and $\lambda = \lambda(\varepsilon) = \min\{2\delta, \varepsilon/4\tilde{K}\rho(\delta)\}$ with $\rho(\delta)$ defined by (2.38), we have (by (5.11) and (5.12))

$$\|J_\lambda(s)U(s,r)y-U(s,r)y\| < \frac{\varepsilon}{4} + \frac{\varepsilon}{4} + \frac{\varepsilon}{4} < \varepsilon \tag{5.13}$$

for all $y \in Y$.

Since $U(s,r)Y$ is a bounded subset of D and $J_{\lambda(\varepsilon)}(s)$ is compact, $J_{\lambda(\varepsilon)}(s)$ $U(s,r)Y$ is relatively compact. By (5.13) so it is $U(s,r)Y$. The proof is complete.

Lemma 5.2 gives an estimate of $J_\lambda(s)$ in terms of $U(t,s)$. A converse estimate is the following one (under Hypotheses of Theorem 5.1)

$$\|U(t,s)x-x\| \le (2+\frac{t-s}{\lambda})\|J_\lambda(s)x-x\| + (t-s)L(\|J_\lambda(s)x\|)\rho(T) \tag{5.14}$$

for all $0 \le s \le t \le T$, $x \in \bar{D}$ and $\lambda > 0$ with L bounded on bounded subsets of $\bar{D}$ (the same L as in (2.11)).

Let us prove (5.14). Let $x \in \bar{D}$, $\tilde{x} \in D(A(s))$ and $\tilde{y} \in A(s)\tilde{x}$. Then (5.4) and nonexpansivity of $x \to U(t,s)x$ lead us to

$$\|U(t,s)x-x\| \le 2\|x-\tilde{x}\| + \|U(t,s)\tilde{x}-\tilde{x}\| \le 2\|x-\tilde{x}\|$$
$$+ (t-s)(\|\tilde{y}\|+L(\|\tilde{x}\|)\rho(T)) \tag{5.15}$$

In particular, for $\tilde{x} = J_\lambda(s)x$ and $\tilde{y}=A_\lambda(s)x \in A(s)J_\lambda(s)x$, (5.15) yields (5.14).

§ 6. Other types of t-dependence of A(t)

Conditions of Kato, Crandall-Pazy

In this section we present other type of conditions on A(t) which are sufficient to guarantee that the DS-approximate solution u_n by given by (1.6) exists, it is convergent and the operator U(t,s) associated with A(t) via (3.58) is an evolution operator in the sense of Definition 3.3. We start with Hypotheses of Crandall-Pazy

H(6.1) For every $t \in [0,T]$, A(t) is ω-dissipative.

H(6.2) $\overline{D(A(t))}=\bar{D}$ is independent of t (so we can choose e.g. $\bar{D}=\overline{D(A(o))}$).

H(6.3) $R(I-\lambda A(t)) \supset \bar{D}$, for all $o<\lambda<\lambda_o$, $o \leq t \leq T$ where λ_o is sufficiently small.

The t-dependence of $J_\lambda(t) = (I-\lambda A(t))^{-1}$ (or of $A_\lambda(t)=\lambda^{-1}(J_\lambda(t)-I)$ is restricted by one of the following two conditions.

H(6.4) There is a continuous function $f\colon [0,T] \to X$ and a monotone increasing function $L\colon [0,\infty[\to [0,\infty[$ such that

$$\|J_\lambda(t)x-J_\lambda(\tau)x\| \leq \lambda\|f(t)-f(\tau)\|L(\|x\|) \tag{6.1}$$

for $0<\lambda<\lambda_o$, $t,\tau \in [0,T]$, $x\in\bar{D}$ or equivalently

$$\|A_\lambda(t)x-A_\lambda(\tau)x\| \leq \|f(t)-f(\tau)\| \; L(\|x\|) \tag{6.1)'$$

H(6.5) There is a continuous function $f\colon [0,T] \to X$ which is of bounded variation on $[0,T]$ and an increasing function $L\colon [0,\infty[\to [0,\infty[$ such that

$$\|J_\lambda(t)x-J_\lambda(\tau)x\| \leq \lambda\|f(t)-f(\tau)\| \; L(\|x\|) \; (1+\|A_\lambda(\tau)x\|) \tag{6.2}$$

for $0<\lambda<\lambda_o$, $t,\tau \in [0,T]$, $x\in\bar{D}$
or equivalently

$$\|A_\lambda(t)x-A_\lambda(\tau)x\| \leq \|f(t)-f(\tau)\| \; L(\|x\|)(1+\|A_\lambda(\tau)x\|) \tag{6.2)'$$

Condition H(6.5) weakens (6.1) to (6.2) but adds the requirement that f be of bounded variation. In particular, if $f(t)=ty$ for some $y\in X$, $y\neq o$ (6.1) becomes a Lipschitz condition with respect to t which was used by Kato [1] (under the additional assumption that X* is uniformly convex).

We will prove that in some particular cases (6.1) implies Condition H(2.1) in Section 2.

Before going on to state the main result we have to introduce "the generalized domain" for semigroup generator essentially due to Crandall [2] in nonlinear case (and to Westphal [1] in linear case).

Let $A:D(A)\subset X\to 2^X$ be a ω-dissipative (possible multivalued) operator satisfying the range condition

$$\overline{D(A)}\subset R(I-\lambda A),\quad 0<\lambda<\lambda_0 \tag{6.3}$$

Set

$$\hat{D}(A)=\{x\in\overline{D(A)},\ \lim_{\lambda\downarrow 0}\|A_\lambda x\|\equiv |Ax\|<+\infty\} \tag{6.4}$$

In Chapter 2 one gives equivalent definitions for $\hat{D}(A)$. We mention merely that

$$\hat{D}(A)=\{x\in\bigcap_{0<\lambda<\lambda_0}R(I-\lambda A),\ \|A_\lambda x\|\leq K(x),\ 0<\lambda<\lambda_0\} \tag{6.5}$$

for some constants $K(x)$, independent of λ.

Indeed, $J_\lambda x-x=\lambda A_\lambda x$ and therefore the boundedness of $A_\lambda x$ implies $x=\lim_{\lambda\downarrow 0}J_\lambda x\in\overline{D(A)}$, so the sets defined by (6.4) and (6.5) coincide.

Moreover, since

$$\|A_\lambda x\|\leq(1-\lambda\omega)^{-1}\|y\|,\quad y\in Ax,\ x\in D(A)$$

It follows $|Ax\|\leq\|y\|$, $\forall y\in Ax$ hence

$$|A\|\leq\inf\{\|y\|;\ y\in Ax\}\equiv|Ax|,\quad x\in D(A)$$
$$D(A)\subset\hat{D}(A)\subset\overline{D(A)},\ \|AJ_\lambda x\|\leq\|A_\lambda x\|,\quad \lambda>0,\ x\in\overline{D(A)} \tag{6.6}$$

Finally, taking into account that (Appendix)

$$(1-\lambda\omega)\|A_\lambda x\|\leq(1-\mu\omega)\|A_\mu x\|,\quad 0<\mu<\lambda \tag{6.7}$$

one has

$$(1-\lambda\omega)\|A_\lambda x\|\leq\lim_{0}(1-\lambda\omega)\|A_\lambda x\|=|Ax\|$$

that is

$$\|A_\lambda x\|\leq(1-\lambda\omega)^{-1}|Ax\|,\ |Ax\|=\sup_{0<\lambda<\lambda_0}(1-\lambda\omega)\|A\,x\| \tag{6.8}$$

for $x\in D(A)$. (For $x\in\bigcap_{0<\lambda<\lambda_0}R(I-\lambda A)\setminus\hat{D}(A)$, $|Ax\|=+\infty$ so (6.8) is also true, but not useful here).

We nox can derive some consequences of H(6.4) and H(6.5). The first one is that either H(6.4) or H(6.5) implies $\hat{D}(A(t))$ - independent of t.

Indeed, triangular inequality, (6.1)' and (6.2)' yield respectively

$$\begin{aligned} |A(t)x\| &\le |A(\tau)x\| + \|f(t)-f(\tau)\|\ L(\|x\|) \\ |A(t)x\| &\le |A(\tau)x\| + \|f(t)-f(\tau)\|\ L(\|x\|)\ (1+/A(\tau)x//) \end{aligned} \tag{6.9}$$

Clearly, (6.9) shows that $\hat{D}(A(\tau)) \subset \hat{D}(A(t))$ for all $t,\tau \in [o,T]$; hence $\hat{D}(A(\tau)) = \hat{D}(A(t))$ for all $t,\tau \in [o,T]$. For simplicity, set

$$\hat{D} = \hat{D}(A(t)),\ \forall\ t \in [o,T] \tag{6.10}$$

Next result gives the relationship of Crandall-Pazy conditions above with Kato's continuity assumption and with our condition H(2.1) in § 2.

<u>Proposition 6.1.</u> 1) Let $A(t)$ be single-valued and satisfy H(6.1), H(6.2) and H(6.3). Moreover, let $D(A(t))=D$ be independent of $t \in [o,T]$. If

$$\|A(t)x-A(\tau)x\| \le \|f(t)-f(\tau)\| L(\|x\|)1+\|A(\tau)x\| \tag{6.11}$$

for $x \in D$ and $t,\tau \in [o,T]$, then (6.2) holds for $x \in \bar{D}$ and a suitable L.

2) Conversely, if X^* is uniformly convex and $R(I-\lambda A(t))=X$ for $o<\lambda<\lambda_o$, $o \le t \le T$, then (6.2) implies (6.1) with $\tilde{L} = L$ and (6.1) implies

$$\|A(t)x-A(\tau)x\| \le \|f(t)-f(\tau)\|\ L(\|x\|) \tag{6.12}$$

for $x \in D$, $t,\tau \in [o,T]$.

3) Condition (6.12) implies H(2.1) in § 2, i.e.

$$\langle A(t)x-A(\tau)y,\ x-y\rangle_i \le \omega\|x-y\|^2 + \|f(t)-f(\tau)\|\ L(\|x\|)\|x-y\| \tag{6.13}$$

for $x,y \in D$ and $t,\tau \in [o,T]$, $o \le t \le \tau \le T$

Proof. 1) Let (6.11) be true. Take $x \in \bar{D}$. Then

$$\begin{aligned} &\|J_\lambda(t)x-J_\lambda(\tau)x\| \le \|J_\lambda(\tau)(I-\lambda A(\tau))J_\lambda(t)x-J_\lambda(\tau)\ (1-\ \lambda A(t))\ J_\lambda(t)x\| \\ &\le (1-\lambda\omega)^{-1}\ \lambda\|A(\tau)J_\lambda(t)-A(t)J_\lambda(t)x\| \le \\ &\quad 2\lambda\|f(t)-f(\tau)\|\ \tilde{L}(\|J_\lambda(t)x\|(1+\|A(t)J_\lambda(t)x\|) \end{aligned} \tag{6.14}$$

where we have used $(1-\lambda\omega)^{-1} \le 2$. Now let $y \in D$ be fixed. We have

$$\begin{aligned} &\|J_\lambda(t)x-y\| = \|J_\lambda(t)x-\ J_\lambda(t)(1-\lambda A(t)y)\| \le \\ &(1-\lambda\omega)^{-1}\ (\|x\|+\|y\| + \lambda\|A(t)y\|) \le \tilde{K}(1+\|x\|) \end{aligned} \tag{6.15}$$

with $\tilde{K}$ independent of t. This is because (6.11) implies the boundedness of $\|A(t)y\|$ on $[o,T]$.

Since $A(t)J_\lambda(t)x=A_\lambda(t)x$, (6.14) leads to (6.2).

2) In this case X is reflexive. If $x \in D$, then $J_\lambda(t)x \to x$ as $\lambda \downarrow o$, for each $t \in [o,T]$. On the other hand, $A_\lambda(t)x = A(t)J_\lambda(t)x$ and $\|A_\lambda(t)x\| \le (1-\lambda\omega)^{-1}|A(t)x\|$, $x \in D$, that is $A_\lambda(t)x$ and $A_\lambda(\tau)x$ contain weakly convergent subsequences (to $A(t)x$ and $A(\tau)x$ respectively). We may assume (relabeling if necessary) that even $A_\lambda(t)x \rightharpoonup A(t)x$ and $A_\lambda(\tau)x \rightharpoonup A(\tau)x$ as $\lambda \downarrow o$. Consequently

$$\begin{aligned} \|A(t)x-A(\tau)x\| &\le \liminf_{\lambda \downarrow o} \|A_\lambda(t)x - A_\lambda(\tau)x\| \\ &\le \|f(t)-f(\tau)\|\, L(\|x\|)(1+|A(t)x\|), \quad x \in D \end{aligned} \tag{6.16}$$

where (6.2)' has been used.

In view of (6.6), $|A(t)x\| \le \|A(t)x\|$ for $x \in D = D(A(t))$, hence (6.16) gives (6.11). Similarly one proves that (6.1)' yields (6.12).

3) Let (6.12) be fulfilled. We have

$$\begin{aligned} &\langle A(t)x-A(s)y, x-y \rangle_i = \langle A(t)x-A(t)y+A(t)y-A(s)y,\ x-y \rangle_i \\ &\le \langle A(t)x-A(t)y,\ x-y \rangle_i + \|A(t)y-A(s)y\|\ \|x-y\| \end{aligned} \tag{6.17}$$

The ω-dissipativity of $A(t)$, (6.12) and (6.17) lead to (6.13). This completes the proof.

Next we give the main result of this section, namely.

<u>Theorem 6.1</u>. (Crandall-Pazy). Let $\{A(t),\ o \le t \le T\}$ satisfy H(6.1), H(6.2) and H(6.3). Then there is a sequence of DS-approximate solution u_n to (1.1)+(1.1)' in the sense of (1.2)-(1.7) with $p_k^n = o$. Such a sequence u_n is given by (3.61). If in addition we admit either H(6.4) or H(6.5) (with $x_o \in \hat{D}$), then every sequence u_n of DS-approximate solutions of (1.1)+(1.1)' is convergent on $[s,T]$ to $U(t,s)$ given by (3.58).

Proof. Let us consider the difference scheme corresponding to $s \in [o,T]$, $x_o \in \overline{D(A(s))}$ and

$$\frac{x_k^n - x_{k-1}^n}{t_k^n - t_{k-1}^n} \in A(t_k^n)x_k^n, \quad k=1,2,\dots,n, \quad h_k = t_k^n - t_{k-1}^n = \frac{T-s}{n} = h \tag{6.18}$$

with $t_o^n = s$, $x_o^n \in D(A(s))$ $x_o^n \to x_o$ for $n \to \infty$. Clearly,

$$x_k^n = (I-h_k A(t_k^n))^{-1} x_{k-1}^n = J_{h_k}(t_k^n)x_{k-1}^n, \quad k=1,2,\dots,n \tag{6.19}$$

Now take $\hat{s} \in [o,T[$, $\hat{x}_o \in \overline{D(A(\hat{s}))} = \bar{D}$ and $\hat{x}_o^m \in D(A(s))$ with $\hat{x}_o^m \to \hat{x}_o$. The corresponding difference scheme is the following one

$$\frac{\hat{x}_j^m - \hat{x}_{j-1}^m}{\hat{t}_j^m - \hat{t}_{j-1}^m} \in A(\hat{t}_j^m)\hat{x}_j^m \ , \ j=1,2,\ldots,m, \ \hat{h}_j^m = \hat{t}_j^m - \hat{t}_{j-1}^m = \frac{T-\hat{s}}{m} = \hat{h} \tag{6.20}$$

with $\hat{t}_o^m = \hat{s}$.

It is easily seen that an estimate similar to (2.35) and (2.42) holds. Indeed, let $\bar{u} \in D(A(s))$. Then (with $h_k = h$)

$$\|x_k^n - \bar{u}\| = \|J_h(t_k^n)x_{k-1}^n - \bar{u}\| \le \|J_h(t_k^n)x_{k-1}^n - J_h(t_k^n)\bar{u}\| +$$

$$+ \|J_h(t_k^n)\bar{u} - \bar{u}\| \le \|x_{k-1}^n - u\| + h\|A_h(t_k^n)\bar{u}\|, \ k=1,2,\ldots,n \tag{6.21}$$

If we are in Hypothesis (6.1)',

$$\|A_h(t_h^n)\bar{u}\| \le \|f(t_k^n) - f(s)\| L(\|\bar{u}\|) + \|A_h(s)\bar{u}\| \tag{6.22}$$

where $\|A_h(s)\bar{u}\| \le |A(s)\bar{u}|$ (see (6.6)).

Similarly, Hypothesis (6.2)' yields

$$\|A_h(t_k^n)\bar{u}\| = \|f(t_k^n) - f(s)\| \ L(\|\bar{u}\|)(1 + |A(s)\bar{u}|) + |A(s)\bar{u}| \tag{6.23}$$

Consequently, either H(6.4) or H(6.5) implies

$$\|x_k^n - \bar{u}\| \le \|x_{k-1}^n - \bar{u}\| + h_k(\|f(t_k^n) - f(s)\| L(\|\bar{u}\|) K(s,\bar{u}) + |A(s)\bar{u}|) \tag{6.24}$$

Where $K(s,u)=1$ in the situation H(6.4) and $K(s,u) = 1 + |A(s)u|$ in the situation H(6.5).

Iterating (6.24) one obtains (see also (2.35))

$$\|x_k^n - \bar{u}\| \le \|x_o^n - \bar{u}\| + (t_k^n - s)|A(s)\bar{u}| + \Sigma_{j=1}^k \ h_j^n \|f(t_j^n) - f(s)\| L(\|\bar{u}\|) K(s,\bar{u}) \tag{6.25}$$

In particular, (6.25) shows that there is a constant $M = M(s,u,x_o) > o$ independent of n and k such that

$$\|x_k^n\| \le M, \ k=o,1,\ldots,n, \ n=1,2,\ldots \tag{6.26}$$

Another step of the proof is to show that there is a positive constant $C = C(s,s,x_o,x_o)$ such that

$$\|x_k^n-\hat{x}_j^m\| \le \frac{h}{h+\hat{h}} \|x_k^n-\hat{x}_{j-1}^m\| + \frac{\hat{h}}{h+\hat{h}}\|x_{k-1}^n - \hat{x}_j^m\| + \\ + \frac{h\,\hat{h}}{h+\hat{h}} \|f(t_k^n)-f(\hat{t}_j^m)\| \; C \tag{6.27}$$

for all $n,m \in N$, $k=1,\ldots,n$, $j=1,\ldots,m$, where x_k^n and $\hat{x}_j^m$ are given by (6.18) and (6.20) respectively.

Obviously (6.27) is just (2.45) with $\omega = 0$, $\alpha_{k,j} = \frac{\hat{h}}{h+\hat{h}}$, $\beta_{k,j} = \frac{h}{h+\hat{h}}$ and $\gamma_{k,j} = \frac{h\,\hat{h}}{h+\hat{h}}$.

Since there is no danger of confusion, suppress the superscripts n and m on all letters t,s and h except the x_o^n's and x_o^m's (i.e. we write $t_k^n = t_k$, $h_k^n = h_k$ h, $x_k^n = x_k$ and so on).

According to the definition of J_λ we have

$$J_\lambda(t)(x-\lambda y)=x,\; A_\lambda(t)(x-\lambda y)=y,\; \forall\, x \in D(A(t)),\; y \in A(t)x \\ t \in [o,T] \text{ and } \lambda > o \tag{6.28}$$

Therefore (6.18) and (6.20) imply (with $r = \frac{h\,\hat{h}}{h+\hat{h}}$)

$$J_{r\lambda}(t_k)(x_k - r\lambda \frac{x_k-x_{k-1}}{h}) = x_k \\ J_{r\lambda}(\hat{t}_j)(\hat{x}_j - r\lambda \frac{\hat{x}_j-\hat{x}_{j-1}}{\hat{h}}) = \hat{x}_j \tag{6.29}$$

and so we have

$$\|x_k-\hat{x}_j\| = \|J_{r\lambda}(t_k)(x_k - r\lambda \frac{x_k-x_{k-1}}{h}) - J_{r\lambda}(t_k)(\hat{x}_j - r\lambda \frac{\hat{x}_j-\hat{x}_{j-1}}{\hat{h}})\| + \\ + \|J_{r\lambda}(t_k)(\hat{x}_j - r\lambda \frac{\hat{x}_j-\hat{x}_{j-1}}{\hat{h}}) - J_{r\lambda}(\hat{t}_j)(\hat{x}_j - r\lambda \frac{\hat{x}_j-\hat{x}_{j-1}}{\hat{h}})\| \tag{6.30}$$

It is easy to check that

$$x_k - r\lambda \frac{x_k-x_{k-1}}{h} - (\hat{x}_j - r\lambda \frac{\hat{x}_j-\hat{x}_{j-1}}{\hat{h}}) = (1-\lambda)(x_k-\hat{x}_j) + \\ + \lambda \frac{h}{h+\hat{h}}(x_k-\hat{x}_{j-1}) + \lambda \frac{\hat{h}}{h+\hat{h}}(x_{k-1}-\hat{x}_j) \tag{6.31}$$

Suppose that (6.1)' holds. Then the nonexpansivity of $x \to J_{r\lambda}(t_k)x$, (6.30) and (6.31) lead us to

$$\|x_k-\hat{x}_j\| \le (1-\lambda)\|x_k-\hat{x}_j\| + \frac{\lambda h}{h+\hat{h}}\|x_k-\hat{x}_{j-1}\| + \frac{\lambda\hat{h}}{h+\hat{h}}\|x_{k-1}-\hat{x}_j\| +$$

$$+ \lambda r\ \|f(t_k)-f(\hat{t}_j)\|L(\|\hat{x}_j-r\lambda\frac{\hat{x}_j-\hat{x}_{j-1}}{\hat{h}}\|) \tag{6.32}$$

Rearranging, dividing by λ and then letting $\lambda\downarrow o$, (6.32) yields (6.27) with

$$C = L(\hat{M}) \ge L(\|\hat{x}_j\|) \tag{6.33}$$

where M is the constant in (6.26) corresponding to s, and x_o and L was supposed to be continuous.

Suppose now that (6.2)' holds. In this case choose $\hat{x}_o^m = \hat{x}_o \in \hat{D}$, $\forall m$). Then the fourth term in (6.32) is multiplied by

$$(1+\|A_{r\lambda}(\hat{t}_j)(x_j-r\lambda\frac{\hat{x}_j-\hat{x}_{j-1}}{\hat{h}}\|) \tag{6.34}$$

In view of (6.28) and (6.20),

$$A_{r\lambda}(\hat{t}_j)(\hat{x}_j-r\lambda\frac{\hat{x}_j-\hat{x}_{j-1}}{\hat{h}}) = \frac{\hat{x}_j-\hat{x}_{j-1}}{\hat{h}} \tag{6.35}$$

so we have to prove that there is a constant $C_1(\hat{s},\hat{x}_o) \equiv C_1$ such that

$$\|\frac{\hat{x}_j-\hat{x}_{j-1}}{\hat{h}}\| \le C_1 \tag{6.36}$$

Set $\hat{y}_j = \frac{\hat{x}_j-\hat{x}_{j-1}}{\hat{h}}$. Hence $\hat{y}_j \in A(\hat{t}_j)\hat{x}_j$, $\hat{x}_j-\hat{h}\ \hat{y}_j = \hat{x}_{j-1}$, and by (6.28), $A_{\hat{h}}(\hat{t}_j)(\hat{x}_j-\hat{h}\ \hat{y}_j) = \hat{y}_j$. Therefore, $\hat{y}_j = A_{\hat{h}}(\hat{t}_j)\hat{x}_{j-1}$, so $\|\hat{y}_j\| \le |A(\hat{t}_j)\hat{x}_{j-1}\| = a_{j-1}$ (see (6.8)). Thus, it suffices to estimate a_j. In view of (6.2)'

$$|A(\hat{t}_j)\hat{x}_{j-1}\| \le |A(\hat{t}_{j-1})\hat{x}_{j-1}\| + \|f(\hat{t}_j) -$$

$$- f(\hat{t}_{j-1})\|L(\|\hat{x}_{j-1}\|)(1+|A(\hat{t}_{j-1})\hat{x}_{j-1}\|) \tag{6.37}$$

On the other hand, on the basis of (6.19), $\hat{x}_{j-1} = J_{\hat{h}}(\hat{t}_{j-1})\hat{x}_{j-2}$. This fact along with (6.6)-(6.8) yields:

$$|A(\hat{t}_{j-1})\hat{x}_{j-1}\| = |A(t_{j-1})J_{\hat{h}}(\hat{t}_{j-1})\hat{x}_{j-2}\| \le \|A_{\hat{h}}(\hat{t}_{j-1})\hat{x}_{j-2}\|$$

$$\le |A(\hat{t}_{j-1})\hat{x}_{j-2}\| \equiv a_{j-2}$$

Consequently, (6.37) implies

$$a_{j-1} \leq a_{j-2} + \|f(\hat{t}_j)-f(\hat{t}_{j-1})\| \quad (1+a_{j-2})L(M), \; j \geq 2 \tag{6.38}$$

where (6.26) has also been used.

Denote $b_{j-1} = L(M)\|f(\hat{t}_j)-f(\hat{t}_{j-1})\|$. Then (6.38) can be written as below

$$a_{j-1} \leq (1+b_{j-1})a_{j-2}+b_{j-1} \leq a_{j-2}(\exp b_{j-1})+b_{j-1} \tag{6.38)'$$

It is now easy to check that (6.38)' gives

$$a_k \leq (a_o + \prod_{i=1}^{k} b_i) \exp \prod_{i=1}^{k} b_i) \tag{6.39}$$

Since $a_o = |A(\hat{t}_1)\hat{x}_o\|$ with $\hat{x}_o \in \hat{D}$ (in this case $\hat{x}_o^m = \hat{x}_o$) we can estimate a_o by a constant independent of m (see (6.9) with $t = \hat{t}_1^m$, $x = \hat{x}_o$ and $\tau = o$). Finally, taking into account that f is bounded variation on $[o;T]$, (6.39) provides a bound for a_k, so (6.36) holds. The main remark now is to observe that (6.25) and (6.27) are similar to (2.42) and (2.45 respectively. It follows that Hypotheses of the theorem lead us to (2.40) which (in view of Theorem 3.1 (see (3.2)) completes the proof.

§ 7. Range condition. Tangential condition

This section is mainly concerned with the relationship between "tangential conditions" (1.11) and (1.12). For the sake of simplicity we start with the following elementary result.

Lemma 7.1. Let $v\colon [o,T] \to X$ be a function and let $D\colon]o,T[\to 2^X$ be a multivalued function. Then the statement (i), (ii) and (iii) below are equivalent

(i) $\lim\limits_{h \downarrow o} \frac{1}{h} d[v(h); D(h)] = o, \quad h > o$

(ii) There is a function $y\colon]o,T[\to X$ such that

$$\lim_{h \downarrow o} \frac{1}{h} \|v(h)-y(h)\| = o, \; y(h) \in D(h), \qquad h \in]o,T[$$

(iii) There is a function $r\colon]o,T] \to X$ such that $r(h) \to o$ as $h \downarrow o$ and $v(h)+hr(h) \in D(h)$ for all $h \in]o,T[$

Proof. Inasmuch as $d[v(h); D(h)] = \inf_{y\in D(h)} \|v(h) - y\|$, it follows that for each $h\in[0,T]$, there is $y(h)\in D(h)$ such that

$$\|v(h)-y(h)\| \leq d[v(h) ; D(h)] + h^2 \tag{7.1}$$

With

$$r(h) = \frac{1}{2}(y(h)-v(h)) \tag{7.2}$$

the proof is complete. It is now easy to prove

Proposition 7.1. Let $\mathfrak{F} = \{A(t); 0 \leq t \leq T\}$ be a family of single valued operators from $D((A(t))$ into X with $D(A(t))=\overline{D(A(t))}$ and $(t,x)\to A(t)x$ continuous $(x \in D(A(t)))$. Then

$$\lim_{h\downarrow 0} h^{-1}d|x+hA(t)x; D(A(t+h))|=0, \quad t\in[0,T], \; x\in D(A(t)) \tag{7.3}$$

implies

$$\lim_{h\downarrow 0} h^{-1} d[x;R(I-hA(t+h))] = 0, \quad t \in [0,T], \; x \in D(A(t)) \tag{7.4}$$

If in addition $\mathfrak{F}$ is ω-dissipative (i.e. (2.11) holds) then (7.3) is equivalent to (7.4).

Proof. According to Lemma 7.1, (7.3) means that there is $r(h)\in X$ such that $y(h) = x+hA(t)x+hr(h)\in D(A(t+h))$, with $r(h)\to 0$ as $h\downarrow 0$. Consequently

$$\frac{1}{h} d[x;R(I-hA(t+h))] \leq \frac{1}{h}\|x-(y(h)-hA(t+h)y(h))\| =$$

$$\|A(t+h)y(h)-A(t)x-r(h)\| \to 0 \text{ as } h\downarrow 0$$

so (7.4) is a consequence of (7.3). Let us show the converse implication. Indeed (involving again Lemma 7.1) (7.4) means the existence of $r(h)\in X$ with $r(h)\to 0$ as $h\downarrow 0$ such that $x+hr(h)\in R(I-hA(t+h))$. In other words, there is $a(h)\in D(A(t+h))$ such that

$$x+hr(h) = a(h)-hA(t+h)a(h) \tag{7.5}$$

It follows

$$h^{-1} d[x+hA(t)x; D(A(t+h))] \leq h^{-1} \|x+hA(t)x-a(h)\|$$

$$= \|A(t)x-r(h)-A(t+h)a(h)\| \tag{7.6}$$

On the other hand, (2.15) (a consequence of (2.11)) yields $(1-h\omega)$ $\|a(h)-x\| \leq \|a(h)-hA(t+h)a(h)-x\| + h\|A(t)x\| + h\|f(t+h)-f(t)\|\, L(\|x\|)$ that is $(1-h\omega)\|r(h)+A(t+h)a(h)\| \leq \|r(h)\| + \|A(t)x\| + \|f(t+h)-f(t)\|$ $L(\|x\|)$. This shows that $\|A(t+h)a(h)\|$ is bounded with respect to h and therefore (by (7.5)) $a(h) \to x$ as $h \downarrow o$. In view of (7.6) we see that (7.3) holds.

Now let us consider the "range conditions" below

(I) $R(I-hA(t+h)) \supset \overline{D(A(t))}$, $o \quad h \leq h_o$, $o \leq t \leq T$ (7.7)
for some small $h_o > o$

(II) $\overline{D(A(t))} = \bar{D}$ is independent of $t \in [o,T]$ and
$R(I-\lambda A(t)) \supset \bar{D}$, $o < \lambda < \lambda_o$, $o \leq t \leq T$ (7.8)
for some small $\lambda_o > o$ (take e.g. $D=D(A(o))$)

(III) $R(I-\lambda A(t)) = X$, $\forall t \in [o,T]$, $\lambda > o$ (7.9)

Clearly, we have: (III) (II) (I) (7.4).
However "Tangential condition" (7.4) is strictly more general than "Range condition" (7.7). An interesting example (arising in PDE) is given in Ch. 3, § 2. Here we give the following simple counter example (due to R.H. Martin Jr.): $x=R^2$, and $A(t) = A$ with

$$Ax = x^{\perp},\ D(A)=S(r)=\{x=(x_1,x_2) \in R^2;\ \|x\|=(x_1^2+x_2^2)^{1/2} = r\} \tag{7.10}$$

where $x^{\perp} = (-x_2,x_1)$ for $x=(x_1,x_2)$. Denote by $<\cdot,\cdot>$ the inner product of R^2. Since $<Ax,x> = o$, Ax is tangent to S(r) at $x \in S(r)$ in classical sense (see Proposition 7.2) so

$$\lim_{h\downarrow o} \frac{1}{h}\, d\,[x+hAx;\ S(r)] = o,\ \|x\| = r \tag{7.11}$$

or equivalently

$$\lim_{h\downarrow o} \frac{1}{h}\, d\,[x;R(I-hA)] = o,\ \|x\| = r \tag{7.12}$$

are fulfilled.

However, (7.7) is not true in this case, since we actually have

$$D(A) \cap R(I-hA) = \emptyset,\ \forall h > o \tag{7.13}$$

Indeed, if we assume the existence of an element $y \in D(A) \cap R(I-hA)$, then there is $x \in D(A)$ such that

$$y = x-hAx = x-hx^{\perp}\ ,\ \|x\| = r,\ \|y\| = r$$

which lead us to the absurdity $r^2=r^2+h^2r^2$, for $h > o$. The fact that (7.11) holds is a consequence of $\langle Ax,x\rangle = o$ and of the following general result.

Proposition 7.2. Let X be a Banach space. Conditions (1) and (2) below are equivalent

(1) $\lim_{h\downarrow o} h^{-1}\ d\,[x+hy;S(r)] = o,\ y \in X$

(2) $\langle x,y\rangle_+ = \lim_{h\downarrow o} \dfrac{\|x+hy\|-\|x\|}{h} = o,\ \|x\| = r, y \in X$

If (in particular) X is real Hilbert space H of inner product $\langle .,.\rangle$ then (1) is equivalent to

(3) $\langle x,y\rangle = o,\ \|x\| = r,\ y \in X$

Proof. In view of Lemma 7.1. Condition (1) means the existence of $g(h) \in X$ with $g(h) \to o$ as $h\downarrow o$, such that $\|x+hy+hg(h)\|=r$, which implies $\|x\|=r$. Moreover

$$|\ \|x+hy\| - \|x\|\ | = |\ \|x+hy\| - \|x+hy+hg(h)\|\ | \le h\,\|g(h)\|$$

and therefore (1) $\Rightarrow$ (2). To prove that (2)$\Rightarrow$(1), choose g(h) such that

$$x+hy+hg(h) = r(x+hy)/\|x+hy\| \qquad (7.14)$$

It is easy to check that

$$\|g(h)\| = \frac{|\ \|x+hy\| - \|x\|\ |}{h}$$

so $g(h)\to o$ as $h\downarrow o$ (on the basis of Condition (2)).

By (7.14) we now conclude that (2)$\Rightarrow$(1).

For Condition (3) we note that $\langle x,y\rangle = \|x\|\ \langle x,y\rangle_+$.

Remark 7.1. If in Condition (1) we replace "lim" by "lim inf" one obtains an equivalent condition. Simialrly, (7.3) and (7.4) are also equivalent with "lim inf" in place of "lim".

Proposition 7.2 is a consequence of the very general result below.

Definition 7.1. A vector y X is said to be "tangent" to D at $x \in D$ (or quasi-tangent to D at x) if

$$\lim_{h\downarrow o} h^{-1}\ d\,[x+hy;\ D] = o \qquad (7.15)$$

One proves that in the case in which D is a submanifold of X, then quasi-tancency means just tangency in a classical sense.

In order to state the result in a precise form, some preliminaries are needed.

Let M be a Hausdorff C^k-manifld (with $k \geq 1$) which is modelled by a Banach space X. Denote by $T_x(M)$ the tangent space of M at $x \in M$.

A vector $y \in T_x(M)$ is said to be quasi-tangent to a (nonempty) subset S of M at $x \in S$ if there is a chart (U, φ) of M at x such that

$$\lim_{h \downarrow o} h^{-1} \; d[\varphi(x)+h\, D(\varphi)_x y;\, \varphi(U \cap S)] = o \qquad (7.16)$$

It is easy to prove that this notion (of quasi-tangency) is independent of the local chart (U,φ) (i.e. (7.16) holds for any other local chart (V, ψ) of M at x).

Moreover we have

<u>Proposition 7.3</u>. Let S be a C^k-submanifold of M ($k \geq 1$) and let $x \in S$ and $y \in T_x(M)$. Then y is quasi-tangent to S at $x \in S$ iff $y \in T_x(S)$.
(The proof of Proposition 7.3 as well as the Definition of quasi-tangency can be found in Montreanu and Pavel [1] or in the book [15], Chapter 4, § 1 by Pavel.

Let A be an open subset of the Banach space X and $g: A \to Y$ a function (Y-a finite dimensional space).

Set

$$D = D_g = \{ x \in A,\ g(x) = o \} \qquad (7.17)$$

The followong result holds (see Pavel and Ursescu [2]).

<u>Theorem 7.1.</u> Suppose that $g: A \subset X \to Y$ (Y- of finite dimension) is continuous on A and Fréchet differentiable at $x \in A$. If the Fréchet derivative $\dot{g}(x): X \to Y$ (of g at x) is surjective, then Conditions (7.18) and (7.19) below are equivalent.

$$\lim_{h \downarrow o} \frac{1}{h} \; d[x+hy, D_g] = o, \ y \in X \qquad (7.18)$$

$$g(x) = o, \ \dot{g}(x)(y) = o \qquad (7.19)$$

where $\dot{g}(x)(y)$ denotes the value of (the linear continuous function) $\dot{g}(x)$ at y

The proof of this theorem can also be found in the author's book [15], ch. 3, § 2.

Chapter 2

Nonlinear Semigroups

This chapter deals with the theory of nonlinear semigroups generated by dissipative operators and with the application of this theory to autonomous differential equations (inclusions). The fundamental result in this theory (on general Banach spaces) is the exponential formula (see (3.10)) of Crandall-Liggett. It is derived from the remarkable estimate (1.16) due to Kobayashi. Among many other general results we have to point out Brezis' characterization of compact semigroups in § 6, and Bruck's results (Theorem 9.3) on the asymptotic behaviour of nonlinear semigroups.

A section (i.e. § 6.2) on linear semigroups, presenting Hille-Yosida Theorem, Pazy's characterization of linear compact semigroups as well as the characterization of compactness solely in terms of the resolvent (Theorem 6.4 due to Vrabie) is included.

§ 1. Discrete schemes in autonomous case.

Let us consider the Cauchy problem $(CP;x_o)$ associated with a nonlinear (possible multivalued) operator $A:D(A)\subset X\to 2^X$ namely

$$u'(t) \in Au(t), \quad o \leq t \tag{1.1}$$

$$u(o)=x_o \in \overline{D(A)} \tag{1.1'}$$

In other words, we shall study the problem (1.1)+(1.1)' in Chapter 1, in the particular case A(t)=1-independent of t (i.e. in the autonomous case). Of course, most of the fundamental results in this chapter will be derived from the corresponding results in Chapter 1.

Fix a positive number T and let n be a positive integer. The DS-approximate solution u_n is also defined by (1.2) in Ch. 1, i.e.

$$u_n(t) = \begin{cases} x_o^n \text{ , for } t = t_o^n = o \\ x_k^n \text{ , for } t \in]t_{k-1}^n, t_k^n] \end{cases} \tag{1.2}$$

$k=o,1,\dots,N_n$, where $t_k^n \in [o,T]$ and $x_k^n \in D(A)$ are defined as below:

$$o=t_o^n < t_1^n < \dots < t_{N_n-1}^n < T \leq t_{T_n}^n \tag{1.3}$$

$$d_n = \max_{1 \le k \le N_n} (t_k^n - t_{k-1}^n) \to 0 \quad \text{as } n \to \infty \tag{1.4}$$

Given $\omega > 0$, one considers $\omega d_n \le \frac{1}{2}$ (see (2.36) in Ch. 1) in order to apply inequality (2.40), Ch. 1, with $h_k^n = t_k^n - t_{k-1}^n$. The elements x_k^n are supposed to satisfy the implicit difference-scheme

$$\frac{x_k^n - x_{k-1}^n}{t_k^n - t_{k-1}^n} - p_k^n \in Ax_k^n, \quad k=1,\dots,N_n \tag{1.5}$$

with $x_o^n \in D(A)$ such that $x_o^n \to x_o$ as $n \to \infty$. The "errors" p_k are subjected to restriction

$$b_n = \sum_{k=1}^{N_n} (t_k^n - t_{k-1}^n) \| p_k^n \| \to 0 \text{ as } n \to \infty \tag{1.6}$$

Remark 1.1. The explanation of the fact that $t_{N_n}^n \ge T$ is given in § 2. If A is m-dissipative, then $t_{N_n}^n = T$ (see § 2).

We already know that if A is m-dissipative (§ 3.6, Ch.1) then we may whoose $p_k^n = o$. The operator A is said to be ω-dissipative if

$$\langle y_1 - y_2, x_1 - x_2 \rangle_i \le \omega \| x_1 - x_2 \|^2, \tag{1.7}$$

for all $x_1, x_2 \in D(A), y_1 \in Ax_1, y_2 \in Ax_2$. (In short, $[x,y] \in A$ means $x \in D(A)$ and $y \in Ax$). Thus, we say that (1.7) holds for all $[x_j, y_j] \in A, j=1,2$. Note that o-dissipativity is simply called dussipativity.

Combining Proposition 2.1 and 2.2 (with f=o) in Chapter 1 we get

Proposition 1.1. The following conditions are equivalent

(i) $A: D(A) \subset X \to 2^X$ is ω-dissipative

(ii) For each $x_1, x_2 \in D(A)$, there is $x^* \in F(x_1 - x_2)$ such that $\langle y_1 - y_2, x^* \rangle \le \omega \| x_1 - x_2 \|^2$, for all $y_j \in Ax_j$, j=1,2

(iii) $(1 - \lambda\omega) \| x_1 - x_2 \| \le \| x_1 - x_2 - \lambda(y_1 - y_2) \|$ (1.8)
for all $\lambda > o$ and $[x_j, y_j] \in A$, j=1,2.

The inequality (1.8) (i.e. ω-dissipativity of A) implies

$$(\lambda + \mu - \lambda\mu\omega) \| x_1 - x_2 \| \le \lambda \| x_2 - \mu y_2 - x_1 \| + \mu \| x_1 - \lambda y_1 - x_2 \| \tag{1.9}$$

for all $\lambda, \mu > o$ and $[x_j, y_j] \in A$, j=1,2 and (1.9) implies

$$(1 - \lambda\omega) \| x_1 - u \| \le \| x_1 - \lambda y_1 - u \| + \lambda |Au| \tag{1.10}$$

for all $\lambda > o$, $u \in D(A)$ and $[x_1, y_1] \in A$ where

$$|Au| = \inf \{ \| v \| ; v \in Au \} \tag{1.11}$$

Remark 1.2. Inequality (1.9) is equivalent to

$$\langle y_1, x_1-x_2\rangle_i + \langle y_2,x_2-x_1\rangle_i \le \omega \|x_1-x_2\|^2, \tag{1.9)'$$

for all $[x_j,y_j] \in A$, j=1,2 (see Appendix, § 2). The operator A satisfying (1.9) is said to be ω-quasi-dissipative (Kobayashi [1]). The ω-dissipativity of A guarantees the convergence of DS-approximate solution (while Tangential conditions or more particular-Range Conditions guarantee the existence of u_n).

In order to give the funamental estimate (i.e. (1.16)) let $x_o \in \overline{D(A)}$ and let $\{\hat t_o^m, \hat t_1^m, \dots, \hat t_{N_m-1}, \hat T\}$ and $\{\hat x_o^m, \hat x_1^m, \dots, \hat x_{N_m}\}$ be the elements of the discrete scheme corresponding to x_o in the sense of (1.2)-(1.6). Therefore

$$o=\hat t_o^m < \hat t_1^m < \dots < \hat t_{N_m-1}^m < T=\hat t_{N_m}^m, \quad \hat h_j^m = t_j^m - \hat t_{j-1}^m \tag{1.12}$$

$$\frac{\hat x_j^m - \hat x_{j-1}^m}{\hat t_j^m - \hat t_{j-1}^m} - \hat p_j^m \in A\hat x_j^m, \ j=1,\dots,\hat N_m, \quad \hat x_o^m \in D(A), \ \hat x_o^m \to \hat x_o \ \text{ as } m\to\infty \tag{1.13}$$

$$\hat d_m = \max_{1\le j\le \hat N_m} \hat h_j^m \to o \ \text{ as } m\to\infty, \ \omega d_m \le \frac{1}{2} \quad b_m = \prod_{j=1}^{\hat N_m} \hat h_j^m \|\hat p_j^m\| \to o \ \text{ as } m\to\infty \tag{1.14}$$

In this case u_m is given by

$$u_m(t) = \begin{cases} \hat x_o^m, & \text{for } t=\hat t_o^m = o \\ \hat x_j^m, & \text{for } t\in]\hat t_{j-1}^m, \hat t_j^m], \ j=1,\dots,N_m \end{cases} \tag{1.15}$$

The fundamental estimate (which plays a crucial role in both the convergence of u_n and the properties of $u(t)=\lim_{n\to\infty} u_n(t)$) can easily be derived from Lemma 2.3 (Ch.1) with $s=\hat s=o$; $\rho=o$. Consequently, the following important result holds (Kobayashi)

Lemma 1.1. Suppose that there exist the DS-approximate solutions u_n and u_m of (1.1) corresponding to x_o and x_o respectively (given by (1.2) and (1.15)). If A is ω-dissipative, then

$$\begin{aligned} \prod_{i=1}^{k} (1-\omega h_i^n) \prod_{q=1}^{j} (1-\omega h_q^m) \|x_k^n - \hat x_j^m\| &\le \|x_o^n - u\| + \|\hat x_o^m - u\| + \\ |Au| \, [(t_i^n - t_j^m)^2 + d_n t_i^n + \hat d_m \hat t_j^m]^{1/2} &+ \prod_{i=1}^{k} h_i^n \|p_i^n\| + \prod_{q=1}^{j} h_q^m \|p_j^m\| \end{aligned} \tag{1.16}$$

for $i=o,\dots,N_n$; $j=o,\dots,\hat N_m$ and every $u\in D(A)$.

Remark 1.3. In the proof of (1.16) one uses only (1.9) and its consequence (1.10). See Remark 2.4 in Chapter 1. Therefore in Lemma 1.1, the hypothesis "A is ω-dissipative" in the sense of (1.7) can be weakened to "A is ω-quasi-dissipative" in the sense of (1.9). However, we did so in Lemma 1.1 because in the applications to PDE we do not know quasi-dissipative operators which are not dissipative.

§ 2. DS-limit solutions. Integral solutions. Generation of nonlinear semigroups.

Suppose that there exists a DS-approximate solution u_n defined by (1.2). Then the ω-dissipativity of A guarantees (in view of the fundamental estimate (1.16)) the convergence of u_n (see the proof of Theorem 3.1, with s=o and $\rho(r)=o$, $r \in [o,+\infty]$). Set $u_n(t)=u_n(t;o,x_o^n)$ and

$$u(t;o,x_o)= \lim_{n\to\infty} u_n(t;o,x_o^n),\ t \geq o, x_o \in \overline{D(A)} \tag{2.1}$$

Since T > o has been arbitrarily fixed, we know (from § 3, Ch. 1) that (2.1) holds uniformly on bounded t-intervals of $R_+ = [o,+\infty[$, and that u is continuous on R_+. Moreover, any other DS-approximate solution $u_n(t)=u_n(t;o,x_o^n)$ (with $x_o^n \in D(A)$, $x_o^n \to \underline{x_o}$ as $n\to\infty$) converges also to u. For each $t \geq o$, define $S(t):\overline{D(A)} \to \overline{D(A)}$ by

$$S(t)x_o=u(t;o,x_o),\ t\geq o,\ x_o \in \overline{D(A)} \tag{2.2}$$

We will prove that S(t) is a semigroup on $\overline{D(A)}$ of (nonlinear) operators in the following sense

$$S(o)=I \text{ (the identity restricted to } \overline{D(A)}) \tag{2.3}$$

$$S(t+s)=S(t)\, S(s),\ t,s \geq o \tag{2.4}$$

$$\lim_{t\downarrow o} S(t)x=x,\quad \forall x \in \overline{D(A)} \tag{2.5}$$

$$\|S(t)x-S(t)y\| \leq e^{\omega t}\|x-y\|,\ \forall t \geq o, x,y \in \overline{D(A)} \tag{2.6}$$

(S(t) with the properties (2.3)-(2.6) is said to be a semigroup of type ω. A semigroup of type o is simply said to be nonexpansive or "contraction semigroup".

Definition 2.1. If D is a nonempty subset of X, a semigroup on D is a function S on $[o,+\infty[$ such that $S(t):D \to D$ for each $t \geq o$ and propertie (2.3)-(2.5) hold (with D in place of $\overline{D(A)}$). If in addition (2.6) is also satisfied for some $\omega \in R$, then S is said to be of type ω on D. We will then write $S \in Q_\omega(D)$. (In most cases we say "the semigroup S(t)" in place of "the semigroup S").

In the case $\omega = o$, S is said to be a nonexpansive semigroup on D or a contraction semigroup on D, or still a "semigroup of contractions" on D. If all of S(t) are linear and D=X, then $S \in Q_o(X)$ is said to be a C_o contraction semigroup or a C_o-semigroup of contractions (in this case (2.6) reduces to $\|S(t)\| \leq 1$).

In order to prove the assetions (2.3)-(2.6) it is convenient to use the notion of integral solution of the problem (1.1)+(1.1)'. According to (3.13) (with f=o) and Theorem 3.2 in Chapter 1, the "DS-limit solution" u given by (2.1), is the unique integral solution to (1.1)+(1.1)', that is u is the unique continuous function on R_+, satisfying $u(o)=x_o$ and the inequality

$$\|x(t)-x\| - \|u(t)-x\| \leq \int_t^{\bar{t}} (\omega\|u(s)-x\| + \langle y,u(s)-x\rangle_+)ds \tag{2.7}$$

for all $o \leq t \leq \bar{t}$ and $[x,y] \in A$.

The inequality (2.6) follows from (3.25) in Ch.1 with s=o, u as in (2.2) with $x_o = x$ and $u(t)=u(t;o,y)=S(t)y$. Since $u(o;o,x_o)=x_o$ (2.2) gives (2.3). Clearly (2.5) is just the continuity of u to the right of t=o. As far as (2.4) is concerned, (i.e. the semigroup property), it is a consequence of the integral solution to (1.1)+(1.1)'. Indeed, $y(t)=u(t+s;o,x_o)$ and $z(t)=u(t;o,z(s;o,x))$ are integral solutions to (1.1) with the same initial condition $y(o)=z(o)=u(s;o,x_o) \in \overline{D(A)}$. Consequently $y(t)=z(t)$, for all $t \geq o$, i.e. $S(t+s)x_o=S(t)(S(s)x_o))$, $x_o \in \overline{D(A)}, t,s \geq o$ which is just (2.4). The existence of DS-approximate solutions u_n in the sense of (1.2)-(1.6) is guaranteed by Tangential condition

$$\lim_{h \downarrow o} h^{-1} d[x;R(I-hA)] = o, \ \forall x \in \overline{D(A)} \tag{2.8}$$

(where "lim" can be replaced by "lim inf").

For the proof we refer to § 3.5 (with f=o, A(t)=A) in Ch. 1. In this case (3.55) in Ch. 1 becomes

$$\lim_{k \to \infty} t_k^n = \sum_{k=1}^{\infty} h_k^n = +\infty \tag{2.9}$$

Consequently, given $T > o$ and $n \in N$ there is a positive integer N_n such that $t^n_{N_n-1} < T \leq t^n_{N_n}$. Following the construction of t_k^n (see (3.53) in Ch.1) it is clear that we may have $T \leq t^n_{N_n}$. However, if A satisfies Range condition

$$R(I-hA) \supset \overline{D(A)}, \quad o < h \leq h_o \tag{2.10}$$

or a stronger one, i.e.

$$R(I-hA)=X, \quad o < h \leq h_o \tag{2.11}$$

(for some sufficiently small $h_o > o$) then we can choose $t^n_{N_n}=T$. Indeed, according to § 3.6 in Chapter 1, in the case of (2.10), for $t > o$ we can choose $t^n_o=o$, $t^n_k=k\frac{t}{n}$

$$x^n_k=(I-\frac{t}{n}A)^{-k}x^n_o, \; k=1,\dots,N_n \tag{2.12}$$

Thus we have

$$u_n(\bar{t})=(I-\frac{t}{n}A)^{-k}x^n_o, \text{ if } (k-1)\frac{t}{n} < \bar{t} \leq k\frac{t}{n} \tag{2.13}$$

and $u_n(o)=x^n_o$. Inasmuch as $t=n\frac{t}{n}$, (2.13) implies (for k=n)

$$u_n(t)=(I-\frac{t}{n}A)^{-n}x^n_o \tag{2.14}$$

Actually, we can replace x^n_o in (2.14) by $x_o \in \overline{D(A)}$ (For t=T, we have $t^n_n=n\frac{T}{n}=T$, i.e. $n=N_n$ and $t^n_{N_n}=T$. If $o < t < T$, then we choose N_n as the positive integer with the property

$$(N_n-1)\frac{t}{n} < T \leq N_n\frac{t}{n} \tag{2.15}$$

Of course, in this case N_n is uniquely determined by n, t and T. If A is dissipative, then Range condition (2.11) is equivalent to

$$R(I-hA)=X, \; \forall\, h > o \quad \text{or to} \quad R(I-A)=X \tag{2.11'}$$

A part of the results of this section can be listed (according to Theorem 3.6 in Ch. 1) in the following theorem due to Y. Kobayashi [1].

Theorem 2.1. (Kobayashi). 1) Let $A:D(A) \subset X \to 2^X$ satisfy Tangential condition (2.8). Then for each $T > o$, there are DS-approximate solutions u_n on $[o,T]$ to the problem (1.1)+(1.1)'.

2) If in addition, A is ω-dissipative then there is a continuous function $u=u(t;o,x_o)$ on $[o,T]$, such that all DS-approximate solutions u_n (given by (1.2)) are convergent to u:

$$u(t;o,x_o)=\lim_{n\to\infty}u_n(t), \text{ uniformly on } [o,T]. \tag{2.16}$$

3) The continuous function u (defined by (2.16)) is the unique integral solution (to (1.1)+(1.1)') on [o,T] in the sense of (2.7) for $o \leq t \leq \bar{t} \leq T$.

4) The family $S=\{S(t);\ S(t):\overline{D(A)} \to \overline{D(A)},\ t \geq o\}$ in (2.2) (in short S(t)) is a semigroup on $\overline{D(A)}$ of type ω (in the sense of (2.2)-(2.6))

5) If A is maximal dissipative and $t \to S(t)x_o$ is right-differentiable at $t_o \geq o$, then $S(t_o)x_o \in D(A)$ and

$$(d^+/dt)S(t)x_o\Big|_{t=t_o} \in AS(t_o)x_o \qquad (2.17)$$

Remark 2.1. The function u defined by (2.16) is also called (by Kobayashi) "DS-limit solution" to (1.1)+(1.1)' on $[o,+\infty[$.

Therefore the notions of "DS-limit solution" and "integral solutions" (in the sense (2.7) of Benilan)are equivalent. The semigroup S(t) defined by the DS-limit solution u to (1.1)+(1.1)', via (2.2) is said to be "the semigroup generated by A". In other words if A is ω-dissipative and satisfies Tangential condition (2.8), then it generates a semigroup S(t) on $\overline{D(A)}$, (of type ω). By Theorem 3.2 in Chapter 1, it follows that every strong solution u to (1.1)+(1.1)' on $[o,+\infty[$ is (the unique) integral solution on $[o,+\infty[$ to (1.1)+(1.1)', i.e. $u(t;o,x_o)=S(t)x_o, t \geq o$.

The uniqueness of the strong solution to the Cauchy problem (1.1)+(1.1)' can be proved directly as in Chapter 1 (see the inequality (3.9)). Clearly, by a strong solution to (1.1)+(1.1)' we mean a continuous function $u:[o,+\infty[\to X$ with the property (1), (2) and (3) in Definition 3.1 (Ch. 1), with $T=\infty$, s=o and A(t) = A, $t \geq o$.

Some other remarks are necessary. There are dissipative operators A satisfying (2.8) and even more than (2.8), i.e.

$$\overline{D(A)}=X,\ R(I-\lambda A)=X,\ \forall \lambda > 0 \qquad (2.18)$$

such the semigroup S genenrated by A via Theorem 3.1 is nowhere differentiable (that is for each $x \in X$, $t \to S(t)x$ is not differentiable at any $t_o \geq o$). Such examples can be found in Crandall and Liggett [1]. However, $u=u(t;o,x_o)= S(t)x_o,\ x_o \in \overline{D(A)},\ t \geq o$ is the unique integral solution to (1.1)+(1.1)', in the sense of (2.7), i.e.

$$\|S(t+h)x_o-x\| - \|S(t)x_o-x\| \leq \int_t^{t+h} (\omega\|S(\tau)x_o-x\| + \langle y,\ S(\tau)x_o-x\rangle_+)dr \qquad (2.19)$$

for all $x_o \in \overline{D(A)}$, $[x,y] \in A$, $t,h \geq o$.

In view of (3.29) (in Ch. 1), (2.19) is equivalent to:

$$\|S(t+h)x_o - x\|^2 - \|S(t)x_o - x\|^2 \leq 2\int_t^{t+h} (\omega\|S(\tau)x_o - x\|^2 + \langle y, S(\tau)x_o - x\rangle_s)d\tau \tag{2.19)'$$

for all $x_o \in D(A)$, $[x,y] \in A$, $t,h \geq o$.

Note also that (2.19)' is equivalent to

$$e^{-2\omega(t+h)}\|S(t+h)x_o - x\|^2 - e^{-2\omega t}\|S(t)x_o - x\|^2$$
$$\leq 2\int_t^{t+h} e^{-2\omega\tau}\langle y, S(\tau)x_o - x\rangle_s \, d\tau \tag{2.19)"$$

for all $x_o \in \overline{D(A)}$, $[x,y] \in A$, $t,h \geq o$ (see Kobayashi [1]). In this book (2.19)" will not be used.

Other aspects on the relationship between $u(t)=S(t)x_o$ and the Cauchy problem (1.1)+(1.1)' are given in § 5. We end this section with the following result essentially due to Benilan.

Theorem 2.2. Let $A:D(A) \subset X \to 2^X$ be m-dissipative, and let $f,g \in L^1_{loc}(o, +\infty;X)$. Denote by $u(t)=u(t;s,x)$, $v(t)=v(t;s,y)$ the integral solutions to the problems

$$u'(t) \in Au(t)+f(t), \quad u(s)=x \in \overline{D(A)}; \tag{2.20}$$

$$v'(t) \in Av(t)+g(t), \quad v(s)=y \in \overline{D(A)}, \tag{2.21}$$

The, the following estimate holds

$$\|u(t)-v(t)\| \leq \|u(\tau)-v(\tau)\| + \int_\tau^t \|f(\theta)-g(\theta)\|d\theta \tag{2.22}$$

for all $o \leq s \leq \tau \leq t$.

Proof. One applies Theorem 4.2 in Ch. 1 with f,g continuous from $[o, +\infty[$ into X. Then, by density arguments one obtains the conlusion of the Theorem.

§ 3. Other properties of nonlinear semigroups

3.1. Lipschitz continuity

Let D be a nonempty subset of the banach space X. Recall that a family of operators $S(t):D \to D$ (with $t \geq o$) is said to be a semigroup on D if the properties (2.3) (2.4) and (2.5) are fulfilled. S(t) is said to be of type ω R if (2.6) is satisfied on D.

<u>Proposition 3.1.</u> Let $S(t)$ be a semigroup on D of type ω and let $x \in D$ be such that

$$\liminf_{t \downarrow 0} \frac{\|S(t)x-x\|}{t} \equiv L(x) < +\infty \tag{3.1}$$

Then $t \to S(t)x$ is Lipschitz continuous on bounded subsets of $[0,+\infty[$, namely

$$\|S(t+h)x-S(t)x\| \leq e^{\omega t}((e^{\omega h}-1)/\omega)L(x), \ \forall\, t,h \geq 0 \tag{3.2}$$

<u>Proof.</u> One applies Theorem 3.3 in Ch. 1 with $s=0$, $L(0,x)=L(x)$ and $V_s^T=0$. In our case $U(t+s,s)x=S(t)x$. Actually the reader can prove (3.2) directly (without following the nonautonomous case in Theorem 3.3, Ch. 1). Indeed, it is clear that in this case the inequality (3.44) in Ch. 1, becomes

$$\|S(mh)x-S(nh)x\| \leq \frac{e^{\omega mh}-e^{\omega nh}}{e^{\omega h}-1} \|S(h)x-x\| \tag{3.3}$$

where the notations are those in the proof of Theorem 3.3 Ch. 1. Since $(e^{\omega h}-1)/h \to \infty$, $mh \to \bar{t}$ and $nh \to t$ as $h \downarrow 0$, (3.3) yields

$$\|S(\bar{t})-S(t)x\| \leq \omega^{-1}(e^{\omega \bar{t}}-e^{\omega t})L(x) \tag{3.4}$$

For $t=t+h$, (3.4) is just (3.2) and the proof is complete. For $\omega=0$, (3.2) becomes

$$\|S(t+h)x-S(t)x\| \leq hL(x), \quad \forall t,h \geq 0 \tag{3.5}$$

i.e. $t \to S(t)x$ is Lipschitz continuous of Lipschitz constant $L(x)$ given by (3.1). Note that (1.16) yileds (see, if necessary, the proof of (3.5)' in Ch. 1)

$$\|S(t)x-S(s)x\| \leq |t-s| \ \|y\| \ \exp(\omega_0(t+s)) \tag{3.6}$$

for each $x \in D(A)$ and every $y \in Ax$ where $\omega_0=\max(0,\omega)$, $t,s \geq 0$. In other words,

<u>Proposition 3.2.</u> If $S(t)$ is the semigroup on $\overline{D(A)}$ generated by the operator A via Theorem 2.1, then for each $x \in D(A)$, the function $t \to S(t)x$ is Lipschitz continuous on bounded intervals of R_+ (namely, (3.6) holds). If $\omega=0$, then $t \to S(t)x$ is Lipschitz continuous on R_+, and $|Ax| = \inf\{\|y\|, \ y \in Ax\}$ is a Lipschitz constant ($\|S(h)x-x\| \leq h|Ax|$, $h \geq 0$, $x \in D(A)$).

<u>Remark 3.1.</u> Better results on Lipschitz continuity of $t \to S(t)x$ will be

given in § 4.

The following results will be used in Section 5.

Proposition 3.3. Let $\omega \in R$ and let A be ω-dissipative satisfying Tangential condition (2.8). Let S(t) be the semigroup generated by A (via Theorem 2.1). Then for each $x \in D(A)$, $[u,v] \in A$ and $x^* \in F(x-u)$, we have

$$\liminf_{t \downarrow 0} \left\langle \frac{S(t)x-x}{t}, x^* \right\rangle \leq \omega \|x-u\|^2 + \langle v, x-u \rangle_s \tag{3.7}$$

or equivalently; for each $x \in \overline{D(A)}$ and $[u,v] \in A$ there is $x^* \in F(x-u)$ such that

$$\liminf_{t \downarrow 0} \left\langle \frac{S(t)x-x}{t} - v, x^* \right\rangle \leq \omega \|x-u\|^2 \tag{3.7)'$$

Proof. It is easy to check that

$$2\langle S(t)x-x, x^*\rangle \leq \|S(t)x-u\|^2 - \|x-u\|^2, \quad x^* \in F(x-u)$$

and

$$\liminf_{t \downarrow 0} \frac{1}{t} \int_0^t \langle v, S(\tau)x-u \rangle_s \, d\tau \leq \langle v, x-u \rangle_s \tag{3.8}$$

Combining these two inequalities with (2.19)' (i.e. with)

$$\|S(t)x-u\|^2 - \|x-u\|^2 \leq 2\omega \int_0^t \|S(\tau)x-u\|^2 \, d\tau +$$

$$+ 2 \int_0^t \langle v, S(\tau)x-u \rangle_s \, d\tau \tag{3.9}$$

for $x \in \overline{D(A)}$, $[u,v] \in A$ and $t \geq 0$, one obtains (2.26). The equivalence of (3.7) and (3.7)' follows from the existence of $x^* \in F(x-u)$ such that $\langle v, x-u \rangle_s = \langle v, x^* \rangle$ (Proposition 2.1, Ch. 1). This completes the proof.

3.2. The relationship of the semigroup $S_A(t)$ and $I(-tA)^{-1}$ The exponential formula of Crandall-Liggett

The fundamental result in the theory of nonlinear semigroups on general Banach space is given by

Theorem (of Crandall-Liggett). If A is ω - dissipative satisfying the range condition (2.10), then the semigroup S(t) generated by A (via (2.16)) is given by the following "exponential formula"

$$S(t)x = \lim_{n\to\infty} \left(I-\frac{t}{n}A\right)^{-n} x \equiv e^{tA}x, \quad x \in \overline{D(A)}, \ t \geq 0 \tag{3.10}$$

Proof. Indeed, in this case a DS-approximate solution u_n of the problem

$$u'(t) \in Au(t), u(o) = x \in \overline{D(A)}, \quad t \geq o \tag{3.11}$$

can be chosen as in (2.14). See also Remark 3.4 and Theorem 5.1. In many problem in S(t) (e.g. in the characterization of compactness of S(t)) the following result of Brezis [4] is particularly important

Theorem 3.1. Let S(t) be the semigroup generated by the dissipative operator A via the exponential formula of Crandall-Liggett. Then for all $\lambda > o, t > o$ and $x \in \overline{D(A)}$,

$$||J_\lambda x-x|| \leq \frac{2}{t}(1+\frac{\lambda}{t}) \int_o^t ||S(s)x-x||ds \tag{3.12}$$

and

$$||S(t)x-x|| \leq (2+\frac{t}{\lambda}) \; ||J_\lambda x-x|| \tag{3.13}$$

Proof. The inequality (3.13) follows from (5.14) in Ch. 1, with s=o, $U(t,o)x=S(t)x, A(t)=A$ and $\rho=o$. To get (3.12) we first observe that (with the remarks above) (5.7) in Chapter 1 becomes

$$||J_\lambda x-x|| \leq \frac{2}{t} \int_o^t ||S(\tau)x-x||d\tau + \frac{\lambda}{t} ||S(t)x-x|| \tag{3.14}$$

On the other hand

$$||S(t)x-x|| \leq \frac{2}{t} \int_o^t ||S(\tau)x-x||d\tau, \; t > o, \; x \in \overline{D(A)} \tag{3.15}$$

Clearly, (3.14) and (3.15) give (3.12). It remains to check (3.15). To this goal, we have

$$||S(t)x- \frac{1}{t} \int_o^t S(\tau)xd\tau|| = ||\frac{1}{t} \int_o^t S(t)x-S(\tau)x)d\tau ||$$

$$\leq \frac{1}{t} \int_o^t ||S(t-\tau)x-x||d\tau = \frac{1}{t} \int_o^t ||S(\tau)x-x||d\tau$$

which yields (3.15). This completes the proof. For $\lambda = t$, (3.12) becomes

$$||J_t x-x|| \leq \frac{4}{t} \int_o^t ||S(\tau)x-x||d\tau, \quad t > o, x \in \overline{D(A)} \tag{3.16}$$

which shows that

$$\lim_{\lambda \downarrow o} J_\lambda x=x, \qquad x \in \overline{D(A)} \tag{3.16'}$$

Corollary 3.1. In the hypotheses of Theorem 3.1,

$$\limsup_{t\downarrow o} \frac{\|S(t)x-x\|}{t} = \lim_{t\downarrow o} \frac{\|J_t x-x\|}{t}, \quad x\in \overline{D(A)} \tag{3.17}$$

proof. We know that $t\to\|A_t x\|$ (where $A_t x=t^{-1}(J_t x-x)$, $J_t=(I-tA)^{-1}$) is an increasing function, for each $x\in \overline{D(A)}$ (see Appendix). Consequently $\lim_{t\downarrow o} \|A_t x\| = b(x)$ exists. Obviously, there are only two possibilities: either $b(x)$ is finite or $b(x) = +\infty$.

Let us prove (3.17) in the frist case (i.e. $b(x) < +\infty$) a simple combination of (3.16)' and (3.13) gives

$$\|S(t)x-x\| \le tb(x), \quad \forall t>o \tag{3.18}$$

Set $a(x)= \limsup_{t\downarrow o} \frac{\|S(t)x-x\|}{t}$. By (3.18), $a(x)\le b(x) < +\infty$. Now, since $a(x)< +\infty$, on the basis of (3.5) with $L(x)\le a(x)$ we have

$$\|S(t)x-x\| \le ta(x), \; \forall t>o \tag{3.19}$$

This inequality and (3.12) imply

$$\|J_\lambda x-x\| \le (t+\lambda)a(x), \; \forall t>o, \lambda>o \quad \text{that is } \|J_\lambda x-x\| \le \lambda a(x), \lambda>o \tag{3.20}$$

hence $b(x)\le a(x)$, so we have $a(x)=b(x)$ (if $b(x)< +\infty$). Let us observe that (3.20) holds in the hypothesis $a(x)< +\infty$. Therefore, if $b(x)=+\infty$, then $a(x)$ cannot be finite. We conclude that $b(x)=+\infty$ implies $a(x)=+\infty$, which completes the proof of (3.17).

Remark 3.2. Formula (3.17) does not imply that $\frac{\|S(t)x-x\|}{\|J_t x-x\|}$ converges to 1 as $t\downarrow o$ ($x \in\overline{D(A)}$). Indeed, let us consider the example:

$$X=R= \,]-\infty,+\infty[\, , \quad Ax = \begin{cases} \frac{1}{x}, & x>o \\ \emptyset, & x\le o \end{cases} \tag{3.21}$$

that is $D(A)=]o,+\infty[$. It is clear that A is m-dissipative and $J_t r= \frac{r+\sqrt{r^2+4t}}{2}$, $r\in R$, $t>o$, so $J_t o= \sqrt{t}$. Moreover, in this case

$$S(t)x = \sqrt{2t+x^2}, \; x\in [o,+\infty[\; = \overline{D(A)}, \; t\ge o$$

hence $S(t)o= 2t$ and therefore

$$\frac{\|S(t)o\|}{\|J_t o\|} = 2$$

In the case of a real Hilbert space H(of inner product $<,>$ and norm $\|.\|$) an estimate better than (3.16) holds, namely

Theorem 3.2. Let A be a maximal dissipative operator acting on H.

Then

$$\|J_t x-x\| \le \frac{2}{t}\int_0^t \|S(\tau)x-x\|d\tau \ , \ \forall x\in \overline{D(A)},\ t>o \tag{3.22}$$

Moreover, the constant 2 is the best possible.

Proof. In this case (3.9) becomes

$$\|S(t)x-u\|^2 \le \|x-u\|^2 + 2\int_0^t \langle v,S(\tau)x-u\rangle d\tau \tag{3.23}$$

for all $x\in\overline{D(A)}$, $[u,v]\in A$ and $t\ge o$. But $\langle v,S(\tau)x-u\rangle = \langle v,S(\tau)x-x+x-u\rangle$, therefore

$$\|S(t)x-u\|^2 \le \|x-u\|^2+2t\langle v,x-u\rangle+2\ \ \|v\|\int_0^t \|S(\tau)x-x\|d\tau$$

For $u=J_t x$ and $v=A_t x$ we have

$$\|S(t)x-J_t x\|^2 \le \|x-J_t x\|^2+2\|J_t x-x\|^2+2\|J_t x-x\|\ \frac{1}{t}\int_0^t \|S(\tau)x-x\|d\tau$$

which obviously implies (3.22). In order to prove the last assertion, suppose that (3.22) holds with a constant $c>o$ in place of 2. Recall that for $x\in D(A)$, $\|S(\tau)x-x\|\le\tau\|z\|$, for all $z\in Ax$ (by (3.6), with $\omega_o=o=s$. Consequently, (3.22) yields

$$\|J_t x-x\| \le \frac{c}{t}\ \|z\|\ \frac{t^2}{2},\ \text{i.e.}\ \|A_t x\|\le\frac{c}{2}\|z\|,\ \forall z\in Ax$$

Choosing $z=A_o x$ (where $\|A_o x\| = \inf\{\|z\|,\ z\in Ax\}$) and letting $t\downarrow o$ we get $\|A_o x\|\le\frac{c}{2}\ \|A_o x\|$, $x\in D(A)$, so $c\le 2$.

Remark 3.3. In the case $A=-\partial\varphi$ (the subdifferential of a lower semicontinuous convex function $\varphi: H\to\ ]-\infty,+\infty]$)the following estimate holds

$$\|J_t x-x\| \le (1+\frac{1}{\sqrt{2}})\|S(t)x-x\|,\ x\in\overline{D(A)},\ t>o \tag{3.24}$$

(where the best constant are not known).

For a general dissipative operator an inequality of the form (3.24) does not hold (For example, in the case $H=R^2$ and A - a rotation of $\frac{\pi}{2}$, (3.24) fails) see Brezis [4].

Remark 3.4. It is natural to think of $S(t)$ (given by (3.10)) as e^{tA}. This is because if A is a number (and X=R) then

$$e^{At}=\lim_{n\to\infty}(1-\frac{t}{n}A)^{-n} \text{ and } u(t)=xe^{At}\equiv S(t)x$$

is the unique solution to the problem $u'(t)=Au(t)$, $t\in R, u(o)=x, x\in R$. Moreover, the "inversion" formula

$$\frac{1}{\lambda-A} = \int_0^\infty e^{-\lambda t}e^{tA}dt,\ \lambda>\mathrm{Re}(A) \tag{3.25}$$

holds. This formula suggests the operator version

$$R(\lambda:A)\equiv(\lambda I-A)^{-1}x=\int_0^\infty e^{-\lambda t}S(t)x\,dt,\ x\in X,\lambda>0 \tag{3.26}$$

Indeed, if A is a linear m-dissipative and densely defined operator acting in the Banach space X, then (3.26) holds for all $\lambda>0$. If A is ω-dissipative then (3.26) holds for $\lambda>\omega$. See the proof of Theorem 6.6.

§ 4. A generalized domain for semigroup generators

We have already see that if the operator A acting in the Banach space X is dissipative and satisfies the range condition (2.10), the the semigroup S(t) generated by the exponential formula (3.10) leaves $\overline{D(A)}$ invariant. In general, S(t) will not leave D(A) invariant. In this section we assign a "generalized domain" (following Crandall [2]) $\hat{D}(A)$ to A with the properties

$$D(A)\subset\hat{D}(A)\subset\overline{D(A)},\quad S(t)\hat{D}(A)\to\hat{D}(A),\ \forall\ t\geq 0 \tag{4.1}$$

In words, we assign a set $\hat{D}(A)$ to A with the property that S(t) leaves $\hat{D}(A)$ invariant and that $\hat{D}(A)$ is larger than D(A) and included in $\overline{D(A)}$. By an example, one shows that in general the above inclusion holds strictly. Recall that if A is ω-dissipative then $\lambda\to(1-\lambda\omega)\|A_\lambda x\|$ is nonincreasing (Appendix). For simplicity, assume that $\omega=0$. Consequently, $\lim_{\lambda\downarrow 0}\|A_\lambda x\|$ with $x\in\overline{D(A)}$ exists. Set

$$|Ax\| = \lim_{\lambda\downarrow 0}\|A_\lambda x\|,\quad x\in\overline{D(A)} \tag{4.2}$$

and define (see (6.4)in Ch. 1)

$$\hat{D}(A) = \{x\in\overline{D(A)};\ \ |Ax\|<+\infty\}=\{x\in\overline{D(A)};\ \|A_\lambda x\| \text{ is bounded as } \lambda\downarrow 0\} \tag{4.3}$$

Since $\|A_\lambda x\|\leq\|y\|$, for all $x\in D(A)$ and $y\in Ax$, it follows

$$|Ax\|\leq\|y\|;\ \forall x\in D(A),y\in Ax \tag{4.4}$$

hence, for $0<\lambda\leq h_0$ (see (2.10))

$$\|A_\lambda x\|\leq|Ax\|\leq|Ax| = \inf\{\|y\|,\ y\in Ax\},\quad x\in D(A) \tag{4.4'}$$

For every $x\in D(A)$, $J_\lambda x\in D(A)$ and $A_\lambda x\in AJ_\lambda x$. This remark (along with (4.4)) yields

$$\|AJ_\lambda x\| \le \|A_\lambda x\|,\ \forall x\in D(A) \tag{4.5}$$

According to (4.4)' we have the inclusions in (4.1). In order to give some characterizations of D(A) set:

$$D_L(A)=\{x\in D(A);\ t\to S(t)x \text{ is Lipschitz continuous on } [o,+\infty[\ \} \tag{4.6}$$

$$D_a(A)=\{x\in D(A);\ \exists\ \{x_n\}\subset D(A), \text{ such that } x_n\to x \text{ and } \sup|Ax_n|<\infty\} \tag{4.7}$$

We are now in the position to give

Theorem 4.1 Let $A:D(A)\subset X\to 2^X$ be a dissipative operator acting in the real Banach space X. Suppose in addition that A satisfies Range condition (2.10). Then

(1) $\hat{D}(A)=D_a(A)=D_L(A)$ and Inclusions in (4.1) hold.

(2) $x\to|Ax\|$ is lower semicontinuous on $\hat{D}(A)$.

Proof (1). The fact that $\hat{D}(A)$ equals $D_L(A)$ uis a direct combination of (3.17) in Corollary 3.1 with Proposition 3.1. In view of the semigroup property $S(t+s)=S(t)S(s))$ it is clear that $S(t):D_L\to D_L$. Now let x be in $\hat{D}(A)$, that is $\|A_\lambda x\| \le \lim_{\lambda\downarrow o} \|A_\lambda x\| = |Ax\| < \infty$ for $\lambda > o$. Set $x_n=J_{1/n}x$. Then $x_n\in D(A)$ and $x_n\to x$ as $n\to\infty$ (see (3.16)' or directly: $\|x_n-x\| \le \frac{1}{n}\|A_{1|n}x\| \le \frac{1}{n}|Ax\|$). Moreover, since $A_{1|n}x\in Ax_n$ it follows $|Ax_n|\le\|A_{1|n}x\| \le |Ax\| < \infty$, for all positive integers n. Therefore $\hat{D}(A)\subset D_a(A)$. Conversely, let $x\in D_a(A\ )$ and let x_n be as indicated in (4.7). We have (see also(4.4)')

$$\|A_\lambda x\| \le \|A_\lambda x-A_\lambda x_n\| + \|A_\lambda x_n\| \le \frac{2}{\lambda}\|x_n-x\| + |Ax_n| \tag{4.8}$$

for $\lambda>o$ and n=1,2,...

Inasmuch as $\sup/Ax_n|<+\infty$ and $x_n\to x$, (4.8) implies $x\in\hat{D}(A)$. We now conclude that $\hat{D}(A)=D_a(A)$. On the basis of (4.4)', we see that $D(A)\subset\hat{D}(A)$. Part (2) is a consequence of $|Ax\| = \sup\|A_\lambda x\|$ and of the continuity of $x\to A_\lambda x$. This completes the proof.

Remark 4.1. If A is ω-dissipative with $\omega>o$, then the condition "$t\to S(t)x$ is Lipschitz continuous on $[o,+\infty[$" in (4.6) must be replaced by "$t\to S(t)x$ is locally Lipschitz continuous on R_+". This is easily seen from (3.2).

Example 4.1. The example below shows $\hat{D}(A)$ need not equal D(A) even if A is linear and densely defined. Set $X=C_o(R_+)$ $(R_+=[o,+\infty[)$ the set of all real-valued continuous functions on R_+ tending to zero at infinity, under the maximum norm. Define

$$D(A)=\{ f\in C_o(R_+);\ f'\in C_o(R_+)\},\ Af=f',\ f\in D(A) \tag{4.9}$$

where f' denotes the strong derivative of f. It is easy to check that

$$R(I-\lambda A)=C_o(R_+),\ \forall\lambda>0 \tag{4.10}$$

Indeed, if g $C_o(R_+)$, then the unique solution f to the problem $f_\lambda - \lambda f_\lambda' = g$ in $C_o(R_+)$, is given by

$$f_\lambda(x)=(J_\lambda g)(x)=((I-\lambda A)^{-1}g)(x)=\frac{1}{\lambda}\int_x^\infty e^{\frac{x-s}{\lambda}} g(s)ds =$$

$$= \int_o^\infty e^{-s} g(x+\lambda s)ds,\quad g\in C_o(R_+),\quad x\in R_+ \tag{4.11}$$

Clearly $\|J_\lambda g\| \le \|g\| \int_o^\infty e^{-s}ds=\|g\|$, for all $g\in C_o(R_+)$, that is A is dissipative and therefore (in view of (4.10)) A is m-dissipative. In this case the semigroup S(t) generated by A via the exponential formula ig given by

$$S(t)f)(x)=f(t+x),\ t,x\ge o,\ f\in C_o(R_+) \tag{4.12}$$

(i.e. translations) and $D(A)=X=C_o(R_+)$. Obviously, Lipschitz continuity on R_+ of t S(t) is just Lipschitz continuity of f on R_+. This remark along with Theorem 4.1 yields

$$\hat{D}(A)=\{ f\in C_o(R_+);\ f \text{ is Lipschitz continuous on } R_+\} \tag{4.13}$$

hence $\hat{D}(A)$ is strictly larger than D(A) in this case (and $\overline{D(A)}=C_o(R_+)$ is strictly larger than $\hat{D}(A)$). Indeed let f_o be the function $f_o(x)= \int_x^{+\infty}\sin t^2dt$, $x\ge o$. Since the Fresnel integral $\int_o^\infty \sin t^2 dt$ u is convergent, it follows $f\in C_o(R_+)$. Moreover, $f_o' \bar{\in} c_o(R_+)$ $(f_o'(x)=-\sin x^2)$ hence $f\in\hat{D}(A)$ but $f_o \bar{\in} D(A)$. It is clear that the function $g(x)= \sqrt{1-x}$ if $o \le x < 1$ and $g(x)=o$ if $x \ge 1$ is in $C_o(R_+)=\overline{D(A)}$, but $g \bar{\in} \hat{D}(A)$. It is also interesting to point out that $|Af\| = \lim\|A_\lambda f\|$ is just the least Lipschitz constant L for f, where $A_\lambda f = (J_\lambda f-f)/\lambda$, therefore

$$(A_\lambda f)(x) = \int_o^\infty e^{-s}\ \frac{f(x+ s)-f(x)}{\lambda}\ ds \tag{4.13'}$$

for all $x\in R_+$, $f\in C_o(R_+)$ and $\lambda > 0$. Indeed, if we have $|f(t+x)-f(x)| \le tL$, $\forall t$, $x\ge o$ then $\limsup \frac{\|S(t)f-f\|}{t} \ge a(f) \le L$. According to (3.17), $\lim\|A_\lambda f\| = a(f)\le L$ and (in view of Proposition 3.1) a(f) is a Lipschitz constant for t S(t)f, hence for f on R_+. Thus $|Af\| = \lim\|A f\|=a(f)=L$ (the least Lipschitz constant for f on R_+). Consequently, we have proved (via semigroup theory) the following classical result:

Proposition 4.1. Let $A_\lambda : C_o(R_+) \to C_o(R_+)$ be defined as in (4.13). Then for each $f \in C_o(R_+)$ the following properties hold

(1) $\lambda \to \|A_\lambda f\|$ is nonincreasing on R_+

(2) f is Lipschitz continuous on R_+ if and only if $\lim \|A_\lambda f\| \equiv |A\| < +\infty$. Moreover, in the case $|Af\| < +\infty$, $|Af\|$ is the least LIpschitz constant for f on R_+.

(3) If f is differentiable in R_+ and $f' \in C_o(R_+)$, then
$A_\lambda f \to f$ as $\lambda \downarrow o$ (in $C_o(R_+)$)

Of course, by using (4.13) one can prove directly Proposition 4.1 (without semigroup approach!) In this case we have

$$Af = \lim_{t \downarrow o} \frac{S(t)f-f}{t}, \text{ in } X=C_o(R_+), \forall f \in D(A) \tag{4.14}$$

i.e. A is the infinitesimal generator of S(t).

Clearly, in this case D(A) is given by (4.9). In general,

$$D(A)= \{f \in X; \lim_{t \downarrow o} \frac{S(t)f-f}{t} \text{ exists}\} \tag{4.14)'$$

A C_o semigroup S on X is uniquely determined by its infinitesimal generator A. Indeed let S and $\bar{S}$ be two C_o semigroupos on X having the same infinitesimal generators A. Set

$$u(s)=S(s)\bar{S}(t-s)f, \ f \in D(A), \ o \leq s < t \quad \text{Then}$$

$$\frac{du(s)}{ds} = S(s)(-A)S(t-s)f+S(s)A\bar{S}(t-s)f=o$$

which implies that u is constant, so u(o)=u(t), i.e. $\bar{S}(t)f=S(t)f$, $\forall t > o$ and $f \in D(A)$). Since $\overline{D(A)}=X$ it follows $\bar{S}(t)=S(t)$ on X. Now let S be a C_o contraction semigroup on X and let A be its infinitesimal generator. Then A is a densely defined m-dissipative operator. Moreover, the semigroup S generated by A via the exponential formula $\lim_{n\to\infty} (I-\frac{t}{n}A)^{-n}f$ is just S, i.e.

$$S(t)f= \lim_{n\to\infty} (I-\frac{t}{n}A)^{-n}f, \ f \in X, \ t \geq o$$

(See Hille-Yosida Theorem in § 6).

Remark 4.2. If X* is uniformly convex and A id m-dissipative then $\hat{D}(A)=D(A)$. Indeed, $x \in \hat{D}(A)$ is equivalent to the boundedness of $\|A_\lambda x\|$ so we may assume (relabeling if necessary) that $A_\lambda x$ is weakly convergent to an element $y \in X$ (this is because X is reflexive). On the other hand, A is demiclosed $[J_\lambda x, A_\lambda x] \in A$, and $J_\lambda x \to x$ as $\lambda \downarrow o$. It follows that $x \in D(A)$

(and $y \in Ax$). See Appendix. In particular, for every maximal dissipative operators A acting in a Hilbert space, $\hat{D}(A)=D(A)$. In other words, in this case $\|A_\lambda x\| \to +\infty$ for all $x \bar{\in} \overline{D(A)}$. Moreover if X and X* are uniformly convex, then

$$\lambda\|A_\lambda x\| = \|J_\lambda x-x\| \geq d[x,\overline{D(A)}] > 0, \ \forall \lambda > 0, x \bar{\in} \overline{D(A)} \qquad (4.14)''$$

hence $\lambda\|A_\lambda x\|^2 \to +\infty$ as $\lambda \downarrow 0$, for $x \bar{\in} \overline{D(A)}$.

<u>Remark 4.3</u> It is well known that in the linear case the infinitesimal generator A of a contraction semigroup S (t) on X is densely defined (i.e. $\overline{D(A)}=X$). In nonlinear case, $\overline{D(A)}$ need not equal X. The first example in this direction was given by Webb namely:

Set $X=C([0,1])$-the space of all real-valued continuous functions $x:[0,1] \to R$, under the maximum norm. Define the function $f:R \to R$ by $f(r)=r$ if $r > 0$ and $f(r)=2r$ if $r \leq 0$. For each $t \geq 0$, set

$$(S(t)x)(s)=f(t+f^{-1}(x(s))), \ \forall s \in [0,1], \ x \in C([0,1]) \qquad (4.15)$$

It is easy to check that S(t) is a contraction semigroup on C([0,1]). Inasmuch as $f^{-1}(r)=r$ if $r \geq 0$ and $f^{-1}(r)=\frac{r}{2}$ if $r \leq 0$, it is readily seen that for each $s \in [0,1]$,

$$\frac{(S(t)x)(s)-x(s)}{t} = \begin{cases} 1, & \text{if } x(s) \geq 0, \ \forall\, t > 0 \\ 2, & \text{if } x(s) < 0, \quad (0 < t < -x(s)/2) \end{cases} \qquad (4.16)$$

for every $x \in C([0,1])= C$. It follows that

$$D(A)=\{x \in C;\ x(s) \geq 0,\ \forall s \in [0,1]\} \cup \{x \in C; x(s) < 0,\ \forall s \in [0,1]\} \qquad (4.17)$$

and therefore

$$\overline{D(A)}= \{ x \in C;\ x \text{ does not change the sign on } [0,1] \} \qquad (4.17)'$$

so C([0,1]) is strictly larger than $\overline{D(A)}$. However, if X* is uniformly convex then the infinitesimal generator of S(t) is also densely defined as in linear case. This aspect was carried out by Baillon [1]. If X is nonreflexive, there are examples of contraction semigroups S(t) on X for which the infinitesimal generator (in the sense of (4.14) and (4.14)-is empty i.e.(S(t)f-f)/t do not converge (in any sense as $t \downarrow 0$ to an element "Af" of X, for any $f \in X$. See Crandall-Liggett [1]

<u>Definition 4.1.</u> A (nonempty) subset D of $\overline{D(A)}$ is said to be "flow-invariant" or simply - "invariant" with respect to the differential equation (inclusion) (1.1) if every integral (strong) solution $u=u(t;t_0;x_0)$ to

this equation) starting in D(i.e. $u(t_o)=x_o \in D$) remains in D as long as it exists (i.e. $u(t;t_o,x_o) \in D$, $\forall t \geq t_o, t$ in the domain of u).

On the basis of Theorem 2.1, we can say that if A is ω-dissipative then Tangential condition (2.8) is sufficient for the invariance of $\overline{D(A)}$ with respect to (1.1). This is because the integral solution $u(t;t_o,x_o) = S(t-t_o)x_o$ to (1.1) has the property in Definition 4.1 (i.e. $S(t):\overline{D(A)} \to \overline{D(A)}$).

We will see in the next section that if in addition A is single-valued and continuous on $\overline{D(A)}$, then (2.8) is also necessary for the invariance of $\overline{D(A)}$ (that is, in this case, for the existence of the solution to (1.1)). A proper sense has the invariance of $\hat{D}(A)$ with respect to (1.1). This is because $\hat{D}(A)$ is strictly larger than D(A). We can lalso say that(in general) D(A) is not invariant with respect to (1.1) (since S(t) does not leave D(A) invariant). See also Corollary 5.3.

§ 5. <u>Strong solutions</u>

5.1. <u>A characterization of strong solution under "Range condition"</u>.

We will give other results on the Cauchy problem

$$u'(t) \in Au(t), \text{ a.e. on } R_+, \ u(o)=x \in \overline{D(A)} \qquad (5.1)$$

where $A:D(A) \subset X \to 2^X$ is a ω-dissipative operator satisfying Range condition (2.10) (hence A generates the semigroup S(t) of type ω on $\overline{D(A)}$ given by (3.10)). Recall that by a strong solution to (5.1) on $R_+ = [o,+\infty[$ we mean a function $u:R_+ \to \overline{D(A)}$ which is:-

(1°) Lipschitz continuous on compact subsets of R_+,

(2°) Differentiable almost everywhere on R_+ and (5.1) is satisfied a.e. on R_+.

We have already proved (Theorem 2.1,5)) that if A is maximal dissipative and $u(t)=S(t)x$ is right-differentiable at $t_o \geq o$, then $u(t_o)=S(t_o)x \in D(A)$ and $u'(t_o) \in Au(t_o)$. Unfortunately, if X id not reflexive, the problem (5.1) may not have a strong solution on $[o,T[$ for any $T > o$. This is because if X is not reflexive, then an absolutely (or even Lipschitz) continuous function of a real variable with values in X need not be differentiable anywhere. There are examples of m-dissipative operators A for which $S(t)=u(t)$ is not differentiable anywhere on R_+ (see Crandall-Liggett[1]). They give an example of a dissipative set A satisfying $\overline{D(A)}=X=C([-1,1])$ and $R(I-hA)=X$ for all $h>o$, such that any of the usual generators one would assign to the semigroup $S(t)$ generated by A via (3.10) is empty. Namely, they show that $(S(t)x-x)/t$ do not converge in any sense as $t\downarrow o$ to an element of $C([-1,1])$ for any $x\in C([-1,1])$. We known that $u(t)=S(t)x$ is (the unique) integral solution to (5.1). The relationship of the function $t\to S(t)x$ and the Cauchy problem (5.1) is given by:

Theorem 5.1. (Crandall-Liggett) Let $A:D(A)\subset X\to 2^X$ be ω-dissipative satisfying Range condition $R(I-hA)\supset\overline{D(A)}$ for all sufficiently small positive h. Assume in addition that A is closed and $x\in\hat{D}(A)$. Then Conditions (1°) and (2°) below on a function $u:R_+\to X$ are equivalent

(1°) u is a strong solution to (5.1)

(2°) $u(t) = \lim_{n\to\infty}(I-\frac{t}{n}A)^{-n}x \equiv S(t),\ t\in R_+$

and $t\to S(t)x$ is (strongly) differentiable almost everywhere (a.e.) on R_+.

Remark 5.1. A is said to be closed if its graph (denoted again by A - see Appendix) is closed in $X\times X$, that is: if $x_n\in D(A)$, $x_n\to x, y_n\in Ax_n$, $y_n\to y$ as $n\to\infty$, then $x\in D(A)$ and $y\in Ax$. If A is maximal dissipative then A is closed. However if A is dissipative and $(R(I-hA)\supset\overline{D(A)}$ for all $h>o$, then A need not be closed (even in linear case). Indeed, take $X=R$, $D(A)= = [0,1[$ and $Ax=-x$, $o\le x<1$. Clearly A is ω-dissipative with $\omega=-1$ and $R(I-hA) = [o,1+h[\supset[o,1]$, $\forall\ h>o$ but A is not closed. Now, a comparison between Theorems 2.1 and 5.1 is easy to be done. The hypothesis $x\in\hat{D}(A)$ is necessary for $u(t)=S(t)x$ to be Lipschitz continuous on compact subsets of R_+ (i.e. locally Lipschitz continuous on R_+ - see Remark 4.1).

Proof of Theorem 5.1. The assertion (1°)$\Rightarrow$(2°) it was already proved in Remark 2.1 (A strong solution u on R_+ to (5.1) is the unique integral solution, which is just $S(t)x$). We now prove the assertion (2°)$\Rightarrow$(1°). Take $x\in\overline{D(A)}$ and assume that $t\to S(t)x$ is differentiable at $t_o>o$, i.e.

$$S(t_o-h)x=S(t_o)x-hy+o(h) \tag{5.2}$$

where $y=\frac{d}{dt}S(t)x$ evaluated at $t=t_o$ and $\frac{1}{h}o(h)\to o$ as $h\downarrow o$. Denote (for simplicity) $y=S'(t_o)x$. Inasmuch as $S(t_o-h)x\in\overline{D(A)}\subset R(I-hA)$ for sufficiently small $h>o$ (say $o<h<h_o$), there is $[u_h,v_h]\in A$ such that $S(t_o-h)x=u_h-hv_h$ for each $o<h<h_o$. Therefore (5.2) becomes

$$S'(t_o)x-v_h = \frac{1}{h}(S(t_o)x-u_h)+\frac{1}{h}o(h),\quad o<h<h_o \tag{5.3}$$

On the other hand, with $u=u_n, v=v_h$ and $S(t_o)x$ in place of x, (3.7)' yields

$$<S'(t_o)x-v_h, x^*>\le \omega\|S(t_o)x-u_h\|^2,\quad o<h\le h_o \tag{5.4}$$

for some $x^*\in F(S(t_o)x-u_h)$. A simple combination of (5.3) and (5.4)gives

$$h^{-1}\|S(t_o)x-u_h\|\le\frac{1}{1-h\omega}\cdot\frac{\|o(h)\|}{h} \tag{5.5}$$

for all sufficiently small $h>o$ which shows that $u_h\to S(t_o)x$ as $h\downarrow o$. Moreover, on the basis of (5.5), (5.3) implies $v_h\to S'(t_o)x$.

Since A is closed it follows that $S(t_o)x\in D(A)$ and $S'(t_o)x\in AS(t_o)x$. For this fact x need not be even in $\hat{D}(A)$. However, $t\to S(t)x$ is locally Lipschitz on R_+ iff $\hat{x}$ $D(A)$). The proof of Theorem 5.1 is complete.

Remark 5.2. We say that the semigroup $S(t)$ generated by A has a "smoothing effect" at $t_o>o$ on $x\in\overline{D(A)}$ if $S(t_o)x\in D(A)$.

From the proof of Theorem 5.1 it follows that if $t\to S(t)x$ (with $x\in\overline{D(A)}$) is differentiable at $t_o>o$ and A is closed, then $S(t_o)x\in D(A)$) i.e. $S(t_o)$ has a smoothing effect on x. In short we say that differentiability implies the smoothing effect of $S(t)$ on the initial data (see also Theorem 2.1).

From Theorem 5.1 we can easily derive

Corollary 5.1. In addition to the hypotheses of Theorem 5.1, assume that X is reflexive. Then for every $x\in\hat{D}(A)$, the initial value problem (5.1) has a unique strong solution u on $[o,+\infty[$ given by

$$u(t)=S(t)x=\lim_{n\to\infty}(I-\frac{t}{n}A)^{-n}x \tag{5.6}$$

Moreover, $u(t)\in\hat{D}(A)$ for all $t>o$.

Proof. Since $x\in\hat{D}(A)$, $t\to S(t)x$ is $\hat{D}(A)$-valued, locally Lipschitz continuous, and therefore (on the basis of reflexivity of X) it is a.e. differentiable on R_+. According to Theorem 5.1, $u(t)=S(t)x$ is a strong solution to (5.1) on R_+. The uniqueness of the solution to the problem (5.1) can be proved as in Ch. 1, § 3.2, by using a lemma of Kato. Indirectly, the uniqueness of u is a consequence of the fact that $t\to S(t)x$

is the unique integral solution to (5.1) and that a strong solution is an integral solution.

Remark 5.3. The conclusion of Corollary 5.1. and its proof remain valid if the hypothesis that X is reflexive is weakened to that every (locally) Lipschitz continuous function from R_+ into X is almost everywhere differentiable (e.g. this is the case for $X=l^1$). If X* is uniformly convex and A is m-dissipative, then $\hat{D}(A)=D(A)$ so Corollary 5.1 can be restated as

Corollary 5.2. Let X be a real Banach space with X* uniformly convex and let $A \subset X \times X$ be m-dissipative. Then for every $x \in D(A)$ the initial value problem

$$u'(t) \in Au(t), \text{ a.e. on } R_+, \ u(o)=x$$

has a unique strong solution u given by

$$u(t)=S(t)x= \lim_{n\to\infty} (I-\frac{t}{n}A)^{-n}x, \ t \geq o, \ x \in D(A)$$

Moreover $u(t)=S(t)x \in D(A)$, for all $t > o$ (i.e. $S(t):D(A) \to D(A)$) and $t \to S(t)x$ is Lipschitz continuous on R_+,

$$||S(t)x-S(s)x|| \leq |t-s| \ |Ax| \tag{5.6'}$$

where $|Ax| = \inf \{ \ ||y||, \ y \in Ax \}$

For (5.6)' see Proposition 3.2. let us recall the rpoof of S(t) D(A) $\subset D(A)$, $t \geq o$. Take $t > o$, $x \in D(A)$ and set $y=S(t)x$. We have $||S(s)y-y|| \leq ||S(s)x-x|| \leq s|Ax|$ so according to (3.17),

$$\lim_{\lambda \downarrow 0} {}^{-1} ||J_\lambda y-y|| = \lim_{\lambda \downarrow 0} ||A_\lambda y|| = \lim_{\lambda \downarrow 0} \sup h^{-1} ||S(h)y-y|| \leq |Ax|$$

which implies $||A_\lambda y|| \leq |Ax|$, $\forall \lambda > o$. Inasmuch as $J_\lambda y \to y$ for $\lambda \downarrow o$ and $A_\lambda y \in AJ_\lambda y$ it follows that $y=S(t)x \in D(A)$, q.e.d. (Here we have used the fact that A is demiclosed as indicated in Theorem 2.1 in Appendix).

Remark 5.4. If both X and X* are uniformly convex, then $S(t):D(A_o)$ $D(A_o)$, $\forall \ t > o$ (see Th. 7.1, (5)).

5.2. Strong solutions on closed subsets in continuous case. Flow-invariance.

If in addition to Hypotheses of Theorem 2.1 we assume that A is singl valued and continuous on $\overline{D(A)} \equiv D$, the the semigroup S(t) generated by A is strongly differentiable (i.e. for each $x \in D$ and each $t_o > o$, the function $t \to S(t)x$ is strongly differentiable at t_o). Recall that in this case Tangential condition (2.8) is equivalent to

$$\lim_{h\downarrow o} h^{-1} d(x+hA;D)=o, \ \forall \ x \in D \qquad (5.7)$$

(see Proposition 7.1 in Chapter 1). Consequently, $u(t)=S(t)x_o$ is the unique strong solution to the Cauchy problem

$$u'(t) = Au(t), \ u(o)=x_o \in D, \ t \geq o \qquad (5.8)$$

Let us state the result into a precise form, i.e.

Theorem 5.2. (Martin [1]). Let D be a (nonempty) closed subset of the Banach space X and let $A:D \to X$ be continuous and ω-dissipative. Then for each $x_o \in D$ the Cauchy problem (5.8) has a (unique) solution $u:[o,\infty[\to D$ if and only if (5.7) holds (i.e. for every $x \in D$, Ax is "tangent" to D at x). $S(t):D \to D$ defined by $S(t)x_o=u(t;o,x)$ is a semigroup of type ω on D.

Proof. Necessity. Take $x \in D$ and assume that (5.8) has a solution u with $u(o)=x$ and $u(t) \in D$ for all t o. Then (5.7) holds. Indeed

$$h^{-1}d(x+hAx;D) \leq h^{-1} \ \|x+hAx-u(h)\| =$$

$$\|(u(h)-x)h^{-1}-Ax\| \to \|u'(o+)-Ax\| = \|Au(o)-Ax\| = o \qquad (5.9)$$

where $u'(o+)$ denotes the right derivative $\frac{d}{dt}u(t)$ evaluated at $t=o$.

Sufficiency. Let $x_o \in D$ be fixed. Since A is continuous at x_o, there are $M > o$ and $r > o$ such that

$$\|Au\| \leq M, \ \forall u \in D \cap B(x_o,r) \qquad (5.10)$$

where $B(x_o,r)=\{y\in X; \ \|y-x_o\| \leq r\}$. Choose $T > o$ with the property $T(M+1)<r$. Let n be an arbitrary positive integer. Set $t_o^n = o$ and $x_o^n=x_o$. Inductively define $t_{i+1}^n=t_{i+1} > o$ and $x_{i+1}^n \equiv x_{i+1} \in D$ as follows: if $t_i=T$ set $t_{i+1} = t_i=T$ and if $o < t_i < T$, define d_i as the largest number in $]o,\frac{1}{n}]$ with the properties:

$$t_i+d_i \leq T(o < d_i \leq \frac{1}{n}), \ t_i+d_i \leq T$$

$$\|Au-Ax_i\| \leq \frac{1}{n}, \ \forall \ \ u \in D, \ \|u-x_i\| \leq d_i(M+1) \qquad (5.11)$$

$$\frac{1}{d_i} \ d(x_i+d_iAx_i;D) \leq \frac{1}{2n} \qquad (5.12)$$

In view of the continuity of A on D and of (5.7), d_i is well-defined.

Set $t_{i+1}=t_i+d_i$. By (5.12) there is an element $x_{i+1}\in D$ such that $\frac{1}{d_i}\|x_i+d_iAx_i-x_{i+1}\|\le\frac{1}{n}$. Set

$$\bar{p}_i=(x_{i+1}-x_i-d_iAx_i)/d_i \tag{5.13}$$

It follows

$$x_{i+1}=x_i+(t_{i+1}-t_i)(Ax_i+\bar{p}_i),\quad \|\bar{p}_i\|\le\frac{1}{n} \tag{5.14}$$

It is easy to check that $x_i\in B(x_o,r)$, so according to (5.10), (5.11) and (5.14) we have

$$\|Ax_{i+1}-Ax_i\|\le\frac{1}{n}\ ,\ i=o,1,\ldots \tag{5.15}$$

We now prove that $\lim_i t_i=T$. To this goal, assume by contradiction that $\lim_{i\to\infty} t_i^n=\bar{t}<T$ which means that the series $\sum_{i=o}^{\infty}d_i$ is convergent. Consequently, $\lim_i d_i=o$ and $\lim_i x_i=x^*\in D$ exists (this is because $\|x_{i+1}-x_i\|\le d_i(M+1)$ according to (5.14)). Choose $h^*\in\,]o,\frac{1}{n}]$ such that:

$$\frac{1}{h^*}\,d(x^*+h^*Ax^*;D)\le\frac{1}{4n} \tag{5.16}$$

$$\|Au-Ax^*\|\le\frac{1}{2n},\ \text{if}\ \|u-x^*\|\le 2h^*(M+1),\ u\in D \tag{5.17}$$

Let i_o be a positive integer with the properties:

$$d_i<h^*,\ \|x_i-x^*\|\le h^*(M+1),\ \|Ax_i-Ax^*\|\le\frac{1}{2n},\ \forall\ i\ge i_o \tag{5.18}$$

Let us observe that (5.11) holds with h^* in place of d_i. Indeed, if $\|u-x_i\|\le h^*(M+1)$ (with $u\in D$), the $\|u-x^*\|\le\|u-x_i\|+\|x_i-x^*\|\le 2h^*(M+1)$ for all $i\ge i_o$. Consequently

$$\|Au-Ax_i\|\le\|Au-Ax^*\|+\|Ax^*-Ax_i\|\le\frac{1}{2n}+\frac{1}{2n}\quad,\ \forall\ i\ge i_o$$

Inasmuch as $d_i<h^*$ is the maximal number satisfying (5.11) and (5.12), it follows that (5.12) is not valid if d_i is replaced by h^*, so we have

$$\frac{1}{h^*}\,d(x_i+h^*Ax_i;D)\ge\frac{1}{2n},\ \forall\ i\ge i_o \tag{5.19}$$

Letting $i\to\infty$ in (5.19) we get: $\frac{1}{h^*}d(x^*+h^*Ax^*;D)\ge\frac{1}{2n}$ which contradicts (5.16). Thus we have $\lim_i t_i=T$. Let $\bar{T}=T/2$. Then there is a positive integer $i=N_n$ such that

$$t_{N_n-1}<\bar{T}\le t_{N_n}\quad (n=1,2,\ldots) \tag{5.20}$$

Define $y_n:[o,\bar{T}]\to X$ as below

$$y_n(t)=x_i+(t-t_i)Ax_i+\bar{p}_i),\quad t_i\le t\le t_{i+1} \tag{5.21}$$

for $i=o,1,\dots,N_n-1$. It is easy to check that $y_n(t)\in B(x_o,r)$ for all $t\in[o,\bar{T}]$ and

$$\|y_n(t)-y_n(s)\|\le(M+1)(t-s),\quad t,s\in[o,T] \tag{5.21'}$$

The crucial point is to prove the convergence of y_n. One can follows Martin's original proof (Martin [1]) or a simplified version given by Kenmochi and Takahashi [1] (which is still delicate). However, in our framework here, the convergence of y_n can be immediately derived from Theorem 2.1 of Kobayashi (i.e. from the estimate (1.16). To this goal, one observes that (5.14) can be rewritten as the implicit difference-scheme (1.5), i.e.,

$$\frac{x_i-x_{i-1}}{t_i-t_{i-1}}-p_i\in Ax_i,\quad i=1,\dots,N_n \tag{5.22}$$

with $p_i=\bar{p}_{i-1}+Ax_{i-1}-Ax_i$. By (5.15) and (5.14) we have $\|p_i\|\le\frac{2}{n}$ and therefore (1.6) is satisfied. In this case $d_n=\max_{1\le i\le N_n}(t_i-t_{i-1})\le\frac{1}{n}$, hence (1.4) is also true. Consequently, the step function $u_n:[o,\bar{T}]\to D$ given by $u_n(t)=x_o$ if $t=o$ and $u_n(t)=x_i$ if $t_{i-1}<t\le t_i$, $i=1,\dots,N_n$ is a DS-approximate solution on $o,\bar{T}$ to the problem (5.8). According to Theorem 2.1, the function u given by $u(t)=\lim u_n(t)$, $t\in[o,\bar{T}]$ is continuous and it defines the semigroup $S(t):D\to D$, by $S(t)x_o=u(t;o,x_o)$, $x_o\in D$, $t\ge o$. On the other hand, it is easily seen that

$$\|y_n(t)-u_n(t)\|\le 2(M+1)/n,\quad t\in[o,\bar{T}] \tag{5.23}$$

so

$$\lim_{n\to\infty}y_n(t)=\lim_{n\to\infty}u_n(t)=u(;o,x_o)=S(t)x_o,\quad t\ge o$$

Moreover, by (5.21)' it follows that $y_n(t)\to u(t;o,x_o)$ as $n\to\infty$, uniformly on $[o,\bar{T}]$.

But it is well-known that the limit u of the polygonal lines y_n satisfies

$$u(t)=x_o+\int_o^t Au(s)ds,\quad t\in[o,\bar{T}] \tag{5.24}$$

Let us recall the proof of (5.24). Set $b_n(s)=t_{i+1}$ if $t_{i-1}<s\le t_i$, $i=1,\dots,N_n$ and $b_n(o)=o$. Then

$$y_n(t)=x_o+\int_o^t Ay_n(b_n(s))ds+g_n(t),\quad t\in[o,\bar{T}] \tag{5.25}$$

with

$$g_n(t) = \sum_{j=0}^{i-1} (t_{j+1}-t_j)p_j+(t-t_i)p_i, \quad t_i \leq t \leq t_{i+1} \qquad (5.25$$

hence $\|g_n(t)\| \leq \frac{\bar{T}}{n}$, $t \in [o,\bar{T}]$

By definition, $|b_n(s)-s| \leq \frac{1}{n}$ so $y_n(b_n(s)) \to u(s)$ as $n \to \infty$ uniformly on $[o,\bar{T}]$. Letting $n \to \infty$ in (5.25) one obtains (5.24). In view of the continuity of A on D, (5.24) implies the differentiability of u on [o. Thus, we have proved that us is a strong solution on $[o,\bar{T}]$ to (5.8). The extendability of the (local) solution u from $[o,\bar{T}]$ to the whole semiaxis $R_+ = [o, +\infty[$ can be proved by standard arguments. Indeed, let $[o,t_{max}[$ be the maximal domain of u. If $t_{max} = +\infty$ we have nothing to pr It $t_{max} = t_m < +\infty$, then we proceed as follows. From

$$u'(t+h)-u'(t)=Au(t+h)-Au(t), \quad o \leq t < t+h < t_m \qquad (5.26$$

we derive (inasmuch as A is ω-dissipative)

$$\frac{d}{dt} \|u(t+h)-u(t)\| \leq \omega \|u(t+h)-u(t)\|, \text{ a.e. on } [o,t_m-h[$$

which yields (see (3.7) in Ch. 1)

$$\|u(t+h)-u(t)\| \leq \|u(h-u(o)\| e^{\omega t}, \quad t \in [o,t_m-h[\qquad (5.26$$

It follows from (5.26) that $\lim_{t \to t_m} u(t)$ exists, which implies that u can be extended to $[o,t_{max}]$ as a solution of (5.8). Now, we can extend u the right of t_{max} (by the above proof with $u(t_{max}) \in D$ in place of x_o. This contradiction shows that $t_{max} = +\infty$. The uniquenes of the solution on R_+ to (5.8) has already been proved (Corollary 5.1). Thus, the proof is complete.

Remark 5.6. Let $A:D \to X$ be continuous, dissipative and satisfying tangential condition to D in the sense of (5.7), with D closed. Then i follows

$$< Ax_1-Ax_2, x^* > \leq o, \quad \text{for all } x_1,x_2 \in D \qquad (5.27$$

and for all $x^* \in F(x_1-x_2)$ (F the duality mapping of X). Indeed, let S(t be the semigroup on D generated by A via Cauchy probelm (5.8), i.e. $S(t)x=u(t;o,x_o)$, $x_o \in D, t \geq o$, where $u=u(t;o,x_o)$ is the solution to (5.8 Then for every $x^* \in F(x_1-x_2)$ we have

$$< \frac{S(t)x_1-x_1}{t} - \frac{S(t)x_2-x_2}{t}, x^*> = t^{-1} < S(t)x_1-S(t)x_2,x^*>$$

$$-t^{-1} \|x_1-x_2\|^2 \leq o, \quad x_1,x_2 \in D \qquad (5.28)$$

(since $\|S(t)x_1-S(t)x_2\| \leq \|x_1-x_2\|$, $\|x^*\| = \|x_1-x_2\|$ and $<x_1-x_2,x^*> = \|x_1-x_2\|^2$. But

$$\lim_{t \downarrow o} \frac{S(t)x_j-x_j}{t} = Ax_j, \ j=1,2$$

and therefore (5.28) implies (5.27). The remark is interesting since (by definition) the dissipativity of A means (5.27) only for some $x^* \in F(x_1-x_2)$.

Theorem 5.2 leads to a characterization of the flow-invariant sets D with respect to (5.8) (see Definition 4.1), namely

Corollary 5.3. Let U be an open subset of X and D a closed (nonempty) subset of U. Suppose that $A:U \to X$ is continuous and dissipative. Then Tangential condition (5.7) is necessary and sufficient for the flow-invaraince of D with respect to the equation (5.8).

Flow invariance (theory and applications) is treated by the author in his book [15].

§6. GENERATION OF COMPACT SEMIGROUPS

6.1. Nonlinear compact semigroups

Let S be a semigroup of type ω on a nonempty subset D of the Banach space X (i.e., $S \in Q_\omega(D)$, see Definition 2.1).

Definition 6.1. (1) S is said to be ultimately compact if there exists $t_0 > 0$ such that the operator $S(t_0) : D \to D$ is compact. (2) S is said to be compact is $S(t)$ is compact for every $t > 0$ (sometimes we say " $S(t)$ is compact" instead of "the semigroup S is compact").

Clearly, if for some $t_0 > 0$, $S(t_0)$ is compact, then (in view of $S(t) = S(t-t_0)S(t_0)$, $t \geq t_0$) $S(t)$ is compact for every $t > t_0$. Therefore, $S(t)$ is ultimately compact if it is compact for all $t \geq t_0 > 0$ (for some $t_0 > 0$).

Note that, if $D = X$, $S(0) = I$ is compact iff X is finite dimensional.

If S is compact, then it is obviously ultimately compact. The converse assertion need not be true. Indeed, let us consider the semigroup arising in the case of the delay ordinary differential equation

$$u'(t) = au(t-r), \quad t \geq 0, \quad r > 0, \quad a \in R \tag{6.1}$$

with the initial datum condition

$$u(s) = f(s), \quad -r \leq s \leq 0, \quad f \in C([-r,0];R). \tag{6.2}$$

Obviously, the solution u of the problem (6.1) and (6.2) on $[0,r]$ is given by

$$\begin{aligned} u(t) &= u(0) + a\int_0^t u(y-r)\,dy = u(0) + a\int_{-r}^{t-r} u(s)\,ds \\ &= f(0) + a\int_{-r}^{t-r} f(s)\,ds, \quad \text{for all} \quad t \in [0,r], \end{aligned} \tag{6.3}$$

i.e., u is "the extension" of f to $[0,r]$ as defined by (6.3). The extension of u to $[r,2r]$ is given by

$$u(t) = u(r) + a\int_r^t u(y-r)\,dr, \quad r \leq t \leq 2r.$$

In such a manner, one constructs the solution u of (6.1) and (6.2) on R_+ (this is the so-called method of steps). Le us define the following (C_0) semigroup on $C([-r,0];R)$. For $0 \leq t \leq r$ and $f \in C([-r,0];R)$ set

$$(S(t)f)(s) = \begin{cases} f(t+s), & \text{if } s \in [-r,-t] \\ f(0) + a\int_{-r}^{s+t-r} f(y)\,dy, & \text{if } s \in [-t,0] \end{cases} \tag{6.4}$$

and thus $S(t)$ is defined for $t \in [0,r]$. For $t > r$, $S(t)$ is determined by the semigroup property. Clearly, for $0 \leq t < r$, $S(t)$ is not compact (in $C([-r,0];R)$). For $t = r$,

$$S(r)f)(s) = f(0) + a \int_{-r}^{s} f(y)\,dy, \quad -r \le s \le 0, \; f \in C([-r,0];R), \tag{6.4$'$}$$

so $S(r)$ is compact (anD so is $S(t)$ for all $t \ge r$). Therefore, $S(t)$ defined viA (6.4) is ultimately compact (with $t_0 = r$), but not compact. Note that $u(t) = (S(t)f)(0)$ is just the solution of (6.1) and (6.2) on $[0,r]$. Of course, the above conclusions remain valid if $C([-r,0];R)$ is replaced by $C([-r,0];R^n)$ (where R^n is the Euclidian n-dimensional space).

The infinitesimal generator A of S is given by

$$(Af)(s) = \begin{cases} f'(s), & \text{if } s \in [-r,0[\\ af(-r), & \text{if } s = 0 \end{cases} \tag{6.5}$$

with $D(A) = \{f \in C^1([-r,0];R),\, f'(0) = af(-r)\}$.

Similar properties hold for the semigroup associated with the functional differential equation

$$u'(t) = F(u_t), \quad t \in R_+, \quad u_0 = f \in C([-r,0];R^n), \tag{6.5$'$}$$

where u_t is defined by $u_t(x) = u(t+s)$, $s \in [-r,0]$, $u \in C([-r,+\infty[;R^n) \equiv C$ and $t \ge 0$. Suppose that $F : C \to R^n$ is a bounded linear operator. For $t \ge 0$ and $f \in C$, set $S(t)f = u_t$, where u is the unique solution to (6.5)'. Then $S(t)$ is a (C_0) semigroup on C and its infinitesimal generator A is given by $Af = f'$ for $f \in D(A) = \{g \in C;\, g'(0) = F(g)\}$. Therefore, the solution u to (6.5)' is given by $u(t) = (S(t)f)(0)$, $t \in R_+$ (See Hale [1]).

In what follows, we will use the results in Chapter 1, §5 in order to give a characterization of compact semigroups.

In this case, Proposition 5.1 (Ch. 1) may be restated as:

Proposition 6.1. If the semigroup $S \in Q_\omega(D)$ (see Definition 2.1) is compact, then the functions $t \to S(t)x$ are equicontinuous at $t = t_0$ for each $t_0 > 0$ with respect to x in bounded subsets Y of D.

For Theorem 5.1 in Chapter 1, we derive the following characteriation (due to Brezis [4], 1974) of the compactness of S.

Theorem 6.1. (Brezis) Let $S \in Q_\omega(\overline{D(A)})$ be the semigroup generated by A via the exponential formula (3.10) of Crandall-Liggett. Then $S(t)$ is compact for every $t > 0$ if and only if Conditions (I) and (II) below hold

(I) $J_\lambda = (I - \lambda A)^{-1} : R(I - \lambda A) \to D(A)$ is compact (for every $\lambda > 0$).

(II) For every bounded subset Y of $\overline{D(A)}$, the functions $\{t \to S(t)x;\, x \in Y\}$ are equicontinuous at every $t = t_0 > 0$.

(In the linear case, Theorem 6.1 with $\overline{D(A)} = X$ is due to Pazy [1] - see Th. 6.3. In the case $X = H$ – a real Hilbert space and $A = -\partial\varphi$, where φ is the subdifferential of a proper, l.s.c., convex function defined on H, Theorem 6.1 is due to Konishi [2].)

Note that the compactness of J_λ for $\lambda = \lambda_0 > 0$ (e.g., for $\lambda = 1$) implies the compactness of J_λ for every $\lambda > 0$. In other words, Brezis' theorem aserts that: " $S \in Q_\omega(\overline{D(A)})$ is compact iff $J_1 = (I - A)^{-1}$ is compact and the functions $t \to S(t)x$ are equicontinuous on bounded subsets of $\overline{D(A)}$ at every $t = t_0 > 0$ ".

It is easy to check the assertion below.

Proposition 6.2. Let A be dissipative, satisfying Range condition (2.10). Then $(I - A)^{-1}$ is compact iff for each positive number $r > 0$, the level set

$$L_r = \{x \in D(A);\ \| x \| \leq r \quad \text{and} \quad \| y \| \leq r,\ \text{for some } y \in Ax\} \tag{6.6}$$

is precompact in X (see also Remark 6.1.(4)).

Proof. Suppose, for the simplicity of writing, that A is single-valued. If $(I - A)^{-1}$ is compact then for every $r > 0$, $(I-A)^{-1}(I-A)L_r = L_r$ is relatively compact. This is because $(I-A)L_r$ is bounded. Conversely, let L_r be precompact for every $r > 0$, and let M be a bounded subset of $R(I - A)$. Denote by $L(M) = \{x \in D(A);\ x - Ax \in M\}$. Let $m > 0$ be such that $\| x - Ax \| \leq m$ for each $x \in L(M)$. Clearly, $M = \{x - Ax;\ x \in L(M)\}$. Fix $x_0 \in D(A)$. Then, the dissipativity of A yields

$$\| x - x_0 \| \leq \| x - Ax - (x_0 - Ax_0) \| \leq m + \| x_0 - Ax_0 \|, \quad \forall x \in L(M),$$

i.e.,

$$\| x \| \leq m + \| x_0 - Ax_0 \| + \| x_0 \| = \bar{r}, \qquad \forall x \in L(M).$$

We now have

$$(I - A)^{-1}M = L(M) \subset \{x \in D(A);\ \| x \| \leq \bar{r},\ \| Ax \| \leq m + \bar{r}\} \subset L_r$$

with $r = m + \bar{r}$. Since L_r is supposed to be precompact, it follows that $(I - A)^{-1}M$ is also precompact. This completes the proof.

In the case in which $-A$ is the subdifferential $\partial\varphi$ of a proper, l.s.c. convex function φ: $H \to]-\infty, +\infty]$ from the real Hilbert space H into $]-\infty, +\infty]$, we have

Proposition 6.3. In the case $A = -\partial\varphi$, $(I - A)^{-1}$ is compact iff for every $r > 0$, the level set

$$C_r = \{x \in D_e(\varphi);\ \| x \| \leq r \quad \text{and} \quad \varphi(x) \leq r\} \tag{6.7}$$

is compact (C_r is closed). See also Remark 6.1.(4).

Note that $D_e(\varphi)$ is the effective domain of φ, i.e.,

$$D_e(\varphi) = \{x \in D(\varphi);\ \varphi(x) < +\infty\} \tag{6.8}$$

(see Appendix).

In other words, in this case the compactness of C_r given by (6.7) is equivalent to the precompactness of L_r given by (6.6).

Proof of Proposiiton 6.3. Let $x_0 \in D_e(\varphi)$. Then, for each $x \in D(\partial\varphi)$ and $-y \in (\partial\varphi)(x) \equiv \partial\varphi(x)$, we have by definition

$$\varphi(x_0) - \varphi(x) \geq \langle -y, x_0 - x \rangle, \quad \text{i.e.} \quad \varphi(x) \leq \langle y, x_0 - x \rangle + \varphi(x_0), \tag{6.9}$$

where $\langle \cdot, \cdot \rangle$ and $\| \cdot \|$ denote the inner product and the norm of H, respectively.

Consequently, given $r > 0$, there is an $r_1 = r_1(r) > 0$ such that

$$\begin{aligned} L_r &= \{x \in D(A);\ \| x \| \leq r \quad \text{and} \quad \| y \| \leq r \quad \text{for all} \quad y \in Ax\} \\ &\subset \{x \in D_e(\varphi);\ \| x \| \leq r_1 \quad \text{and} \quad \varphi(x) \leq r_1\} = C_{r_1}. \end{aligned}$$

Therefore, the compactness of C_{r_1} implies the precompactness of L_r, i.e., the compactness of $J_\lambda = (I - A)^{-1}$. Conversely, we now prove that the compactness of J_λ implies the compactness of C_r.

Inasmuch as $J_\lambda x \in D(A)$ and $A_\lambda x \in AJ_\lambda x$, the first inequality in (6.9) yields

$$\varphi(x) - \varphi(J_\lambda x) \geq \langle -A_\lambda x, x - J_\lambda x \rangle = \frac{1}{\lambda} \| x - J_\lambda x \|^2, \quad \lambda > 0. \tag{6.10}$$

On the other hand, φ is bounded from below by an affine function and, therefore, (6.10) yields (for $x \in C_r$)

$$\frac{1}{\lambda} \| x - J_\lambda x \|^2 \leq r + a \| J_\lambda x \| + b \leq p \| J_\lambda x - x \| + q \tag{6.11}$$

where a, b, p and q are positive constants, independent of λ and $x \in C_r$. According to an elementary result, from (6.11) one obtains

$$\| J_\lambda x - x \| \leq \lambda p + (\lambda^2 p^2 + 4\lambda q)^{\frac{1}{2}}, \quad \forall x \in C_r, \tag{2.12}$$

That is $\lim_{\lambda \downarrow 0} J_\lambda x = x$ uniformly with respect to $x \in C_r$. Since $J_\lambda C_r$ is precompact it follows that C_r is also precompact (Lemma 5.1 in Ch. 1). This completes the proof.

It is interesting (and important in applications to PDE - see Ch. 3) that the semigroup $S(t)$ generated by $-\partial\varphi$ is equicontinuous at $t_0 > 0$ on bounded subsets of $\overline{D(\partial\phi)}$. In other words, we have

Proposition 6.4. Let $A = -\partial\varphi$. Then for every bounded set $Y \subset \overline{D(A)}$, the functions $\{t \to S(t)x;\ x \in Y\}$ are equicontinuous at every $t_0 > 0$.

Proof. The key of the proof is the following inequality

$$\| A_0 S(t)x \| \leq \| A_0 z \| + \frac{1}{t} \| x - z \|, \ \forall x \in \overline{D(A)},\ \forall z \in D(A),\ t > 0, \tag{6.13}$$

where $\| A_0 z \| = \min\{\| y \|;\ y \in Av\}$ (where A_0 is the minimal section of $A = -\partial\varphi$). This is proved in §7.

Let $t_0 > 0$ and $t \geq t_0/2$. Then

$$\| S(t)x - S(t_0)x \| = \| S(t - \frac{t_0}{2})S(\frac{t_0}{2})x - S(\frac{t_0}{2})S(\frac{t_0}{2})x \|, \ \forall x \in \overline{D(A)}. \qquad (6.13)'$$

In this case $S(t) : \overline{D(A)} \to D(A)$, for $t > 0$, so $S(\frac{t_0}{2})x \in D(A)$ (due to the "smoothing effect" of $S(t)$ on x in $\overline{D(A)}$ - see §7).

According to (3.6), (6.13) implies

$$\| S(t)x - S(t_0)x \| \leq | t - t_0 | \| A_0 S(\frac{t_0}{2})x \| \leq | t - t_0 | (\| A_0 z \| + \frac{2}{t_0} \| x - z \|),$$

$$\forall t \geq \frac{t_0}{2}, \quad x \in \overline{D(A)}, \quad z \in D(A). \qquad (6.14)$$

The conclusion follows from (6.14).

A combination of Theorem 6.1, Proposition 6.3 and Proposition 6.4 lead to the following result (due also to Brezis [4]).

Theorem 6.2. (Brezis) The semigroup $S(t)$ generated by $-\partial\varphi$ (via the exponential formula) is compact iff for every $r > 0$, the level set C_r defined by (6.7) is compact in H (or equivalently, if $(I - \partial\varphi)^{-1}$ is compact).

Examples of compact semigroups are given in Ch. 3. Recall that $-\partial\varphi$ is m-dissipative.

Remark 6.1. (1) Let $A : D(A) \subset X \to X$ be a continuous and m-dissipative operator. If X is infinite dimensional, then $(I - A)^{-1}$ is not compact. This is because $I - A$ is continuous and $(I - A)(I - A)^{-1} = I$ - the identity operator on $R(I - A) = X$. According to Theorem 6.1, it follows that $S(t)$ generated by a continuous, m-dissipative and everywhere defined operator (acting in a Banach space of infinite dimension) is not compact. In particular, the semigroup $S(t)x_0 = u(t; 0, x_0)$ generated by A via Theorem 5.2 (of Martin) is not compact if X is of infinite dimension.

(2) If $S \in Q_\omega(D)$ and, in addition, S is defined on all of R and $S(t + s) = S(t)S(s)$ for all $t, s \in R$, then S is said to be a group on D. In this case we will write $S \in G_\omega(D)$. If $S \in G_\omega(X)$, then $S(-t)S(t) = S(0) = I$, for each $t > 0$. Consequently, if X is infinite dimensional, S cannot be compact.

(3) In a Hilbert space H, a compact semigroup S generated by a maximal dissipative set A has the property: for each $x \in \overline{D(A)}$, $t \to S(t)x$ has a precompact range (i.e., $\{S(t)x ;\ t \geq 0\}$ is precompact). The proof is given in §9.

(4) The compactness of C_r given by (6.7) is obviously equivalent to the compactness of

$$\tilde{C}_r = \{x \in H ;\ \| x \|^2 + \varphi(x) \leq r\}, \qquad \forall r > 0. \qquad (6.7)'$$

The precompactness of L_r given by (6.6) is equivalent to the precompactness of

$$\tilde{L}_r = \{x \in D(A);\ \| x \| + | Ax | \leq r\}, \qquad \forall r > 0 \qquad (6.6)'$$

with $| Ax | = \inf\{\| y \| ;\ y \, in \, Ax\}$.

The latter assertion can easily be proved. Indeed, $L_r \subset \tilde{L}_{2r}$, so the only fact we have to check is that "the precompactness of L_r implies the precompactness of $\tilde{L}_r''$. Since L_r is precompact for each $r > 0$, it follows that J_λ is compact for each $\lambda > 0$. In addition,

$$\| J_\lambda x - x \| \leq r, \quad \forall x \in \tilde{L}_r \quad (\text{since } \| A_\lambda x \| \leq | Ax |, \quad x \in D(A)),$$

and thus taking $K_\lambda = J_\lambda \tilde{L}_r$ and $K = \tilde{L}_r$ in Lemma 5.1, Ch. 1, we easily deduce the precompactness of $\tilde{L}_r$, q.e.d.

(5) The fact that conditions (I) and (II) in Theorem 6.1 are, independent is pointed out in Ch. 3, §3. In the "hyporbolic case" (even linear) J_λ is compact and, however, $S(t)$ is not (this is because (II) fails. See Ch. 3, §4. Actually the compact semigroups occur in the study of "parabolic" PDE. See Ex. 4.6, §4, Ch. 3. Examples in this direction can also be found in the forthcoming book of Vrabie [3].

6.2. Linear compact semigroups. Hille-Yosida Theorem.

It is clear that in the linear case condition (II) in Theorem 6.1 (i.e, the equicontinuity of $t \to S(t)x$ at each $t_0 > 0$ on bounded subsets of X) is equivalent to the continuity of $t \to S(t)$ at each $t_0 > 0$ in the uniform operator topology of $B(X)$. Recall that $B(X)$ is the space of all linear bounded operators from X into itself. Consequently, Theorem 6.1 contains the well-known result of Pazy [1, 1968] below.

Theorem 6.3. (Pazy) Let $A : D(A) \subset X \to X$ be a linear, densely defined unbounded, m-dissipative operator. Then the semigroup $S \in Q_0(X)$ generated by A (via the exponential formula) is compact if and only if conditions (I) and (II) below hold.

(I) $(I - \lambda A)^{-1} : X \to D(A)$ is compact for $\lambda > 0$.

(II) $t \to S(t)$ is continuous at every $t_0 > 0$ in the uniform operator topology.

In other words, Theorem 6.1 of Brezis is a nonlinear version of Theorem 6.3 of Pazy.

Remark 6.2. Conditions (I) and (II) above are independent. Indeed, there are examples (some PDE of "hyperbolic type" - see Ch. 3) of operators A with $(I - A)^{-1}$ compact which generates groups. In such a case (I) holds while (II) not.

In what follows we will present two cases in which condition (II) in Theorem 6.3 is automatically satisfied. The first case is " A - self-adjoint acting in a Hilbert space H ", and the second case is " A generateds an analytic C_0 semigroup of contractions. See Corollary 6.1 and Corollary 6.2, repsectively.

Corollary 6.1. Let $A : D(A) \subset H \to H$ be a linear, densely defined (unbounded) m-dissipative operator. Suppose, in addition, that A is self-adjoint. Then the semigroup generated by A is compact iff $(I - A)^{-1}$ is compact.

Proof. The hypotheses imply that $-A$ is the subdifferential of a proper, l.s.c., convex function $f : H \to]-\infty, +\infty[$ (i.e., $A = -\partial f$). Such a function is the following

$$f(x) = \begin{cases} \frac{1}{2} A^{1/2} x, & \text{if } x \in D(A^{1/2}) \\ +\infty, & \text{otherwise} \end{cases}, \qquad \text{(see Appendix)}$$

where $A^{1/2}$ is the square root of A. The conclusion of this corollary follows now from Theorem 6.2.

In the linear case $S \in Q_0(X)$ means that S in a C_0 semigroup. See Definition 2.1. It is know (see e.g., Pazy [4, pp. 60-62], or Goldstein [1, p. 33]) that if $0 \in \rho(A)$ - the resolvent of A - and $S(t)$ is analytic in a sector

$$D = \{z;\ \Theta_1 < arg z < \Theta_2,\ \Theta_1 < 0 < \Theta_2\}$$

containing the nonnegative real axis, then $S(t)$ is differentiable for $t > 0$ (i.e., $S(t) : X \to D(A)$ for $t > 0$ - see Remark 5.2) and there is a constant $C > 0$ such that

$$\| AS(t) \| \leq \frac{C}{t}, \qquad \text{for all } t > 0,$$

where A is the infinitesimal generator of $S(t)$. It is easy to see that

Lemma 6.1. If $S(t) : X \to D(A)$ for $t > 0$ and $\| AS(t) \| \leq \frac{C}{t}$, then $S(t)$ is continuous at each $t_0 > 0$ in the uniform operator topology.

Proof. Let $x \in X$ and $t_0 > 0$. For $t > \frac{1}{2} \cdot t_0$ we have

$$\begin{aligned} \| S(t)x - S(t_0)x \| &= \| S(t - \frac{t_0}{2})S(\frac{t_0}{2})x - S(\frac{t_0}{2})S(\frac{t_0}{2})x \| \\ &\leq | t - t_0 | \| AS(\frac{t_0}{2})x \|, \end{aligned} \tag{6.15}$$

where we have used (3.6) with $S(\frac{t_0}{2})x$ in place of x, $s = \frac{1}{2} \cdot t_0$ and $t - \frac{1}{2} \cdot t_0$ in place of t. It follows from (6.15) that

$$\| S(t) - S(t_0) \| \leq | t - t_0 | \| AS(\frac{t_0}{2}) \| \leq 2C\, | t - t_0 | / t_0, \quad \forall t \geq \frac{1}{2} \cdot t_0$$

which completes the proof.

In particular, if $S(t)$ is analytic in D and $0 \in \rho(A)$, then $S(t)$ is continuous in the uniform operator topology at every $t_0 > 0$. Consequently, from Theorem 6.3 it follows

Corollary 6.2. Let A be the infinitesimal generator of an analytic C_0 contraction semigroup S. If $0 \in \rho(A)$ and $(I - A)^{-1}$ is compact, then S is compact.

Theorems 6.1 and 6.3 give characterizations of the compactness of $S(t)$ (generated by A) in terms of A and of $S(t)$ itself.

It would be interesting to characterize the compactness of $S(t)$ solely in terms of the properties of its generator A. Such a result is known in the case $A = -\partial\varphi$ (Theorem 6.2). In general, for the nonlinear case, there seems to be no such results. In the linear case the compactness of $S(t)$ is characterized by the compactness of $(I - \lambda A)^{-1}$ and by the fact that $\lim_{n\to+\infty} (I - \frac{t}{n}A)^{-n}x = S(t)x$ is uniform with respect to x in bounded subsets of X and t in compact subsets of $[0, +\infty[$; this result has been proved by Vrabie [2]. In the form below, the proof of this result (i.e., Theorem 6.4) is given by the author and is similar to that given by Vrabie.

Let us proceed now to the statement of this result in a precise form, namely

Theorem 6.4. (Vrabie) Let $A : D(A) \subset X \to X$ be a linear, densely defined m-dissipative operator. Then the C_0 semigroup S on X generated by A is compact if and only if

(I)' $(I - \lambda A)^{-1} : X \to D(A)$ is compact for $\lambda > 0$.

(II)' $\lim_{n\to+\infty} (I - \frac{t}{n}A)^{-n}x = S(t)x$ uniformly with respec to x in bounded subsets of X and t in bounded subsets of $[0, +\infty[$.

Remark 6.3. Clearly, Conditions (I)' and (II)' in Theorem 6.4 are sufficient to guarantee the compactness of S even in the nonlinear case. Indeed, if A is m-dissipative (possible multivalued) and

(I) $(I - \lambda A)^{-1}$ is compact for $\lambda > 0$, and

(II) $J^n_{t/n}x \equiv (I - \frac{t}{n}A)^{-n}x \to S(t)x$ as $n \to +\infty$, uniformly for x in bounded subsets of $\overline{D(A)}$ and for t in bounded subsets in $[0, +\infty[$, then $S(t)$ is compact for $t > 0$. (See e.g., Corollary 5.1 in Ch. 1, with $D = \overline{D(A)}$ and $T_n = (I - \frac{t}{n}A)^{-n}$.) The problem is whether or not Condition (II) above is necessary for the compactness of $S(t)$ in the nonlinear case.

For the proof of Theorem 6.4, some preliminaries are needed. It is known that

$$R(\lambda; A)x = (\lambda I - A)^{-1}x = \int_0^\infty e^{-\lambda s}S(s)x\,ds\,, \quad \lambda > 0\,. \tag{6.16}$$

See the proof of Theorem 6.6. Differentiating (6.16) with respect to λ and substituting $s = tv$, one obtains

$$(R(\lambda; A))^{(n)}x = (-1)^n t^{n+1} \int_0^\infty v^n e^{-\lambda tv} S(tv)x\,dv\,. \tag{6.17}$$

On the other hand, from the resolvent identity

$$R(\lambda; A) - R(\mu; A) = (\mu - \lambda)R(\lambda; A)R(\mu; A)\,, \quad \lambda, \mu > 0 \tag{6.17$'$}$$

we derive (by differentiation)

$$(R(\lambda; A))^{(n)} = (-1)^n n!(\lambda I - A)^{-n-1}\,, \quad \lambda > 0,\ n = 1, 2, \ldots\,. \tag{6.18}$$

Combining (6.17) and (6.18) and taking $\lambda = \frac{n}{t}$, we find

$$(I - \frac{t}{n}A)^{-n-1}x = \frac{n^{n+1}}{n!}\int_0^\infty (ve^{-v})^n S(tv)x\,dv\,. \tag{6.19}$$

Inasmuch as

$$\frac{n^{n+1}}{n!}\int_0^\infty (ve^{-v})^n\,dv = 1\,, \tag{6.19$'$}$$

(see (6.24) with $b = 0$) (6.19) can be written in the form

$$(I - \frac{t}{n}A)^{-n-1}x - S(t)x = \frac{n^{n+1}}{n!}\int_0^\infty (ve^{-v})^n [S(tv)x - S(t)x]\,dv\,, \tag{6.20}$$

for all $t \geq 0$, $x \in X$, and $n = 1, 2, \ldots$.

Proof of Theorem 6.4. On the basis of Remark 6.3 and Theorem 6.1, the only fact we have to prove is that the compactness of $S(t)$ implies condition (II)'. Let us fix an arbitrary $\varepsilon > 0$ and a bounded subset M of X (say $\| x \| \leq m$ for all $x \in M$). Since $S(t)$ is supposed to be compact, it is continuous in the uniform operator topology at $t > 0$ (in what follows, we will use only this property of $S(t)$). Given a bounded interval $0 < \alpha < \beta < \infty$, choose $0 < \overline{\alpha} < \alpha$ and $\beta < \overline{\beta} < +\infty$. Let $\delta = \delta(\varepsilon) > 0$ be such that

$$\| S(s) - S(w) \| \leq \varepsilon/m, \quad \text{for all } s, w \in [\overline{\alpha}, \overline{\beta}], \ | s - w | \leq \delta(\varepsilon). \tag{6.21}$$

Choose $0 < a < 1 < b < +\infty$, such that $t \in [\alpha, \beta]$ and $v \in [a, b]$ imply $tv \in [\overline{\alpha}, \overline{\beta}]$ and $| tv - t | \leq \beta(b-a) < \delta(\varepsilon)$. (These hold iff $\frac{\overline{\alpha}}{\alpha} \leq a < 1 < b < \frac{\overline{\beta}}{\beta}$ and $b - a < \beta^{-1}\delta(\varepsilon)$.)

In these conditions we have

$$\| S(t)x - S(tv)x \| \leq m \| S(t) - S(tv) \| < \varepsilon \tag{6.22}$$

for all $t \in [\alpha, \beta]$, $v \in [a, b]$ and $x \in M$.

We now break the integral in the formula (6.20) into three integrals I_n^1, I_n^2 and I_n^3 on $[0, a]$, $[a, b]$ and $[b, +\infty[$ respectively. Then, for all $x \in M$ (i.e., $\| x \| \leq m$)

$$\| I_n^1 \| \leq 2m \frac{n^{n+1}}{n!} \int_0^a (ve^{-v})^n dv, \quad \| I_n^3 \| \leq 2m \frac{n^{n+1}}{n!} \int_b^\infty (ve^{-v})^n dv$$

since $\| S(t)x \| \leq \| x \| \leq m$, and

$$\| I_n^2 \| \leq \varepsilon \frac{n^{n+1}}{n!} \int_a^b (ve^{-v})^n dv .$$

Set

$$I_k^a = \int_0^a v^k e^{-nv} dv \quad \text{and} \quad \tilde{I}_k^b = \int_b^\infty v^k e^{-nv} dv, \quad a, b > 0.$$

Integrating by parts, we get

$$\begin{aligned} I_k^a &= -\frac{1}{n} \cdot a^k e^{-na} + \frac{k}{n} \cdot I_{k-1}^a, \quad k = 1, 2, \ldots \\ \tilde{I}_k^b &= \frac{k}{n} \cdot \tilde{I}_{k-1}^b + \frac{b^k}{n} e^{-nb}, \quad k = 1, 2, \ldots . \end{aligned} \tag{6.23}$$

Iterating these formulas for $k = 1, 2, \ldots, n$, we find

$$\frac{n^{n+1}}{n!} \int_0^a (ve^{-v})^n dv = 1 - e^{-na}[1 + na + \frac{(na)^2}{2!} + \ldots + \frac{(na)^n}{n!}] \tag{6.24}$$

$$\frac{n^{n+1}}{n!} \int_b^\infty (ve^{-v})^n dv = e^{-nb}[1 + nb + \frac{(nb)^2}{2!} + \ldots + \frac{(nb)^n}{n!}] \tag{6.25}$$

for all $a, b > 0$, $n = 1, 2, \ldots$. On the basis of $\lim_{n\to\infty} \frac{n^n}{n!} e^{-n} = 0$ and $ze^{-z} < e^{-1}$, $\forall z \neq 1$ it is easy to check that

$$\lim_{n\to+\infty} [1 + nz + \frac{(nz)^2}{2!} + \ldots + \frac{(nz)^n}{n!}]e^{-nz} = \begin{cases} 1, & \text{if } 0 < z < 1 \\ 0, & \text{if } 1 < z \end{cases} \tag{6.25$'$}$$

For example, if $z > 1$, we have $\frac{(nz)^k}{k!} < \frac{(nz)^{k+1}}{(k+1)!}$ for $n > k + 1$, and hence

$$(\frac{n^n}{n!} \cdot e^{-n})(e \cdot z \cdot e^{-z})^n (n+1) \to 0 \quad \text{as} \quad n \to +\infty$$

(since $(n+1)(e \cdot z \cdot e^{-z})^n \to 0$, $\frac{n^n}{n!} e^{-n} \to 0$ as $n \to +\infty$).

It follows that $\lim_{n\to+\infty} I_n^1 = \lim_{n\to+\infty} I_n^3 = 0$, and consequently (6.20) yields

$$\lim_{n\to+\infty} \sup \| (I - \frac{t}{n} A)^{-n-1} x - S(t)x \| < \varepsilon \quad \text{for all } t \in]\alpha, \beta[\quad \text{and} \quad x \in M,$$

where $\varepsilon > 0$ is arbitrary. In other words,

$$J_{t/n}^{n+1} x = (I - \frac{t}{n} A)^{-n-1} x \to S(t)x \quad \text{as} \quad n \to +\infty \tag{6.26}$$

uniformly with respect to t in bounded intervals $[\alpha, \beta] \subset R_+$ $(0 < \alpha < \beta)$ and x in bounded subsets $M \subset X$. It is easy to prove that even

$$(I - \frac{t}{n+1} A)^{-n-1} x \to S(t)x \quad \text{as} \quad n \to +\infty$$

uniformly for $x \in M$ and $t \in [\alpha, \beta]$. Replace t by $\frac{n}{n+1} \cdot t$ in (6.26). Indeed, let $0 < \overline{\alpha} < \alpha < \beta$ and let $\varepsilon > 0$. There is a $\delta = \delta(\varepsilon) > 0$ such that

$$\| S(s) - S(w) \| < \frac{\varepsilon}{2 \cdot m}, \quad \forall s, w \in [\overline{\alpha}, \beta], \quad | s - w | < \delta(\varepsilon). \tag{6.27}$$

Obviously, $\frac{n}{n+1} \cdot t \in [\overline{\alpha}, \beta]$ for all $t \in [\alpha, \beta]$ and $n > \frac{\overline{\alpha}}{\alpha - \overline{\alpha}}$. On the other hand, $| nt/(n+1) - t | < \beta/(n+1) < \delta$, for all $t \in [\alpha, \beta]$ and $n + 1 > \beta\delta^{-1}$. We conclude that

$$\| S(\frac{n \cdot t}{n+1})x - S(t)x \| \le m \| S(\frac{n \cdot t}{n+1}) - S(t) \| < \varepsilon/2 \tag{6.27$'$}$$

for all $t \in [\alpha, \beta]$, $x \in M$ and $n > \max\{\frac{\overline{\alpha}}{\alpha - \overline{\alpha}}, \frac{\beta}{\delta} - 1\}$.

According to (6.26) (with $[\overline{\alpha}, \beta]$ in place of $[\alpha, \beta]$) and to (6.27)', there is a positive integer $n(\varepsilon)$ such that

$$\| (I - \frac{s}{n} A)^{-n-1} x - S(s)x \| < \varepsilon/2 \tag{6.28}$$

for all $n > n(\varepsilon)$, $s \in [\overline{\alpha}, \beta]$ and $x \in M$. In particular, for $t \in [\alpha, \beta]$ and $s = \frac{n}{n+1} \cdot t$, with $n > n(\varepsilon)$, (6.28) implies

$$\| (I - \frac{t}{n+1} A)^{-n-1} x - S(\frac{n \cdot t}{n+1}) x \| < \varepsilon/2, \quad \forall x \in M. \tag{6.28'}$$

On the basis of (6.27)-(6.28)' we conclude that there is a positive integer $n(\varepsilon)$ (which depends also on α, β and M) such that

$$\| (I - \frac{t}{n+1} A)^{-n-1} x - S(t)x \| < \varepsilon,$$

for all $n > n(\varepsilon)$, $t \in [\alpha, \beta]$ and $x \in M$. The proof is complete.

As we have already mentioned, in the proof of the necessity of Condition (II)' in Theorem 6.4, we have used only the continuity of $t \to S(t)$ (at each $t_0 > 0$) in the uniform operator topology. In other words, we have proved.

Theorem 6.5. Let A be the infinitesimal generator of a C_0 contraction semigroup $S(t)$. Then for every $x \in X$

1^0) $(I - \frac{t}{n} A)^{-n} x \to S(t)x$ as $n \to +\infty$, uniformly (in t) on bounded subsets of $R_+ = [0, +\infty[$.

2^0) If $t \to S(t)$ is continuous at every $t > 0$ in the uniform operator topology, then $(I - \frac{t}{n} A)^{-n} x \to S(t)x$ as $n \to +\infty$, uniformly with respect to t in bounded subsets of $[0, +\infty]$ and x in bounded subsets of X.

Proof. The part 2^0) has already been discussed. For the part 1^0), we need not the continuity of $t \to S(t)$ at each $t > 0$ in $B(X)$. It suffices only the property: $\lim_{t \downarrow 0} S(t)x = x$, $\forall x \in X$ of a C_0 semigroup. Indeed, for $M = \{x\}$, the continuity of $t \to S(t)x$ on $[\overline{\alpha}, \beta]$ implies

$$\| S(t)x - S(tv)x \| < \varepsilon, \quad t \in [\alpha, \beta], \quad v \in [a, b],$$

i.e., (6.22), as well as

$$\| S(\frac{n \cdot t}{n+1})x - S(t)x \| < \varepsilon/2,$$

i.e., (6.27).

The only fact given without proof is the formula (6.16). (We will prove it below.)

We have also proved the most part of the following celebrated theorem of Hille-Yosida.

Theorem 6.6. (Hille-Yosida) (I) A linear (unbounded) operator $A : D(A) \subset X \to X$ is the infinitesimal generator of a C_0 contraction semigroup $S(t)$ on X $(t \geq 0)$ is and only if

(1) A is closed and $\overline{D(A)} = X$.

(2) Every $\lambda > 0$ belongs to the resolvent set $\rho(A)$ of A and

$$\| (\lambda I - A)^{-1} \| \leq \frac{1}{\lambda}. \tag{6.29}$$

(II) If A is the infinitesmial generator of the C_0 contraction semigroup $S(t)$, then

$$\lim_{n\to+\infty}(I-\frac{t}{n}A)^{-n}x = S(t)x\,,\quad t\geq 0\,,\quad x\in X\,, \tag{6.30}$$

and the limit holds uniformly with respect to t in bounded intervals of $[0,+\infty[$.

Proof. (I) By definition

$$D(A) = \{x\in X;\ \lim_{h\downarrow 0}\frac{S(h)x-x}{h}\ \text{exists}\}\,, \tag{6.31}$$

and

$$Ax = \lim_{h\downarrow 0}\frac{S(h)x-x}{h}\,. \tag{6.32}$$

Necessity. For $x\in X$, one easily proves that

$$x_t = \frac{1}{t}\int_0^t S(s)x\,ds \in D(A)\,.$$

Since $x_t \to x$ at $t\downarrow 0$, it follows that $\overline{D(A)} = X$. To prove that A is closed, let $x_n\in D(A)$, $x_n\to x$ and $Ax_n\to y$ as $n\to+\infty$. Since (for $0\leq s<t$)

$$S(t)x - S(s)x = \int_s^t S(\tau)Ax\,d\tau\,,\quad \forall x\in D(A)\,,$$

it follows

$$S(t)x_n - x_n = \int_0^t S(\tau)Ax_n\,d\tau$$

which yields

$$S(t)x - x = \int_0^t S(\tau)y\,d\tau\,,\qquad t>0\,.$$

Dividing by t and letting $t\downarrow 0$, we conclude that $x\in D(A)$ and $y = Ax$. Hence A is closed.

Inasmuch as $\| S(t)\| \leq 1$, the improper Riemann integral

$$\int_0^\infty e^{-\lambda t}S(t)x\,dt \equiv R(\lambda)x\,,\qquad \lambda>0 \tag{6.33}$$

is well-defined. Moreover, it defines a linear bounded operator $R(\lambda)$ with $\| R(\lambda)\|\leq 1/\lambda$. It is easy to check that $R(\lambda)x\in D(A)$, $\forall x\in X$ and $AR(\lambda)x = \lambda R(\lambda)x - x$ $= \lim_{h\downarrow 0}\frac{S(h)-I}{h}R(\lambda)x$. Thus $(\lambda I - A)R(\lambda) = I$. On the other hand, for $x\in D(A)$, we have (because A is closed)

$$R(\lambda)Ax = \int_0^\infty e^{-\lambda t}S(t)Axdt = A\int_0^\infty e^{-\lambda t}S(t)xdt = AR(\lambda)x\,,$$

and therefore,

$$R(\lambda)(\lambda I - A) = (\lambda I - A)(R(\lambda) = I\,,$$

that is $R(\lambda) = (\lambda I - A)^{-1} \equiv R(\lambda; A)$, $\lambda > 0$.

Sufficiency. Clearly (6.29) means $\| (I - \mu A)^{-1} \| \leq 1$, $\mu > 0$. This fact is equivalent to the fact that A is m-dissipative (hence A is maximal dissipative and, therefore, it is closed). Let

$$S(t)x = \lim_{n\to\infty} (I - \frac{t}{n}A)^{-n}x \tag{6.34}$$

be the C_0 semigroup generated by A via the exponential formula (see (3.10)). Therefore, $\lambda \in \rho(A)$ for all $\lambda > 0$ and (6.29) are sufficient for A to generate a C_0 semigroup S (given by (6.34)). The problem is to prove that the infinitesimal generator of S is just A. Indeed, let $x \in D(B)$ where B is defined by

$$\lim_{t\downarrow 0} \frac{S(t)x - x}{t} = \frac{d^+S(t)x}{dt}\,|_{t=0} = Bx\,.$$

By Theorem 2.1 (part 5) it follows that $S(0)x \in D(A)$ (i.e., $x \in D(A)$ and $Bx = Ax$. Therefore, $D(B) \subset D(A)$ and the restriction of A to $D(B)$ is just B. By the necessity part, $1 \in \rho(B)$. It follows $(I - B)^{-1} = (I - A)^{-1}$ which implies $A = B$. Part (II) has already been proved (see Theorem 6.5). The proof is complete.

Let S be an arbitrary C_0 semigroup on X. Then, there exists two constants $M \geq 1$ and $\omega \geq 0$ such that

$$\| S(t) \| \leq Me^{\omega t}, \qquad \forall t \geq 0. \tag{6.35}$$

From Theorem 6.6 we can derive the following characterization of the infinitesimal generator A of an aribtrary C_0 semigroup S (see e.g., Goldstein [1, p. 19] for the proof).

Theorem 6.7. A is the infinitesimal generator of a C_0 semigroup S iff A is closed, densely defined and $M \geq 1$, $\omega \in R$ exist such that $\lambda \in \rho(A)$ for each $\lambda > \omega$, and

$$\| (\lambda I - A)^{-n} \| \leq M/(\lambda - \omega)^n, \qquad n = 1, 2, \ldots. \tag{6.36}$$

Remark 6.4. Since the resolvent set $\rho(A)$ of A is

$$\rho(A) = \{\lambda \in R,\ I - A : D(A) \to X \text{ is bijective and } (\lambda I - A)^{-1};\ X \to D(A) \text{ is continuous}\},$$

it follows that $\rho(A) \neq \emptyset$ implies A is closed. Therefore, condition (2) in Theorem 6.6 implies " A is closed", i.e., the latter condition is superfluous. We add, however, this condition because in the proof of the necessity we need first to prove (1) and then condition (II).

Moreover, condition (6.29), i.e., $(I - \mu A)^{-1} : X \to D)(A)$ is nonexpansive is equivalent to the fact that A is m-dissipative. Recall that "A is m-dissipative" implies "A is maximal dissipative" which in its turn implies "A is closed". Therefore, Hille-Yosida theorem can be restated in the form Lumer-Phillips Theorem below.

Theorem 6.7. (Lumer-Phillips) A linear operator A is the infinitesimal generator of a C_0 semigroup of contractions on X iff A is densely defined and m-dissipative.

Remark 6.5. The condition $D(A) = X$ for a linear m-dissipative operator implies "A continuous from X into X" (which is not useful in applications to PDE). Indeed $(I - A)^{-1} : X \to X$ with $\| (I - A)^{-1} \| \leq 1$ implies $(I - A)^{-1}$ is bijective and continuous. Consequently, if $D(A) = X$, then $I - A$ is continuous, and therefore, A is continuous.

The importance of C_0 semigroups is illustrated by

Theorem 6.8. Let A be a linear, densely defined, m-dissipative operator. Then the C_0 contraction semigroup $S(t)$ generated by A via the exponential formula $\lim_{n\to+\infty}(I - \frac{t}{n}A)^{-n}x = S(t)x$, $x \in X$, is differentiable (for $x \in D(A)$) and $u(t, t_0, x) = S(t - t_0)x$ is the unique solution to the Cauchy Problem

$$u'(t) = Au(t), \quad t \geq t_0, \quad u(t_0) = x, \quad x \in D(A). \tag{6.37}$$

Proof. It is enough to consider the case $t_0 = 0$. Let $x \in D(A)$ and $t > 0$. Then

$$\lim_{h\downarrow 0} \frac{S(t+h)x - S(t)x}{h} = \lim_{h\uparrow 0} \frac{S(t+h)x - S(t)x}{h} = AS(t)x, \tag{6.37$'$}$$

i.e., $u(t) = S(t)x$ satisfies $u'(t) = Au(t)$ for all $t > 0$ and $u(0) = x$. Of course, by $u'(t)\,|_{t=0}$ we mean the right derivative of u calculated at $t = 0$. The uniqueness is already known (see e.g., Remark 2.1). Note that the dissipativity of A and (6.37) yield

$$\frac{d}{dt} \| u(t; t_0, x) \| \leq 0 \quad \text{a.e. on } [t_0, +\infty[, \quad \forall x \in D(A), \tag{6.38}$$

and hence

$$\| u(t; t_0, x) \| \leq \| x \|, \quad \forall t \geq t_0, \quad x \in D(A),$$

that is "$x = 0$" implies "$u(t, t_0, x) = 0$, $\forall t \geq t_0$" which is equivalent to the uniqueness of the solution to the problem (6.37).

Applications to PDE of this important result are given in Chapter 3.

Remark 6.6. A few comments on (6.37)'. If $h > 0$, $t \geq 0$, then

$$R_h = \frac{1}{h}(S(t+h)x - S(t)x) = S(t)\frac{S(h)x - x}{h} = \frac{S(h) - I}{h} S(t)x. \tag{6.39}$$

Therefore, according to (6.31), (6.32) and to the properties of $S(t)$, we have

$$\lim_{h \downarrow 0} R_h = S(t)Ax = AS(t)x = \frac{d^+}{dt} S(t)x \,, \quad x \in D(A), \quad t \geq 0 \,. \tag{6.40}$$

If $t > 0$, and $0 < h < t$, then

$$R_{-h} = -\frac{1}{h}(S(t-h)x - S(t)x) = S(t-h)\frac{S(h)x - x}{h} \,. \tag{6.41}$$

Set $x_h = \frac{S(h)x - x}{h}$, with $x \in D(A)$. Then $\lim_{h \downarrow 0} x_h = Ax$. Since $\| S(t) \| \leq 1$, we have $\| S(t-h)x_h - S(t)Ax \| \leq \| x_h - Ax \| + \| S(t-h)Ax - S(t)Ax \| \leq \| x_h - Ax \| + \| Ax - S(h)Ax \| \to 0$ as $h \downarrow 0$. Therefore,

$$\lim_{h \downarrow 0} R_{-h} = S(t)Ax = AS(t)x = \frac{d^-}{dt} S(t)x \,, \quad x \in D(A), \quad t > 0 \,. \tag{6.42}$$

In other words, if $t > 0$ and $x \in D(A)$, then

$$\frac{d^-}{dt} S(t)x = \frac{d^+}{dt} S(t)x = \frac{d}{dt} S(t)x = AS(t)x = S(t)Ax \,, \tag{6.43}$$

i.e., $t \to S(t)x$ is differentiable at $t > 0$. In the case $t = 0$, by $\frac{dS(t)x}{dt} |_{t=0}$ we mean the right derivative $\frac{d^+}{dt} S(t)x |_{t=0}$ (and $\frac{d^+}{dt} S(t)x |_{t=0} = Ax$, $x \in D(A)$).

§ 7. Differential inclusions in Hilbert spaces Smoothing effect of semigroups generated by subdifferentials.

Let A be a m-dissipative operator acting in the Banach space X and let $S \in Q_o(\overline{D(A)})$ be the contraction semigroup generated by A via the exponential formula. We have seen that if $x \in \overline{D(A)}$ and if $t \to S(t)x$ is strongly (or even only weakly) right differentiable at $t_o \geq o$ then $S(t_o)x \in D(A)$ and

$$d^+/_{dt}(S(t)x|_{t=t_o} \equiv S'(t_o^+)x \in AS(t_o)x \qquad (7.1)$$

In this case we simply say that $S(t_o)$ has a smoothing effect on $x \in \overline{D(A)}$. If X is a reflexive space, then for every $x \in \overline{D(A)}$ it follows

$$S(t)x \in D(A) \text{ for almost all } t \geq o \text{ (i.e.: a.e. on } R_+) \qquad (7.2)$$

In the case of a Hilbert space H of inner product $\langle .,. \rangle$ and norm $\|.\|$, we can prove stronger properties, e.g.

$$S(t):D(A) \to D(A), \text{ for all } t \geq o \qquad (7.3)$$

(as in the case A-linear!) and even more, if $A=-\partial\phi$, then

$$S(t):\overline{D(A)} \to D(A), \text{ for all } t > o \qquad (7.4)$$

In what follows we consider H to be real for simplicity. Recall that in Hilbert spaces maximal dissipativity is equivalent to m-dissipativity. Therefore a maximal dissipative set $A \subset H \times H$ is precisely a non-empty subset A of $X \times X$ satisfying dissipativity condition

$$\langle y_1-y_2, x_1-x_2 \rangle \leq o, \ [x_i,y_i] \in A, \ i=1,2 \qquad (7.5)$$

and range condition

$$R(I-A)=H \text{ (or equivalently } R(I-\lambda A)=H, \forall \lambda > o) \qquad (7.6)$$

Recall also that (7.5) and (7.6) imply D(A) convex and A closed. For $x \in D(A)$, the subset Ax is convex and closed, so there is a unique element $A_o x \in Ax$ such that

$$\|A_o x\| = \inf \{\|y\|, y \in Ax\}, \ \|A x\| \leq \|A_o x\|, \lambda > o \qquad (7.7)$$

If $x \in D(A)$ then

$$A_\lambda x \to A_o x \text{ as } n \to \infty \tag{7.8}$$

Moreover, if $x_\lambda \in X$ has the properties:
$x_\lambda \to x$ as $\lambda \downarrow o$ and $A_\lambda x_\lambda$ is bounded, then $x \in D(A)$ and $A_\lambda x_\lambda$ contains a sub-sequence

$$A_{\lambda_n} x_{\lambda_n} \to A_o x \text{ as } n \to \infty \ (\lambda_n \to o, \text{see Appendix}) \tag{7.8'}$$

In this case $\hat{D}(A)=D(A)$ and the conclusion of Corollary 5.1 holds. However in Hilbert spaces the solution u to the Cauchy problem

$$u'(t) \in Au(t),\ u(o)=x,\ x \in D(A),\quad t \geq o \tag{7.9}$$

has some additional properties. We know that $u(t,x)=S(t)x$ where S is the semigroup generated by A via the exponential formula (5.6). Next result collects some of the main properties of the solution u to (7.9).

Theorem 7.1. Let $A \subset H \times H$ be a maximal dissipative set. Then for each $x \in D(A)$, there is a unique function $u(t)=u(t;x)$ with the properties:

(1) $u(t) \in D(A)$, $t \geq o$ (i.e. $S(t)D(A) \subset D(A)$, $\forall t \geq o$)

(2) $t \to u(t)$ is Lipschitz continuous on R_+, its strong derivative $u' \in L^\infty(R_+,H)$ in the sense of distributions and $\|u'\|_\infty \leq \|A_o x\|$.

(3) $u'(t) \in Au(t)$ a.e. on $[o,\infty[$ (precisely, for each t_o in which u is differentiable, $u'(t_o) \in Au(t_o)$.

(4) $t \to \|A_o u(t)\|$ is decreasing and $t \to A_o u(t)$ is continuous to the right on R_+

(5) For each $t \in [o,+\infty[$, u has a right derivative

$$\frac{d^+u}{dt} \quad \text{and} \quad \frac{d^+u(t)}{dt} = A_o u(t) \qquad (\forall\ t \geq o)$$

(6) $\|u(t,x)-u(t,y)\| \leq \|x-y\|$, $\quad t \geq o$, $x,y \in D(A)$

Remark 7.1. In the nonlinear case (i.e. in our case here) a solution everywhere differentiable could not be expected. Here is an example: take $H=R$ and $A \subset R \times R$ defined as follows $Ax=-1$, if $x > o$, $Ax=[-1,o]$ if $x=o$, $Ax=o$ if $x < o$.

Obviously, A is maximal dissipative. it is easy to verify that in this case the strong solution to (7.9) is given by:

$$u(t,x)=(x-t)^+ = \begin{cases} o, & \text{if } x \leq t \\ x-t, & \text{if } o \leq t < x \end{cases}$$

if $x > o$. If $x < o$, $u(t,x)=x$ for all $t \geq o$.

Therefore, if $x > o$, the corresponding solution u is not differentiable at $t=x$.

With some additional effort, Theorem 7.1 can be derived from the reuslts in § 5. However, we will give a proof which is independent of that of results in § 5. This is because we want to point out another standard technique in the proof of existence results to (7.9).

Proof of Theorem 7.1. The existence. Let us consider the Yosida approximate equation

$$u_\lambda'(t) = A_\lambda u_\lambda(t),\ u_\lambda(o) = x \tag{7.10}$$

This equation has a unique solution $u_\lambda \in C^1([o,\infty[:H)$ (since $u \to A_\lambda u$ is Lipschitz continuous on H). We fix $h > o$. Then

$$\frac{d}{dt}(u_\lambda(t+h)) = A_\lambda u_\lambda(t+h),\ t \geq o \tag{7.11}$$

Combining (7.10), (7.11) with the dissipativity of A_λ one obtains

$$\frac{d}{dt}\|u_\lambda(t+h)-u_\lambda(t)\|^2 \leq o,\ \forall\ t \geq o$$

This implies

$$\|u_\lambda(t+h)-u_\lambda(t)\| \leq \|u_\lambda(h)-u_\lambda(o)\|,\ t \geq o,\ h \geq o \tag{7.11)'$$

Dividing by $\lambda > o$ and then letting $\lambda \downarrow o$ it follows

$$\|u_\lambda'(t)\| \leq \|u_\lambda'(o)\| = \|A_\lambda u_\lambda(o)\| = \|A_\lambda x\| \leq \|A_o x\| \tag{7.12}$$

Thus, we also have

$$\|A_\lambda u_\lambda(t)\| \leq \|A_o x\|,\ \forall t, \lambda > o \tag{7.13}$$

It is now easy to prove that $u_\lambda(t)$ is uniformly convergent as $\lambda \downarrow o$ on every compact $[o,T]$. Indeed, since $u = J_\lambda y - \lambda A_\lambda y$, $\forall y \in H$, (7.10) yields

$$\begin{aligned} &< u_\lambda'(t)-u_\mu'(t),\ u_\lambda(t)-u_\mu(t) > = < A_\lambda u_\eta(t) - A_\mu u_\mu(t),\ -\lambda A_\lambda u_\lambda(t) + \\ &+ \mu A_\mu u_\mu(t) + J_\lambda u_\lambda(t) - J_\mu u_\mu(t) >,\ \mu > o \end{aligned} \tag{7.14}$$

Inasmuch as $[J_\lambda y, A_\lambda y] \in A$ for all $\lambda > o$ and $y \in X$, (7.14) and (7.13) yield

$$\frac{1}{2}\ \frac{d}{dt}\ \|u_\lambda(t)-u_\mu(t)\|^2 \leq 2\ \|A_o x\|^2(\lambda+\mu),\ t\ \ o, \lambda\ , \mu > o$$

which gives

$$\|u_\lambda(t)-u_\mu(t)\| \leq 2\|A_0 x\| \; (t(\lambda+\mu))^{1/2}, \; t \geq 0, \lambda, \;\; > 0 \tag{7.15}$$

which shows that u (t) is convergent to a function u(t) as $\lambda \downarrow 0$, uniformly on bounded subsets of R_+. Let us fix (arbitrarily) $T > 0$. Since u'_λ is bounded in L (o,T;H) (by (7.12)), we may assume (relabelling if necessary) that u' is weakly star (w*) convergent to an element (say v) in L (|o,T|;H), i.e.

$$\lim_{\lambda \downarrow o} \int_o^T \langle u'(t), f(t) \rangle dt = \int_o^T \langle v(t), f(t) \rangle dt, \; \forall f \in L^1([0,T];H) \tag{7.16}$$

On the other hand

$$\int_o^T \langle u'_\lambda(t), \phi(t) \rangle dt = - \int_o^T \langle u_\lambda(t), \phi'(t) \rangle dt, \; \phi \in C_0^1([0,T];H) \tag{7.17}$$

The uniform convergence of $u_\lambda(t)$ to u(t), (7.16) and (7.17) lead to

$$\int_o^T \langle v(t), \phi(t) \rangle dt = - \int_o^T \langle u(t), \phi'(t) \rangle dt, \;\; \phi \in C_0^1([0,T];H) \tag{7.17)'}$$

which implies $\frac{du}{dt}$ = v in the sense of distributions. Clearly, (7.10) can be written in the form $(u_\lambda(t)-x)' = A_\lambda u_\lambda(t) - A_\lambda x + A_\lambda x$. This fact (in conjunction with the dissipativity of A_λ and $\|A_\lambda x\| \leq \|A_0 x\|$) yields $\|u_\lambda(h)-x\| \leq h\|A_0 x\|$, $\forall \;\; h > 0$. In view of (7.11)' we now have

$$\|u_\lambda(t+h)-u_\lambda(t)\| \leq h\|A_0 x\|, \; \forall \; t \geqslant 0, \; h \geqslant 0, \lambda > 0$$

Letting $\lambda \downarrow 0$ it follows

$$\|u(t+h)-u(t)\| \leq h\|A_0 x\| \tag{7.18}$$

and therefore u is strongly differentiable a.e; on R_+. Denoting by $\frac{du(t)}{dt}$ = u'(t) the strong derivative of u evaluated at t, it follows that

$$\int_o^T \langle v(t)-u'(t), \; \phi(t) \rangle dt = 0, \; \forall \; \phi \in C_0^1([0,T];X) \tag{7.18)'}$$

Since $u_\lambda(t) \to u(t)$ and $A_\lambda \; u_\lambda(t)$ is bounded, we have $u(t) \in D(A)$ for all $t \in [0,T]$. We have actually (relabeling if necessary) $u'_\lambda \rightharpoonup v$ in L^2 (o,T;X), and

$$\int_o^T \langle v(t)-u'(t), \; g(t) \rangle dt = 0, \; \forall g \in C_0(]0,T[;X) \tag{7.18)''}$$

which yields u'(t)=v(t)a.e. on]o,T[(hence $u' \rightharpoonup u'$ in $L^2(0,T;X)$) (see also Appendinx). Now we need to define the realization $\mathcal{A}$ of A in

$L^2(o,T;H)$, namely

$$D(\mathcal{A}) = \{ u \in L^2(o,T;H);\ u(t) \in D(A) \text{ a.e. on } [o,T] \} \tag{7.19}$$

For $u \in D(\mathcal{A})$, set

$$\mathcal{A}u = \{ w \in L^2(o,T;H),\ w(t) \in Au(t), \text{ a.e. on } [o,T] \} \tag{7.19'}$$

It is easy to check that $\mathcal{A}$ is also maximal monotone in $L^2(o,T;H)$ and $(\mathcal{A}_\lambda w)(t) = A_\lambda w(t)$, $\forall t \in [o,T]$, $\forall w \in L^2(o,T;H)$. Let us point out what we known about u : $u \to u$ in $L^2(o,T;H)$, $u_\lambda' \rightharpoonup u'$ (weakly) in $L^2(o,T;H)$, $u' = \mathcal{A}_\lambda u$ (with $(\mathcal{A}_\lambda u)(t) = A_\lambda u(t)$, so $\mathcal{A}_\lambda u_\lambda$ is bounded in $L^2(o,T;H)$. It follows that $u \in D(\mathcal{A})$ (we know already that actually $u(t) \in D(A)$ for all $t \geq o$) and $u' \in \mathcal{A}u$ (i.e. $u'(t) \in Au(t)$ a.e. on $[o,T]$).

Thus (1), (2) and (3) are verified. We now prove (4). We know that any weak cluster point $y(t)$ of $A_\lambda u_\lambda(t)$ is in $Au(t)$ ($y(t) \in Au(t)$). Or, by (7.13) and the definition of A_o,

$$\|A_o u(t)\| \leq \|y(t)\| \leq \liminf_{\lambda \to 0} \|A_\lambda u_\lambda(t)\| \leq \|A_o x\| \tag{7.20}$$

that is

$$\|A_o S(t)x\| \leq \|A_o x\|, \ \forall t \geq o,\ x \in D(A) \tag{7.21}$$

(since $u(t,x) = S(t)x$). Therefore, if $o \leq t_o$, t and $x \in D(A)$ we have $S(t_o)x \in D(A)$ and consequently (by (7.21) with $S(t_o)x$ in place of x)

$$\|A_o S(t+t_o)x\| = \|A_o S(t)S(t_o)x\| \leq \|A_o S(t_o)x\| \tag{7.22}$$

i.e.

$$\|A_o u(t+t_o)\| \leq \|A_o u(t_o)\| \tag{7.22'}$$

hence $t \to \|A_o u(t)\|$ is decreasing. It remains to prove the continuity (to the right) of this function. To this goal, let $t_n \downarrow o$ as $n \to \infty$ and $y_n = A_o u(t_o + t_n)$. The properties $y_n \in Au(t_o+t_n)$, $u(t_o+t_n) \to u(t_o)$, A demiclosed and y_n bounded ($\|y_n\| \leq \|A_o u(t_o)\|$ according to (7.22')) imply $y \in Au(t_o)$ (where y is the weak limit of a weakly convergent subsequence (denoted again by y_n) of y_n). But $\|y\| \leq \liminf \|y_n\| \leq \|A_o u(t_o)\|$, and therefore $y = A_o u(t_o)$. Since $\|y_n\| \leq \|A_o u(t_o)\|$ and $y_n \rightharpoonup A_o u(t_o)$ it follows that y_n is strongly convergent to $A_o u(t_o)$ (We have proved that every subsequence of y_n contains a subsequence strongly convergent to $A_o u(t_o)$, hence $y_n \to A_o u(t_o)$). It is known that 6) is a consequence of dissipativity of A. Let us prove Property 5). Take an arbitrary $t_o \geq o$. There is $t > o$ such that u is differentiable at t_o+t. Fix such a $t>0$. Therefore $u'(t_o+t)\in Au(t_o+t)$. Let $v(h)=S(h)S(t+t_o)x=u(t_o+t+h)$ be the solution to (7.9) corresponding to $v(o)=S(t+t_o)x\in D(A)$. According to (7.18), $\|v(h)-v(o)\|\le h\|A_ov(o)\|$, i.e. $\|u(t_o+t+h)-u(t_o+h)\|\le h\|A_ou(t_o+t)\|$; $\forall h>0$ which yields $\|u'(t_o+t)\|\le\|A_ou(t_o+t)\|$. This one and $u'(t_o+t)\in Au(t_o+t)$ give $u'(t_o+t)=A_ou(t_o+t)$. We conclude that $u'(t_o+t)=A_ou(t_o+t)$, for almost all $t\ge 0$. Integrating over $[o,t]$ we get

$$t^{-1}(u(t_o+t)-u(t_o)) = \frac{1}{t}\int_o^t A_ou(t_o+s)ds \tag{7.23}$$

On the basis of the continuity of $t\to A_ou(t)$ (to the right), (7.23) implies the existence of d^+u and $d^+u(t_o)=A_ou(t_o)$. The proof is complete.

<u>Remark 7.2.</u> Property (5) in Theorem 7.1 has the following physical interpretation: the system described by (3) "tends" to minimize its speed. More precisely, among all possible speeds $v\in Au(t)$ at the time t, the system chooses that of the least norm $v=u'(t)=A_ou(t)$.

Theorem 7.1 remains valid in the more general case of a Banach space X, whose dual X* is uniformly convex. We have already mentioned (at the beginning of this section) that the differentiability of $t\to S(t)x$ at $t=t_o\ge 0$ implies the smoothing effect of $S(t_o)$ on $x\in\overline{D(A)}$ (in the sense that $S(t_o)x\in D(A)$). It is interesting that in the case $A=-\partial j$ this smoothing effect (i.e. $S(t)\overline{D(A)}\subset D(A)$, $t>0$) holds wihtout additional hypotheses on S(t). Precisely, the following result due to Brezis [3] holds

<u>Theorem 7.2.</u> Let $j:H\to]-\infty,+\infty]$ be a propoer lower semicontinuous convex function (l.s.c.) and S(t) the semigroup generated by $A=-\partial j$ via the exponential formula. Then

$$S(t):\overline{D(A)}\to D(A),\ \forall t>0 \tag{7.24}$$

$$\|A_oS(t)x\|\le\|A_oy\|+\frac{1}{t}\|x-y\| \tag{7.25}$$

for all $x\in\overline{D(A)}$, $y\in D(A)$ and $t>0$.

Proof. Set

$$j_\lambda(z) = \min_{x\in H}\left\{j(x)+\frac{1}{2\lambda}\|z-x\|^2\right\},\ z\in H,\lambda>0 \tag{7.26}$$

$B=\partial j=-A$. It is known that (see Appendix)

$$B_\lambda = (\partial j)_\lambda = \partial j_\lambda \quad (J_\lambda=(I+\lambda B)^{-1},\ B_\lambda={}^{-1}(I-J_\lambda)) \tag{7.27}$$

that is B_λ is just the subdifferential of j_λ. In this case $A_\lambda = -B_\lambda$, so the approximate equation (7.10) becomes

$$u'_\lambda(t)+B_\lambda u_\lambda(t)=0,\ u_\lambda(0)=x,\ x\in\overline{D(A)}=\overline{D(\partial j)} \tag{7.28}$$

By the definition of ∂j_λ, $j_\lambda(u)-j_\lambda(y) \geq \langle B_\lambda y,u-y\rangle$ for all $u\in H$, where y is an arbitrary (but fixed) element of D(A)=D(B). Define (the convex function)

$$g_\lambda(u)=j_\lambda(u)-j_\lambda(y)-\langle B_\lambda y,u-y\rangle,\ u\in H \tag{7.29}$$

Hence $g_\lambda(u)\geq 0$, $\forall u\in H$. Since j_λ is Frechet differentiable so is g_λ and we have

$$g_\lambda(y)=0,\ \partial g_\lambda(u)=\partial j_\lambda(u)-B_\lambda y,\ u\in H \tag{7.30}$$

Equation (7.28) (with $B_\lambda = \partial j_\lambda$) becomes

$$u'_\lambda(t)+\partial g_\lambda(u_\lambda(t))+B_\lambda y=0,\ u_\lambda(0)=x\in\overline{D(A)},\ t\geq 0 \tag{7.31}$$

In view of (7.31) and

$$g_\lambda(y)-g_\lambda(u_\lambda(t))\geq\langle\partial g_\lambda(u_\lambda(t)),\ y-u_\lambda(t)\rangle,\ g_\lambda(y)=0$$

it follows

$$g_\lambda(u_\lambda(t))\leq\langle u'_\lambda(t)+B_\lambda y,y-u_\lambda(t)\rangle,\ \forall t\geq 0 \tag{7.32}$$

Let us fix an (arbitrary) $T>0$. Integrating (7.32) over $[0,T]$ and observing that

$$\langle u'_\lambda(t),y-u_\lambda(t)\rangle = -\frac{1}{2}\langle u_\lambda(t)-y,u_\lambda(t)-y\rangle',$$

we get

$$\int_0^T g_\lambda(u_\lambda(t))dt\leq\frac{1}{2}\|x-y\|^2-\frac{1}{2}\|u_\lambda(T)-y\|^2+\int_0^T\langle B_\lambda y,y-u_\lambda(t)\rangle dt \tag{7.33}$$

Taking into account that ∂g_λ is the Frechet derivative of g_λ it follows $(g_\lambda(u_\lambda(t)))'=\langle g_\lambda(u_\lambda(t)),u'_\lambda(t)\rangle$. Therefore multiplying (7.31) by $tu'_\lambda(t)$, integrating over $[0,T]$ and observing that

$$\int_0^T\langle tB_\lambda y,u'_\lambda(t)\rangle dt=\int_0^T\langle tB_\lambda y,(u_\lambda(t)-y'\rangle dt=$$

$$T\langle B_\lambda y,u_\lambda(T)-y\rangle-\int_0^T\langle B_\lambda y,u_\lambda(t)-y\rangle dt$$

we obtain

$$\int_0^T t\,\|u'_\lambda(t)\|^2dt+Tg_\lambda(u_\lambda(T))-\int_0^T g_\lambda(u_\lambda(t))dt=-T\langle B_\lambda y,u_\lambda(T)-y\rangle$$

$$+\int_0^T <B_\lambda y, u_\lambda(t)-y> dt \tag{7.34}$$

Combining (7.33), (7.34) and $g_\lambda(u_\lambda(T)) \geq o$, it follows

$$\int_0^T t\|u'_\lambda(t)\|^2 dt \leq -\frac{1}{2}\|u_\lambda(T)-y\|^2 - T<B_\lambda y, u_\lambda(T)-y> + \tfrac{1}{2}\|x-y\|^2 \tag{7.35}$$

On the other hand, according to Theorem 7.1 (4) $t \to \|u'_\lambda(t)\| = \|B_\lambda u_\lambda(t)\|$ is a decreasing function, so $\|u'_\lambda(T)\| \leq \|u'_\lambda(t)\|$ for all $o \leq t \leq T$, and therefore (7.35) leads to

$$\frac{T^2}{2}\|u'_\lambda(T)\| \leq \frac{1}{2}\|x-y\|^2 + \frac{T^2}{2}\|B_\lambda y\|^2 \tag{7.36}$$

which implies (for $y \in D(B)=D(A)$)

$$\|u'_\lambda(T)\| = \|B_\lambda u_\lambda(T)\| \leq \|B_o y\| + \frac{1}{T}\|x-y\| \tag{7.37}$$

for all $T > o$, $x \in \overline{D(A)}$ and $y \in D(A)$ that is $B_\lambda u_\lambda$ is bounded with respect to λ. We will prove that $u_\lambda(T) \to S(T)x$. To this aim, take an arbitrary $x_o \in D(A)$ and denote by $\bar{u} = u_\lambda(t,o,x_o)$ the solution to (7.28) corresponding to $\bar{u}_\lambda(o)=x_o$. From the proof of Theorem 7.1 we know that $\bar{u}_\lambda(T) \to S(T)x_o$ Moreover, the dissipativity of $-B_\lambda = A_\lambda$ yields

$$\|u_\lambda(t,o,x)-\bar{u}_\lambda(t;o,x_o)\| \leq \|x-x_o\|, \quad t \geq o. \qquad \text{We now have}$$

$$\|u_\lambda(T)-S(T)x\| \leq \|u_\lambda(T)-\bar{u}_\lambda(T)\| + \|\bar{u}_\lambda(T)-S(T)x_o\| + \|S(T)x_o - S(T)x\|$$

hence

$$\limsup_{\lambda \downarrow o} \|u_\lambda(T)-S(T)x\| \leq 2\|x-x_o\|, \ \forall x_o \in D(A)$$

which yields $u_\lambda(T) \to S(T)x$. We conclude from (7.37) that $S(T)x \in D(A)$ and $B_\lambda u_\lambda(T) \to B_o u(T) = B_o S(T)x$ (see (7.20)). Moreover, since $B_o u(T) = A_o u(T$ and $\|A_o u(T)\| \leq \liminf_{\lambda \downarrow o}\|B_\lambda u_\lambda(T)\|$, the inequality (7.37) implies

$$\|A_o S(T)x\| \leq \|A_o y\| + \frac{1}{T}\|x-y\|, \forall T > o$$

that is (7.25). The proof is complete.

Combining Theorems 7.1 and 7.2 we derive

<u>Theorem 7.3.</u> Let $f: H \to]-\infty, +\infty]$ be l.s.c. and let $S(t)$ be the semigroup generated by $-\partial f$. Then, for every $u_o \in \overline{D(A)}$, the function $u(t) = S(t)u_o$ has the properties: $t \to u(t)$ is continuous from $[o,+\infty[$ into H

(1) $u(t) \in D(A) \subset D_e(f)$, $\forall t > o$ (with $A = -\partial f$), $u(o)=u_o$ and

$$\|A_o S(t)u_o\| = \|A_o u(t)\| \leq \|A_o y\| + \frac{1}{t}\|u_o - y\|, \ \forall t > o, y \in D(A) \tag{7.38}$$

(2) u is Lipschitz continuous on $[d,+\infty[$ for every $d>o$ namely

$$\|S(t)u_o-S(s)u_o\| < |t-s|\|A_oS(d)u_o\|\ ,\ t\ ,\ s>d \qquad (7.38)'$$

(3) u is a.e. differentiable on $]o,+\infty[$, with

$$u'(t)+(\partial f)_o(u(t))=o,\ \text{a.e. on }]o,+\infty[,u'\in L^\infty(d,\infty;H) \qquad (7.39)$$

and

$$\|u'\|_{L^\infty(d,\infty;H)}\le\|A_oy\|+\frac{1}{d}\|u_o-y\|\equiv M_y,\ y\in D(A) \qquad (7.40)$$

(4) The right derivative $\frac{d^+u(t)}{dt}$ exists on $]o,+\infty[$ it is continuous and

$$\frac{d^+u(t)}{dt} + (\partial f)_o(u(t))=o,\ \text{for all } t>o \qquad (7.41)$$

(5) $t\to f(u(t))$ is a decreasing, Lipschitz continuous convex function on $[d,+\infty[$ for every $d>o$, and

$$\frac{d^+}{dt} f(u(t))=-\ \|\frac{d^+u}{dt}(t)\|^2,\ \forall\ t>o \qquad (7.42)$$

(6) If $u_o\in D(\partial f)$ then $u(t)=S(t)u_o\in D(\partial f)$ for all $t\ge o$ and

$$\|S(t)u_o-S(s)u_o\|\le|t-s|\ \|A_ou_o\|,\ \forall\ t,s\ge o \qquad (7.43)$$

with $u'(o)$ the right derivative of u at $t=o$

$$\frac{d^+u(t)}{dt} + (\partial f)_o(u(t))=o,\ \text{for all } t\ge o \qquad (7.43)'$$

and Properties in (5) hold on $[o,+\infty[$

(7) If $f(z)=o$, and $f(x)\ge o$ for all $x\in H$, then:

(7.1°) $\quad\int_o^\infty f(u(t))dt\le\frac{1}{2}\ \|u_o-z\|^2$, i.e. $f(u)\in L^1(R_+),u_o\in\overline{D(A)}$

(7.2°) $\quad\int_o^\infty t\|u'(t)\|^2dt\le\|u_o-z\|^2$, (I.e. $\sqrt{t}u'\in L^2(R_+;H)$), $u_o\in\overline{D(A)}$

hence

$$\int_d^\infty \|u'(t)\|^2\ dt\le\frac{1}{d}\|u_o-z\|^2\ (\text{i.e. } u'\in L^2(d,+\infty;H),\ \forall d>o),u_o\in\overline{D(A)}$$

(7.3°) $\quad\int_o^T \|u'(t)\|^2dt+f(u(T))=f(u_o),\ \forall\ T>o$

$T\to f(u(T))$ is decreasing and

$$f(u(T))\le f(u_o),\ \forall T>o,\ \int_o^\infty\|u'(t)\|^2\ dt\le f(u_o)\ (\text{hence } u'\in L^2(R_+;H))$$

(8) $\|S(t)u-S(t)v\|\le\|u-v\|,\forall\ t\ge o,u,v\in\overline{D(A)}$

<u>Proof.</u> Part (1) is just Theorem 7.2. Part (2). Take $d>o$ and $t>d$. Then

$u(t)=S(t-d)S(d)u_o$ with $y=S(d)u_o \in D(A)$ so (7.38)' is just Proposition 3.2 in Ch. 2. Parts (3) and (4) follows from Theorem 7.1 along with $u(t)=S(t-d)y$, $t>d$ and (7.38). Part (5). Since $u'_+ = \frac{d^+u(t)}{dt} \in -\partial f(u(t))$ for all $t \geq d$, we have

$$f(u(t+h))-f(u(t)) \geq -\langle u'_+(t), u(t+h)-u(t)\rangle ,$$
$$f(u(t))-f(u(t+h)) \geq -\langle u'_+(t+h), u(t)-u(t+h)\rangle, t \geq d, h > o \tag{7.44}$$

Inasmuch as $t \to u'_+(t)$ is continuous at $t \geq d$ we conclude that (7.44) yeilds (7.42). Moreover

$$\|u'_+(t)\| \leq \|u'_+(d)\| = \|A_o u(d)\| \leq M_y \text{ (see (7.40)) } \forall\, t \geq d \tag{7.45}$$

According to Lemma 2.2 in Appendix, $t \to f(u(t))$ is absolutely continuous on every compact of $[d,+\infty[$ hence

$$f(u(t_2))-f(u(t_1)) = \left| \int_{t_1}^{t_2} f'(u(t))dt \right| \leq |t_1-t_2| M_y^2, \forall t_1, t_2 \geq d \tag{7.46}$$

(6) For $u_o \in D(A)$ we can take $d=o$ so we get (7.43) and (7.43)'. Finally, properties (7) are proved in Theorem 10.3 (with $f(t)=o$, and j in place of f). For example, by (10.30) in our case here becomes (since $f_n=o$)

$$\int_o^T f(u(t))dt \leq \frac{1}{2} \|u_o-z\|^2, \forall\, T > o \tag{7.47}$$

which is equivalent to (7.1^o.) (this is because $f(u(t)) \geq o$, $\forall\, t \geq o$). Similarly, (10.32) yields (7.2^o).

<u>Remark 7.3.</u> If $u_o \in D_e(f)$, $f(x) \geq o$, $\forall x \in H$ and $f(z) = o$ for some $z \in H$, then (by (7.1^o), (7.3^o))

$$\lim_{t\to\infty} f(u(t))=o \quad \text{and} \quad \int_o^\infty \|u'(t)\|^2\, dt=f(u_o) \tag{7.48}$$

Finally if the level set

$$C_r=\{ x \in H, \|x\|+f(x) \leq r\} \text{ is precompact in H, } \forall r > o \tag{7.49}$$

then by Bruck's theorem (see Corollary 9.2 in Ch. 2)

$$u(t)=S(t)u_o \to a \quad \text{in H with } a \in D(A),\ o \in \partial f(a) \tag{7.50}$$

and therefore $f(x) \geq f(a)$, $\forall x \in H$. Thus we have proved.

<u>Corollary 7.1.</u> Let $f:H \to [o,+\infty]$ be a l.s.c. function with $(\partial f)^{-1}_{a} \neq \emptyset$, such that the level sets (7.49) are precompact. Then for every $u_o \in D(\partial f)$, the solution $u(t)=S(t)u_o$ of (7.39) has the additional property that $u(t) \to a$ (strongly) as $t \to \infty$ with $o \in \partial f(a)$ (i.e. $\inf_{x \in H} f(x)=f(a)$)

§ 8. Some results of nonlinear perturbation theory

8.1. Perturbation of m-dissipativity

Let X be a Banach space and let A and B; $A,B \subset X \times X$ be two nonempty subsets of $X \times X$. Suppose that one of these sets (say A) is m-dissipative. Simple examples show that even if B id m-dissipative, A+B need not be m-dissipative. In what follows we are concerned with several criteria which guarantees that A+B is m-dissipative. In the case A-m-dissipative and B- a perturbation such that A+B is also m-dissipative we say that "m-dissipativity of A is preserved" or that the perturbation B guarantees "the propagation" of m-dissipativity of A to A+B. In view of these notions, we can say that we are concerned with the propagation of m-sissipativity from A to A+B. The main result we are going to prove in this subsection is given by

Theorem 8.1. (Kobayashi) Let $A \subset X \times X$ be dissipative and let $B: B(B) \subset X \to X$ be a continuous operator such that $\overline{D(A)} \subseteq D(B)$ and A+B is dissipative. Then A+B is m-dissipative if and only if A is m-dissipative.

The proof of Theorem 8.1 (due to Kabayashi [1]) is based on

Theorem 8.2. Let $A \subset X \times X$ be a dissipative set. Then Conditions (i) and (ii) below are equivalent.

(i) A is closed and for every $z \in X$, A+z satisfies Tangential condition

$$\lim_{h \downarrow 0} \frac{1}{h} d[x; R(I-h(A+z))] = 0, \quad \forall x \in \overline{D(A)} \tag{8.1}$$

(ii) A is m-dissipative

In particular, Theorem 8.1 yields

Theorem 8.1' Let $A \subset X \times X$ be m-dissipative and let $B: \overline{D(A)} \to X$ be continuous, dissipative and satisfying tangential condition to $\overline{D(A)}$, i.e.

$$\lim_{h \downarrow 0} \frac{1}{h} d[x+hBx; \overline{D(A)}] = 0, \quad x \in \overline{D(A)} \tag{8.1)'}$$

Then A+B is m-dissipative.

Proof. The only fact we have to observe is that A+B is dissipative. Indeed, in this case the dissipativity of B i.e. the existence of $x^* \in F(x_1 - x_2)$, such that

$$\langle Bx_1 - Bx_2, x^* \rangle \le 0, \quad \forall x_1, x_2 \in D(A) \tag{8.1)"}$$

and (8.1)', imply (8.1)" <u>for all</u> $x^*\in F(x_1-x_2)$ (see Remark 5.5). Consequently A+B is dissipative. Indeed, for every $x_1,x_2\in D(A)$, there is $x_o^*\in F(x_1-x_2)$ such that

$\langle y_1-y_2,x_o^*\rangle\le o$, $\forall\ [x_j,y_j]\in A$, $j=1,2$. Therefore $\langle y_1+Bx_1-(y_2+Bx_2),x_o^*\rangle\le o$, which means just the dissipativity of A+B. Recall that for $z\in X$, $(A+z)x\equiv Ax+z$, $x\in D(A)$.

<u>Proof of Theorem 8.2.</u> Assume that (ii) holds. Then A is maximal dissipative (hence A is closed). Moreover, by definition, $R(I-hA)=X$, $h>o$, so $R(I-h(A+z))=X$ for all $h>o$ and every $z\in X$. Therefore (ii)$\Rightarrow$(i). We now prove that (i)$\Rightarrow$(ii). First we prove that (i) implies the maximal dissipativity of A on $\overline{D(A)}$ (i.e. that A is maximal dissipative on $\overline{D(A)}\times X$ - see Appendix). Indeed, let $x\in\overline{D(A)}$ and $z\in X$ be such that $A\cup[x,z]$ is dissipative. Condition (8.1) with $-z$ in place of z is equivalent to the existence of $r(h)\to o$ as $h\downarrow o$, $x_h\in D(A)$ and $y_h\in Ax_h$ such that $x-(x_h-h(y_n-z))=hr(h)$ (see § 7 in Ch. 1). On the other hand, the dissipativity of $A\cup[x,z]$ means

$$\|x_h-x\|\le\|x_h-x-h(y_h-z)\| \tag{8.1)'''}$$

which yields $\|y_n-z-r(h)\|\le\|r(h)\|\to o$ as $h\downarrow o$. Therefore $y_h\to z$ and $x_h\to x$ as $h\downarrow o$. Since A is supposed to be closed, it follows $x\in D(A)$ and $z\in Ax$ so A is maximal dissipative on $\overline{D(A)}$. We now can prove that $R(I-A)=X$. To this goal, take an arbitrary $y\in X$. We will prove the existence of $x_o\in D(A)$, such that $y\in x_o-Ax_o$. For this fact, set $C=A-I+y$. Clearly C is maximal dissipative on $\overline{D(C)}=\overline{D(A)}$ (since A is so). It is easy to see that (8.1) implies

$$\lim_{h\downarrow o}\frac{1}{h}d[x;R(I-hC)]=o,\ \forall\ x\in\overline{C(D)}=\overline{D(A)} \tag{8.2}$$

Indeed, we have seen that (8.1) implies $x-x_h+hy_n-hz=hr(h)$ (for every $z\in X$). Replace here z by $x-y$. Hence $(1-h)x-x_h+hy_n+hy=hr(h)$, $h>o$.

Dividing by $1-h$ and denoting $\frac{1}{1-h}=t$ (i.e. $h=\frac{t}{1+t}$) we get

$$x-(1+t)x_t+ty_t+ty=tr(t) \tag{8.3}$$

where $y_t\in Ax_t$ (with $x_t=x_h,y_t=y_h)r(t)=r(h)),r(t)\to o$ as $t\downarrow o$

We now have to observe that (8.3) is equivalent to (8.2) (see ch. 1, § 7). According to Theorem 2.1, the ω-dissipative set C generates a semigroup S(t) (of type $\omega=-1$) with $\|S(t)x-S(t)y\|\le e^{-t}\|x-y\|$, $x,y\in\overline{D(A)}$, $t\ge o$

Since $S(t):\overline{D(A)}\to\overline{D(A)}$, by Banach fixed point principle, there is a unique

$x_t \in \overline{D(A)}$ such that $S(t)x_t = x_t$. The semigroup property yields $S(t)S(s)x_t = S(s)S(t)x_t = S(s)x_t$, therefore $S(s)x_t = x_t$, $\forall t,s \geq 0$, so $x_t = x_s$ for all $t,s \geq 0$. In other words, x_t is independent of t, i.e. $x_t = x_0 \in \overline{D(A)}, \forall t \geq 0$. This means $S(t)x_0 = x_0$, $\forall t \geq 0$, so $t \to S(t)x_0$ is differentiable. This fact and the maximal dissipativity of C on $\overline{D(A)} = \overline{D(C)}$ imply (in view of Theorem 2.1, 5)) $x_0 \in D(C) = D(A)$ and $\frac{d}{dt}(S(t)x_0) = \frac{d}{dt}(x_0) = 0 \in Cx_0$, i.e. $0 \in Ax_0 - x_0 + y$. Therefore (i) implies $R(I-A) = X$, that is (ii). The proof is complete.

Remark 8.1. It is clear that (8.1) is equivalent to

$$\lim_{h \downarrow 0} \frac{1}{h} d[x+hz; R(I-hA)] = 0, \ \forall x \in \overline{D(A)}, \ z \in X \tag{8.4}$$

Indeed, we have seen above that (8.1) is equivalent to the existence of $x_h \in D(A)$, $y_h \in Ax_h$ and $r(h) \to 0$ as $h \downarrow 0$ such that

$$x - (x_h - hy_h - hz) = hr(h) \tag{8.5}$$

which can be written as

$$x + hz - (x_h - hy_h) = hr(h) \tag*{(8.5)'}$$

Clearly (8.5)' is equivalent to (8.4).

Theorem 8.2 remains valid if in (8.1) we replace "lim" by "lim inf" (this is easy to check, on the basis of the proof of Theorem 8.2).

Corllary 8.1. Let $A \subset X \times X$ be dissipative satisfying Tangential condition (8.1). Then $\bar{A}$ (the closure of A) is m-dissipative.

Proof. Obviously $R(I-hA) \subset R(I-h\bar{A})$, $h > 0$ hence $d[x+hz; R(I-h\bar{A})] \leq d[x+hz; R(I-hA)]$. Since A satisfies (8.4), (i.e. (8.1)) then $\bar{A}$ satisfies also (8.1). Thus, according to Theorem 8.2, $\bar{A}$ is m-dissipative.

For the proof of Theorem 8.1 we need also the following lemma

Lemma 8.1. Let A be m-dissipative and let $B: \overline{D(A)} \subset X \times X$ be continuous. Then for every $z \in X$,

$$\lim_{h \downarrow 0} \frac{1}{h} d[x+hz; R(I-h(A+B))] = 0, \ \forall x \in \overline{D(A)} \tag{8.6}$$

Proof. Let $x \in \overline{D(A)}$ and $z \in X$. Set

$$x_h = J_h(x+hz+hBx), \text{ where } J_h = (I-hA)^{-1}, \ h > 0$$

Clearly $x_h \in D(A)$, and $x_h \to x$ as $h \downarrow 0$. Moreover, there is $y_h \in Ax_h$ such that $x + hz + hBx = x_h - hy_h$. We now have

$$(x_h - hy_h - hBx_h - x - hz)/h = Bx - Bx_h \equiv r(h) \to 0 \text{ as } h \downarrow 0 \tag{8.7}$$

But (8.7) is equivalent to (8.6), so the proof is complete.

Proof of Theorem 8.1. If A is m-dissipative then A+B is closed and satisfies (8.6) (i.e. (8.1)) with A+B in place of A. According to Theorem 8.2, A+B is m-dissipative. Conversely if A+B is m-dissipative, then A=(A+B)-B is dissipative, closed and satisfies (by Lemma 8.1) the tangential condition (8.1). Consequently A is m-dissipative. The proof is complete.

In particular, if $A \subset X \times X$ is m-dissipative and $B:X \to X$ is continuous and dissipative, the A+B is dissipative and therefore(by Theorem 8.1) is m-dissipative. In other words, an immediate consequence of Theorem 8.1 is teh following result due to Barbu [1].

Theorem 8.3. (Barbu) Let $A \subset X \times X$ be m-dissipative. If $B:X \to X$ is a continuous every where defined dissipative operatoir, then A+B is m-dissipative.

The conclusion of this theorem was first obtained in the case A-linear, by Webb [1]. Finally, for A=O one obtains the result of Martin i.e.

Theorem 8.3' (Martin) Any continuous everywhere defined dissipative operator $B:X \to X$ is m-dissipative.

Note that if B is not everywhere defined (i.e. if X is strictly larger than D(B)) then the conclusion of Martin's theorem need not be true. Here is an elementary example: $X=]-\infty,+\infty[$, $Bx=-\sqrt{x}$, $D(B)=[0,+\infty[=R^1_+$. Clearly B is continuous and dissipative, but $R(I-B)=[0,+\infty[$.

8.2. On the equation $y \in dx-Ax-Bx$.

In applications to boundary value problems (B.V.P.) for elliptic equations of the form $u''(t)+Au(t) \ni 0$ with A dissipative, the equation

$$y \in dx-Ax-Bx, \; d > 0, \; x \in D(A) \cap D(B), \; y \in X \tag{8.8}$$

plays a crucial role (see e.g. Barbu [2], Pavel [4,5,6,8,12].

The basic hypotheses consist of

(H(8.1)) $D(A) \cap D(B) \neq \emptyset$ A,B-m-dissipative.

In these hypotheses for each $\lambda > 0$, $A_\lambda +B$ is m-dissipative (since B is m-dissipative and A_λ is continuous and dissipative). Consequently, for every $\lambda > 0$, there exists a unique $x_\lambda=x_\lambda(d, \lambda ,y) \in D(B)$ such that

$$y \in dx_\lambda -A_\lambda x_\lambda -Bx_\lambda \tag{8.9}$$

A first result on (8.9) is given by

Lemma 8.2. Let (H(8.1)) be satisfied. Then the solution x_λ of (8.9) is bounded i.e. $\|x_\lambda\| \leq b$, with $b > 0$ independent of $\lambda > 0$.

Proof. let x_1 $D(A) \cap D(B)$ and $z \in Bx_1$. There is $z_\lambda \in Bx_\lambda$ such that

$$y \in dx_\lambda \ -A_\lambda \ x_\lambda \ -z_\lambda \tag{8.10}$$

Let $x^* \in F(x_\lambda - x_1)$ with the property $\langle Bx_\lambda - Bx_1,\ x^* \rangle \leq o$. Then $\langle A_\lambda \ x_\lambda \ - A_\lambda \ x_1 ,x^* \rangle \leq 0$. Accordingly, we have

$$d \langle x_\lambda - x_1, x^* \rangle \ = d \ \|x_\lambda - x_1\|^2 \ = \langle dx_\lambda \ - dx_1, -x^* \rangle =$$

$$\langle y + A_\lambda \ x_\lambda \ + z_\lambda \ - dx_1, x^* \rangle \leq \langle y + A_\lambda \ x_1 + z - dx_1, x^* \rangle \leq$$

$$(\|y\| \ + \ \|A_\lambda x_1\| + \ \|z\| \ + d\|x_1\|)\|x_\lambda - x_1\| \tag{8.11}$$

Since $\|A_\lambda x_1\| \leq |Ax_1|$, (8.11) implies the boundedness of x_λ with respect to $\lambda > 0$. We now can prove

Theorem 8.4. Suppose that the dual X* of X is uniformly convex and that (H(8.1)) holds. If $A_\lambda \ x_\lambda$ is bounded then x_λ is strongly convergent to the (unique) solution $x \in D(A) \cap D(B)$ of (8.8). If B is single-valued, then Bx_λ contain a subsequence which converges to Bx.

Proof. Since $z_\lambda = dx_\lambda - A_\lambda x_\lambda - y \in Bx_\lambda$, $A_\lambda \ x_\lambda \in AJ_\lambda \ x_\lambda$ and $x_\lambda = J_\lambda \ x_\lambda$ - $- \lambda A_\lambda \ x_\lambda$ we have

$$\begin{aligned} o \ \leq \ \langle z_\mu \ \ - z_\lambda,\ F(x_\lambda - x_\mu) \rangle = -d\|x_\lambda - x_\mu\|^2 \ + \\ \langle A_\lambda \ x_\lambda - A_\mu x_\mu,\ F(x_\lambda - x_\mu) \rangle \leq -d \ \|x_\lambda - x_\mu\|^2 \\ + \langle A_\lambda \ x_\lambda \ - A_\mu \ x_\mu, F(x_\lambda - x_\mu) - F(J_\lambda \ x_\lambda - J_\mu \ x_\mu) \rangle \end{aligned} \tag{8.12}$$

where we have used the dissipativity of B and A. The boundedness of $A_\lambda \ x_\lambda$ implies $\|J_\lambda \ x_\lambda - x_\lambda\| \ \leq \lambda \|A_\lambda \ x_\lambda\| \to 0$ as $\lambda \downarrow 0$. The uniform continuity of F in bounded sets and (8.12) yield $\lim_{\lambda,\mu \downarrow 0} \|x_\lambda - x_\mu\|$ o that is x_λ is strongly convergent to an element x. But $J_\lambda \ x_\lambda \to x$ and $A_\lambda \ x_\lambda \in AJ_\lambda \ x_\lambda$ with $A_\lambda \ x_\lambda$ bounded imply $\lim J_\lambda \ x_\lambda = x \in D(A)$. Similarly, since $z_\lambda \in Bx_\lambda$ and z_λ is bounded we have $x \in D(B)$. Moreover, every weak cluster point of $A_\lambda x_\lambda (z_\lambda)$ belongs to Ax(Bx respoectively). Therefore, letting $\lambda \downarrow 0$ (or a subsequence of λ) in (8.9), one obtains (8.8). The uniqueness of solution to (8.8) holds if X* is strictly convex (because in this case A,B dissipative implies A+B dissipative). The proof is somplete.

In the case of a Hilbert space H (of inner product $\langle , \rangle$) one can prove that the converse assertion of Theorem 8.4 is also true. Therefore we have

Theorem 8.5. Let $A,B \subset H\times H$ be maximal dissipative with $D(A)\cap D(B)\neq \emptyset$. Let $d > o$ and $y \in H$. Then the equation (8.8) admits a (unique) solution x iff $A_\lambda x_\lambda$ in (8.9) is bounded in λ.

Proof. The only fact we have to prove is that if (8.8) has a solution x then $A_\lambda x_\lambda$ is bounded. Indeed, (8.8) means the existence of $u \in Ax$ such that $z=y-dx+u \in -Bx$. This and (8.10) give

$$o \leq \langle -z_\lambda -z, x_\lambda - x\rangle = -d\|x_\lambda - x\|^2 + \langle A_\lambda x_\lambda - u, x_\lambda - x\rangle$$

hence

$$o \leq \langle A_\lambda x_\lambda - u, J_\lambda x_\lambda - \lambda A_\lambda x_\lambda - x\rangle \qquad \text{since } x_\lambda = J_\lambda x_\lambda - \lambda A_\lambda x_\lambda)$$

Inasmuch as $A_\lambda x_\lambda \in AJ_\lambda x_\lambda$ it follows $o \leq \langle A_\lambda x_\lambda - u, -\lambda A_\lambda x_\lambda\rangle$, i.e. $\|A_\lambda x_\lambda\| \leq \|u\|$, $\forall \lambda > o$ which completes the proof.

Obviously Theorem 8.5 can be restated as

Theorem 8.6. Let $A,B \subset H\times H$ be maximal dissipative with $D(A)\cap D(B)\neq \emptyset$. Then $A+B$ is maximal dissipative iff for each $y \in H$, $A_\lambda x_\lambda$ in (8.9) is bounded in λ.

We end this subject with

Theorem 8.7. Let $A,B \subset X\times X$ be m-dissipative with $D(A)\cap D(B)\neq \emptyset$ and X^*-uniformly convex. If

$$\langle z, F(A_\lambda x)\rangle \geq o, \ \forall \lambda > o, \ x \in D(B) \text{ and } z \in Bx \tag{8.13}$$

then $A+B$ is m-dissipative.

Proof. With the notation in (8.10) we have $A_\lambda x_\lambda = dx_\lambda - z_\lambda - y$ with $z_\lambda \in Bx_\lambda$ Consequently

$$\|A_\lambda x_\lambda\|^2 = \langle A_\lambda x_\lambda, F(A_\lambda x_\lambda)\rangle = \langle -y+dx_\lambda, F(A_\lambda x_\lambda)\rangle -$$

$$\langle z_\lambda, F(A_\lambda x_\lambda)\rangle \leq (\|y\| + d\|x_\lambda\|)\|A_\lambda x_\lambda\| \tag{8.14}$$

which (in view of the boundedness of x_λ - Lemma 8.2) proves the boundedness of $A_\lambda x_\lambda$.

8.3. Perturbation of maximal monotonicity.

Let X be a Banach space and let X^* be its dual. Recall that a nonempty subset $A \subset X\times X^*$ is said to be monotone if

$$\langle y_1 - y_2, x_1 - x_2\rangle \geq o, \text{ for all } x_j \in D(A) \text{ and } y_j \in Ax_j, \ j=1,2 \tag{8.15}$$

A is said to be maximal monotone if A is monotone and it is not properly contained in any other monotone set of $X \times X^*$.

Some properties of monotone sets are given in the Appendix. Clearly, in the case of H "monotonicity" is just "accretivity". Recall without proof the following two results (see Appendix)

<u>Theorem 8.8.</u> Let X be reflexive and let $A \subset X \times X^*$ be maximal monotone. If $B: X \to X^*$ is monotone, hemicontinuous and bounded on bounded subsets, then A+B is maximal monotone.

<u>Theorem 8.9.</u> Let X be reflexive. If A and B are maximal monotone sets in $X \times X^*$ with

$$(\mathrm{Int}\, D(A)) \cap D(B) \neq \emptyset \tag{8.16}$$

then A+B is maximal monotone.

In the case of the real Hilbert space H of inner product $\langle\ ,\ \rangle$ and norm $\|.\|$ the condition (8.16) can eb weakened to $o \in \mathrm{Int}(D(R)-D(B))$ (where Int D means the interior of D in the strong topology). COnsequently one has

<u>Theorem 8.10.</u> Let B and W be maximal monotone (i.e. m-accretive) in $H \times H$. If

$$o \in \mathrm{Int}\ (D(B)-D(A)) \tag{8.17}$$

then A+B is maximal dissipative in $H \times H$.

Proof. We have to prove that R(I+A+B)=H. To this goal, let f be an arbitrary element of H. For every $\lambda > o$ denote by u_λ the solution of the equation

$$f \in u_\lambda + Au_\lambda + B_\lambda u_\lambda \tag{8.18}$$

The existence and uniqueness of u_λ is guaranteed by the fact that in this case (i.e. X=H) "maximal accretive" is equivalent (by Minty's theorem), to "m-accretive" (so we can apply the results in the previous subsection 8.2). In view of Theorem 8.5 it is necessary (and sufficient) to prove that

$$\sup_{\lambda > o} \|B_\lambda u_\lambda\| < +\infty \tag{8.19}$$

By hypothesis (8.17), there is $r > o$ such that every $y \in H$ with $\|y\| \leq r$ belongs to D(B)-D(A), i.e. $y=v-w$, $v \in D(B)$, $w \in D(A)$.

On the other hand, the accretivity of B_λ gives $\langle B_\lambda u_\lambda - B_\lambda v, u_\lambda - v\rangle \geq o$, and therefore

$$\langle B_\lambda u_\lambda, y\rangle = \langle B_\lambda u_\lambda, v\rangle - B_\lambda u_\lambda, w\rangle \leq \langle B_\lambda u_\lambda, u_\lambda - w\rangle + \|B_o v\|\ \|u_\lambda - v\|,\ \forall \lambda > o$$

By (8.18), there is $a_\lambda \in Au_\lambda$ such that $B_\lambda u_\lambda = f - u_\lambda - a_\lambda$ and thus

$$<B_\lambda u_\lambda, y> \le <f-u_\lambda - a_\lambda, u_\lambda - w> + \|B_o v\| \ \|u_\lambda - v\| \tag{8.20}$$

where $<-a_\lambda, u_\lambda - w> \le - <A_o w, u_\lambda - w> \le \|A_o w\| \ \|u_\lambda - w\|$ (since A is accretive). Thus, in view of the (8.20) and boundedness of u_λ we conclude that there exists a constant $K > o$ (independent of λ and y) such that $<B_\lambda u_\lambda, y> \le K$, $\forall y \in H$ with $\|y\| \le r$. This fact implies the boundedness of $B_\lambda u_\lambda$ i.e. (8.19). The proof is complete.

Remark 8.2. In the case $A = \partial f$ and $B = \partial g$ ($f,g:H \to]-\infty,+\infty]$ are lower semicontinuous convex functions (l.s.c.)), the condition (8.17) can be weakened to $o \in Int(D_e(f)-D_e(g))$ (See Attouch [1]). Here $D_e(f)$ is the effective domain of f (defined by (6.8)). Consequently we have

Theorem 8.10'. Let $f,g:H \to]-\infty, +\infty]$ be l.s.c. with $o \in Int(D_e(f)-D_e(g))$. Then $\partial f + \partial g$ is maximal monotone (and consequently $\partial f + \partial g = \partial(f+g)$)

Note that if $D_e(f) \cap D_e(g) \neq \emptyset$, we always have

$$\partial f + \partial g \subset \partial(f+g) \tag{8.20)'$$

Since $\partial(f+g)$ is maximal monotone it follows that

$$\partial f + \partial g = \partial(f+g) \tag{8.20)''$$

iff $\partial f + \partial g$ is maximal monotone.

8.4. Continuous perturbation of compact semigroups generators. The case of subgradients.

Let $A:D(A) \subset X \to 2^X$ be m-dissipative.

In order to avoid confusion denote by S_A the semigroup generated by A via the exponential fromula of Crandall-Liggett. Roughly speaking we are interested in finding perturbation operators B such that the propagation of compactness of S_A to S_{A+B} holds. In this direction we give

Theorem 8.11. Suppose that the semigroup S_A generated by m-dissipative set A is compact. Let $B:\overline{D(A)} \to X$ be a continuous operator which maps bounded sets into bounded sets, such that A+B is dissipative. Then (A+B is m-dissipative and) the semigroup S_{A+B} generated by A+B is also compact.

Proof. We know that "A m-dissipative, $B:\overline{D(A)} \to X$ continuous and A+B dissipative" imply A+B m-dissipative (Theorem 8.1). We have to prove that if in addition S_A is compact and B is bounded on bounded sets of $\overline{D(A)}$, then S_{A+B} is also compact. To this goal, set (for $x,y \in \overline{D(A)}$)

$$u(t;t_o,x)=S_{A+B}(t-t_o)x, v(t;t_o,y)=S_A(t-t_o)y,\ t_o \leq t \qquad (8.21)$$

Then $\tilde{u}$ and v are the integral solutions to the problems $\tilde{u}'(t) \in A\tilde{u}(t)+B\tilde{u}(t), \tilde{u}(t_o)=u(t_o;o,x); v'(t) \in Av(t), v(t_o)=y,\ t_o \leq t$ respectively, where $\tilde{u}(t)=u(t;t_o,u(t_o;o,x))=S(t-t_o)u(t_o;o,x)=u(t;o,x)=S(t-t_o)S(t_o)x, t_o \leq t, x \in \overline{D(A)}$ $(S=S_{A+B})$. On the basis of (2.22) with $f(t)=B\tilde{u}(t)=Bu(t;o,x), g(t)=o, s=t_o=\tau$ and $\tilde{u}$ on place of u it follows

$$\|u(t;o,x)-v(t;t_o,y)\| \leq \|u(t_o;o,x)-y\| + \int_{t_o}^{t} \|Bu(s;o,x)\|ds \qquad (8.22)$$

for all $x,y \in \overline{D(A)}$ and $t \geq t_o \geq o$

Let $t > o$ be fixed. We are now prepared to prove that $S_{A+B}(t)$ is compact. Indeed, let M be a bounded subset of $\overline{D(A)}$. We will prove that $S_{A+B}(t)M$ is precompact. First, the nonexpansivity and semigroup properties of S_{A+B} imply the existence of a constant $C = C(t,M) > o$, such that

$$\|u(s;o,x)\| \leq C, \text{ for all } x \in M \text{ and } s \in [o,t] \qquad (8.23)$$

and therefore there exists $C(t,M) \equiv C > o$ such that $\|Bu(s;o,x)\| \leq C$ for all $x \in M$, $s \in [o,t]$. Take (arbitrarily) $\varepsilon \in]o,t[$. For $t_o=t-\varepsilon$ and $y=u(t_o;o,x)$, (8.22) (where $u(t;o,x)=S_{A+B}(t)x$ yields

$$\|S_{A+B}(t)x-S_A(\varepsilon)u(t-\varepsilon;o,x)\| \leq \int_{t-\varepsilon}^{t} \|Bu(s;o,x)\|ds \leq \varepsilon C \qquad (8.24)$$

for all $x \in M$ and $\varepsilon \in]o,t[$. Since $u(t-\varepsilon;o,M)=\{u(t-\varepsilon;o,x); x \in M\}$ is a bounded subset of $\overline{D(A)}$ and $S_A(\varepsilon)$ is compact, $K_\varepsilon=S_A(\varepsilon)u(t-\varepsilon;o,M)$ is precompact.According to Lemma 5.1 in Chapter 1, (8.24) implies the precompacteness of $K=S_{A+B}(t)M$, q.e.d.

Corollary 8.2. Let A be m-dissipative with $\overline{D(A)} = X$ and let $B:X \to X$ be a linear bounded dissipative operator. If S_A is compact then so is S_{A+B}.

Proof. In this case A+B is m-dissipative and B is bounded on bounded sets. On the basis if Theorem 8.11, S_{A+B} is compact.

Remark 8.3. In the case A-linear, Corollary 8.2. was given by Pazy [4, p. 79]. It can be restated as below.

- Let A be the infinitesimal generator of a compact C_o semigroup S. If B is a bounded operator, then A+B is the infinitesimal generator of a compact C_o semigroup S_{A+B}.

The result below shows that the condition "B is bounded on bounded sets of $\overline{D(A)}$" can be avoided in the case $A=-\partial f$. Let us state the result in a precise form

Theorem 8.12. Let $f,g:H\to]-\infty, +\infty]$ be two lower semicontinuous convex functions from the real Hilbert space H into R such that $o\in \text{Int}(D_e(f)-D_e(g))$ (see (6.8). Set $A=-\partial f$ and $B=-\partial g$. If S_A is compact and $g:H\to [o, +\infty]$ (i.e. e.g. is positive-valued), then S_{A+B} is also compact.

Proof. According to Theorem 8.10', $\partial f+\partial g=\partial(f+g)$. Clearly $D_e(f+g)=D_e(f)\cap D_e(g)$. Consequently, for every $r>o$

$$C_r=\{x\in D_e(f+g);\|x\|\le r,\ f(x)+g(x)\le r\}\subset\{x\in D_e(f);\|x\|\le r, f(x)\le r\}\equiv C_r(f) \tag{8.25}$$

According to Theorem 6.2, $C_r(f)$ is precompact and therefore so is C_r. Invoking again Theorem 6.2, the precompactness of the level sets C_r implies the compactness of the semigroup generated by $-\partial(f+g)=A+B$ (i.e. S_{A+B} is compact) q.e.d.

§ 9. Some results on the asymptotic behaviour of nonlinear semigroups

Let D be a subset of the Banach space X and let S be a semigroup on D (see Definition 2.1). Mainly, in this section we are concerned with the behaviour of the function $t\to S(t)x$ on $[o,+\infty[$ e.g. the boundedness of $S(t)x$ on $[o,+\infty[$, $\lim\limits_{t\to\infty} S(t)x$ a.s.o.

Let us define some notions

Definition 9.1. (1) The semigroup S on D is said to be bounded if for every $x\in D$, there exists $M(x)>o$, such that $\|S(t)x\|\le M(x)$, $\forall\ t\ge o$ (i.e. the function $t\to S(t)x$ is bounded on $R_+=[o,+\infty[$). (2) S has a fixed point x_o if $S(t)x_o=x_o$, for all $t\ge o$.

Clearly if a nonexpansive semigroup S on D has a fixed point, then it is bounded. (Or, if there is x_o such the $\|S(t)x_o\|\le M(x_o)$, then $\|S(t)x\|\le M(x), t\ge o, \forall x\in D$). For the converse assertion we give

Theorem 9.1. Let D be a closed convex subset of a uniformly convex space X and let S be a nonexpansive semigroup on D. Then S is bounded iff S has a fixed point (Or, there is a bounded trajectory $\{S(t)x,\ t\ge o\}$,

$x \in D$, iff S has a fixed point).

The proof of this theorem can be found in the book [15] of the author. Note that Theorem 9.1 is due to Browder. In the proof one makes use of the following well-known result.

Theorem (Browder-Kirk). Let D be a closed convex bounded subset of the uniformly convex space X. If $B:D \to D$ is nonexpansive (i.e. $\|Bx-By\| \leq \|x-y\|$, $\forall x,y \in D$), then B has a fixed point (The set of all fixed points of B is closed and convex).

Proposition 9.1. (1) Let D be a closed subset of the general Banach space X and let $A:D \to X$ be a continuous and dissipative operator satisfying tangential condition (5.7). Then for every $x \in D$, the function $t \to \|AS(t)x\|$ is nondecreasing on R_+ (here S is the semigroup generated by A via Theorem 5.2). (2) If in addition, $\|Ax\| \to +\infty$ as $\|x\| \to +\infty$, then S is bounded (hence, if X is uniformly convex then S has a fixed point $x_o \in D$, with $Ax_o = o$). (3) If in addition to the hypotheses of (1), D is bounded and X is uniformly convex, then there exists $x_o \in D$ such that $Ax_o = o$.

The proof of this resut (essentially due to Martin) can also be found in the book [15] by the author. Note that Parts (2) and (3) follow easily from (1) and Thoerem 9.1. Indeed, if S is bounded, then there is $x_o \in D$ such that $S(t)x_o = x_o = u(t;o,x_o)$. Since $u'(t) = Au(t)$, $u(o) = x_o$, $t \geq o$, it follows $o = Ax_o$. Clearly in the hypotheses of (1), S has a fixed point $x_o \in D$ if $Ax_o = o$. For some generalizations see § 10.4.

If S is a semigroup on D and $x \in D$, we say that $S(t)x$ has a precompact range if the set K_x below (i.e. the trajectory corresponding to x)

$$K_x = \{ S(t)x, \quad t \geq o \} \tag{9.1}$$

is precompact in X (or that Trajectory $S(t)x$ is precompact). A first result in this direction is the following one (Dafermos-Slemrod [1])

Proposition 9.2. Let D be a closed subset of the Banach space X, and let S be a nonexpansive semigroup on D (i.e. $S \in Q_o(D)$. Definition 2.1) Then the set

$$PR = \{ x \in D, \ S(t)x \text{ has a precompact range} \} \tag{9.2}$$

is closed.

Proof. Let $x_n \in PR$ be such that $x_n \to x_o \in D$, and let $\epsilon > o$. There is a positive integer $n(\varepsilon)$ such that

$$\|S(t)x_n - S(t)x_o\| \leq \|x_n - x_o\| < \varepsilon \ , \ \forall n \geq n(\varepsilon) \tag{9.3}$$

Since $K_\varepsilon = k_{x_{n(\varepsilon)}}$ is precompact it follows from (9.3) and Lemma 5.1 in Chapter 1, that $K_{x_o} = \{S(t)x_o, \ t \geq o\}$ is also precompact, i.e. $x_o \in PR$. In what follows we will present some results on asymptotic behaviour of nonlinear contraction semigroups on a real Hilbert space H of inner product $<,>$ and norm $\|.\|$. (Some of these results remain valid in more general situations, e.g. in uniformly convex spaces).

Denote by S the semigroup generated by the maximal dissipative set $A \subset H \times H$ via the exponential formula.

Recall that $\overline{D(A)}$ is convex in this case (Appendix).

Proposition 9.3. Let $A \subset H \times H$ be maximal dissipative with $(I-A)^{-1}$ compact and let $x \in \overline{D(A)}$. If $S(t)x$ is bounded on R_+ (i.e. K_x given by (9.1) is bounded) then it has a precompact range (i.e. K_x is precompact). Equivalently, if $A^{-1}o \neq \emptyset$, then S is bounded and therefore $S(t)x$ has a precompact range for every $x \in \overline{D(A)}$. (An extension to Banach spaces is given in § 10.4).

Proof. Let us assume that K_x is bounded. By Proposition 9.2 it is enough to prove the precompactness of K_x fo $x \in D(A)$. But in this case $S(t)x \in D(A)$, $\forall t \geq o$ and according to Theorem 7.1, $t \to \|A_o S(t)x\|$ is nonincreasing (hence $\|A_o S(t)x\| \leq \|A_o x\|$, $\forall t \geq o$). In other words $(I-A_o)K_x$ is bounded and therefore $(I-A)^{-1}(I-A_o)K_x = K_x$ is precompact (we have used the definition of $(I-A)^{-1}$, i.e. $(I-A)^{-1}(x-y) = x$ for all $x \in D(A)$ and $y \in Ax$. Thus $(I-A)^{-1}(x - A_o x) = x$ for $x \in D(A)$, because $A_o x \in Ax$). Now, if $a \in A^{-1}o$ then $S(t)a = a$, $\forall \ t \geq o$. Hence, for every $x \in \overline{D(A)}$, K_x is bounded (since $\|S(t)x - S(t)a\| = \|S(t)x - a\| \leq \|x-a\|$, $t \geq o$) and therefore it is precompact.

Corollary 9.1. If the semigroup S generated by the maximal dissipative set $A \subset H \times H$ (via the exponential formula) is compact and $A^{-1}o \neq \emptyset$, then for every $x \in \overline{D(A)}$, $S(t)x$ has a precompact range.

Proof. Indeed, if S is compact then $(I-A)^{-1}$ is necessarily compact (Thoerem 6.1). Actually, Corollary 9.1 can be proved directly, observing that $S(t)x = S(\varepsilon)S(t-\varepsilon)x$, $\forall t, \varepsilon > o$.

Remark 9.1. (1) In terms of solution to Cauchy problem $u' = Au$, Proposition 9.3 asserts that if $(I-A)^{-1}$ is compact, then every bounded solution is precompact (or every bounded trajectory is precompact). (2) If A is maximal dissipative, then it follows from Theorem 9.1 that S is bounded if and only if $A^{-1}o \neq \emptyset$. Indeed, $\overline{D(A)}$ is closed and convex. Therefore if S is bounded, then there is $x_o \in \overline{D(A)}$ such that $S(t)x_o = x_o$, $\forall t \geq o$. This implies (in view of the maximal dissipativity of A and differentiability of $S(t)x_o$), $x_o \in D(A)$ and $o \in Ax_o$.

Now let us introduce the ω-limit-sets (for $x \in \overline{D(A)}$)

$$\omega(x) = \{ z \in \overline{D(A)};\ \exists\, t_n \to +\infty \text{ such that } S(t_n)x \to z \} \tag{9.4}$$

Celarly, If $S(t)x$ has a precompact range, $\omega(x)$ is nonempty. If $\lim_{t\to\infty} S(t)x = z$, then $\omega(x) = z$ and $\omega(x) \in A^{-1}o$. Indeed, $S(t+s)x = S(t)x$, $\forall t,s \geq o$ so $z = S(s)z$, $\forall s \geq o$. This implies $z \in D(A)$ and $z \in A^{-1}o$. In other words, "if A is maximal dissipative and $A^{-1}o = \emptyset$, then $\lim_{t\to\infty} S(t)x$ does not exist" (this holds in every Banach space). A simple property of $\omega(x)$ (which holds also in every Banach space) is given by

<u>Proposition 9.4.</u> (1) $z \in \omega(z)$ for every $z \in \omega(x)$ $(x \in \overline{D(A)})$. (2) If $z_n \in \omega(z_n)$ and $z_n \to z$ as $n \to \infty$, then $z \in \omega(z)$.

Proof. (1) Let $z \in \omega(x)$ and let $t_n \to +\infty$ be such that $S(t_n)x \to z$ as $n \to \infty$. Without loss of generality we may assume that $s_n \equiv t_{n+1} - t_n \to +\infty$ as $n \to \infty$. Then we have

$$\begin{aligned} \|S(t_{n+1}-t_n)z - z\| &\leq \|S(t_{n+1}-t_n)z - S(t_{n+1})x\| + \\ \|S(t_{n+1})x - z\| &\leq \|z - S(t_n)x\| + \|S(t_{n+1})x - z\| \end{aligned} \tag{9.5}$$

which shows that $S(s_n)z \to z$ as $n \to \infty$ (i.e. $z \in \omega(z)$).

(2) By definition, $z_n \in \omega(z_n)$ means the existence of $s_m = s_m(n)$, with $s_m \to +\infty$ as $m \to \infty$ and $\lim_{m\to\infty} S(s_m)z_n = z_n$ for every fixed n. There are two positive integers $p(n) \geq n$ and $q(n) \geq n$ such that

$$\|S(s_m)z_n - z_n\| \leq \frac{1}{n},\ \forall\ m \geq p(n) \quad \text{and} \quad s_m \geq n$$

for all $m \geq q(n)$. Set $t_n = s_{m(n)}$ where $m(n) = \max(p(n), q(n))$. Then $t_n \to +\infty$ as $n \to \infty$ and $\|S(t_n)z_n - z_n\| \leq \frac{1}{n}$. Now we have

$$\|S(t_n)z - z\| \leq 2\|z_n - z\| + \frac{1}{n}, \text{ hence } z \in \omega(z), \text{ q.e.d.}$$

Proposition 9.4 suggests that the set K (below) of all limit points of S is closed

$$K = \bigcup_{x \in \overline{D(A)}} \omega(x) \tag{9.6}$$

Theorem 9.2. In every Banach space X the set K given by (9.6) is closed. If X is a Hilbert space H and Trajectories $K_x, x \in C$ (with C-a dense subset of D(A)) are precompact, then K is convex.

Proof. Let $z_n \in K$ with $z_n \to z$ as $n \to \infty$. This means that there is $x_n \in \overline{D(A)}$ such that $z_n \in \omega(x_n)$, therefore $z_n \in \omega(z_n)$. It follows from Proposition 9.4 that $z \in \omega(z)$, hence $z \in K$. The proof of the convexity of K is based on Lemma 9.1 below and convexity of $\overline{D(A)}$.

Lemma 9.1. Let $T:D \to D$ be a nonexpansive operator on a closed convex subset D of H. If $x,y \in D$ are such that

$$\|Tx-Ty\| = \|x-y\|, \text{ then } T\left(\frac{x+y}{2}\right) = \frac{1}{2}(Tx+Ty)$$

Proof. Recall that given $u,v \in H$, then $c = \frac{u+v}{2}$ if and only if $\|c-u\| = \|c-v\| = \frac{1}{2}\|u-v\|$. Set $u=Tx$ and $v=Ty$. it is easy to check that $c=T(\frac{x-y}{2})$ satisfies $\|c-u\| = \|c-v\| = \frac{1}{2}\|u-v\|$. Then it follows $c=\frac{1}{2}(Tx+Ty)$, q.e.d.

Proof of Theorem 9.2. (continuation). Let $z_1, z_2 \in K$. We will prove that $\frac{1}{2}(z_1+z_2) \in K$ (which implies convexity of K). Let us note that the precompactness of K_x with x in a dense subset C of D(A), implies the precompactness of K_x for all $x \in D(A)$ and consequently - the precompactness of K_x for all $x \in \overline{D(A)}$ (see Proposition 9.2 and its proof).

Let $x_1, x_2 \in \overline{D(A)}$ and $t_n, s_n \to +\infty$ be such that $z_1 = \lim_{n\to\infty} S(t_n)x_1$, $z_2 = \lim S(s_n)x_2$. We may assume (by selecting appropriate subsequences and relabeling if necessary) that $s_n \geq t_n$ and $\lim S(s_n-t_n)x_2 = z_3$ exists. (This is because K_x is precompact). Then $\lim S(t_n)z_3 = z_2$. We may also assume that $p_n = t_{n+1} - t_n \to +\infty$ as $n \to \infty$. Then (see (9.5))

$$\lim_{n\to\infty} S(p_n)z_1 = z_1 \quad \text{and} \quad \lim_{n\to\infty} S(p_n)z_2 = z_2 \tag{9.7}$$

Fix (arbitrarily) $t > 0$. There is $n=n(t)$ such that $p_n > t$. Since $S(p_n) = S(p_n-t)S(t)$, the nonexpansivity of $S(p_n-t)$ yields $\|S(t)z_1 - S(t)z_2\| = \|z_1-z_2\|$.

According to Lemma 9.1. (with $T=S(p_n)$) we have

$$\lim_{n\to\infty} S(p_n)\frac{z_1+z_2}{2} = \lim_{n\to\infty} \frac{1}{2}(S(p_n)z_1 + S(p_n)z_2) = \frac{z_1+z_2}{2}$$

i.e. $\frac{1}{2}(z_1+z_2) \in K$, q.e.d.

In general, without the compactness of $(I-A)^{-1}$ we do not know if the trajectory $\{S(t)x,\ t \geq o\}$ is precompact. However, in the case $A = \partial\varphi$ there is a result (due to R. Bruck [1]- of weak convergence without compactness hypothesis, namely

Theorem 9.3. (Bruck). Let $-A = \partial\varphi$ be the subdifferential of a lower semicontinuous convex (proper) function $\varphi : H \to]-\infty, +\infty]$. If $A^{-1}o \neq \emptyset$ then for every $x \in \overline{D(A)}$, there exists $a \in A^{-1}o$ (so a is a minimum point of φ) such that

$$S(t)x \rightharpoonup a \quad \text{as } t \to +\infty \tag{9.8}$$

The proof of this surprising result relies on the following criterion of weak convergence (due to Z. Opial [1])

Lemma 9.2. Let $\mathbf{E}$ be a nonempty subset of a real Hilbert space H, and let $\{u_n\}_{n\in N} \subset H$ be a sequence with the properties

(1) For every $e \in \mathbf{E}$, $\|u_n - e\|$ is convergent

(2) Every weakly convergent subsequence u_{n_k}

has its weak limit $g \in \mathbf{E}$ (i.e. if $u_{n_k} \rightharpoonup g$ as $k \to \infty$, then $g \in \mathbf{E}$). Then u_n is weakly convergent to an element

$u \in \mathbf{E}$ ($u_n \rightharpoonup u \in \mathbf{E}$ as $n \to \infty$)

Proof. Let us consider two subsequences u_{n_k} and u_{m_k} of u_n, such that $u_{n_k} \rightharpoonup e_1$ and $u_{m_k} \rightharpoonup e_2$ as $k \to \infty$. We will prove that $e_1 = e_2$. By hypothesis, e_1, e_2 $\mathbf{E}$ and $\|u_n - e_i\| \to L_i$ as $n \to \infty$ $(i=1,2)$ But

$$\|u_n - e_1\|^2 - \|u_n - e_2\|^2 = \langle e_2 - e_1,\ 2u_n - e_1 - e_2 \rangle \tag{9.9}$$

For $n = n_k$ ($n = m_k$), $k \to \infty$, (9.9) yields respectively

$\|e_1 - e_2\|^2 = L_1^2 - L_2^2 = -\|e_1 - e_2\|^2$, and therefore $e_1 = e_2 = e \in \mathbf{E}$

It follows that u_n itself is weakly convergent to e as $n \to \infty$, q.e.d.

Proof of Theorem 9.3. One applies Opila's lemma with $\mathbf{E} = A^{-1}o$. Therefore, let $b \in A^{-1}o$. We know that $t \to \|S(t)x - b\|$ is nonincreasing on R_+ (this follows from (2.19) with x in place of x_o, b in place of x and $y = o \in Ab$). It follows also directly, since in this case $u(t) = S(t)x$ satisfies $u'(t) \in Au(t)$ a.e. on R_+ hence $\langle (u(t)-b)',\ u(t)-b \rangle \leq o$ a.e. on R_+ a.s.o)

On the other hand, by (7.25) with y=o we have

$$\|A_oS(t)x\| \le \frac{1}{t}\,\|x-b\|, \ \forall t > o,$$

Let $t_n \to +\infty$, be such that $S(t)x \rightharpoonup a$ as $n \to \infty$. Since $A_oS(t_n)x \in AS(t_n)x$ and $A_oS(t_n)x \to o$ as $n \to \infty$ it follows $o \in Aa$, i.e. $a \in E = A^{-1}o$. Thus, by Lemma 9.2, $S(t)x \rightharpoonup a$ as $t \to \infty$, q.e.d.

Corllary 9.2. If $A = -\partial\varphi$ (as in Th.9.3), $A^{-1}o \neq \emptyset$ and the level sets $C_r = \{ x \in D_e(\varphi);\ \|x\| \le r$ and $\varphi(x) \le r\}, r > o$, are precompact, then for each $x \in \overline{D(A)}$, there exists $b \in A^{-1}o$ such that

$$S(t)x \to b \text{ (strongly) as } t \to \infty \tag{9.10}$$

(i.e. $S(t)x$ converges strongly as $t \to \infty$ to an equilibrium point $b \in D(A)$, $o \in Ab$, hence $f(b) \le f(x)$, $\forall x \in H$, i.e. b is a minimum point of f).

(A generalization of this result to Banach spaces is given by Theorem 10.9)

Proof. In this case $(I-A)^{-1}$ is compact and therefore (by proposition 9.3), $S(t)x$ has a precompact range. It follows that in this case the convergence in (9.8) is strong. Since $o \in \partial\varphi(b)$ we have $\varphi(x) - \varphi(b) \geq o$, $\forall x \in H$, q.e.d.

Remark 9.2. It was proved by Bruck [1] that if φ is even, the convergence in (9.8) is strong. Note that Baillon has given an example in which the convergence (9.8) holds only in the weak topology.

For an arbitrary maximal dissipative operator acting in H there exists an ergodic theorem due to Baillon-Brezis (see Haraux [1]) namely

Theorem 9.4. Let $A \subset H \times H$ be maximal dissipative with $A^{-1}o \neq \emptyset$. Then for every $x \in D(A)$, there exists $a \in A^{-1}o$ such that

$$\frac{1}{t}\int_o^t S(s)x\,ds \rightharpoonup a, \qquad \text{as } t \to \infty \tag{9.11}$$

Recall also without proof the following result (Dafermos and Slemrod [1])

Theorem 9.5. Let $A \subset X \times X$ be m-dissipative (X-a Banach space). If $u(t) = S(t)x$ with $x \in \overline{D(A)}$ has a precompact range then there is a continuou function y such that $\lim_{t\to\infty} \|u(t)-y(t)\| = o$, y is almost periodic and satisfies $y' \in Ay$ in the weak sense (i.e. y is an integral solution to this equation).

Clearly, y is uniquely determined by x and it is called "the almost-periodic part of $S(t)x$".

Several generalizations of some results in this section are given in the next section. Recall also some results on the asymptotic behaviour of $J_t x=(I+tA)^{-1}x$, $x \in X$.

Theorem 9.6. (Reich [1]) Let X^* be uniformly convex and let $A \subset X \times X$ be m-accretive. If $o \in R(A)$, then for each $x \in X$, the (strong) limit $\lim_{t\to\infty} J_t x=y$ exists and $y \in A^{-1}o$.

The current state of this problem as well as many other results of this type can be found in the above paper by Simone Reich and in the corresponding references.

§. 10. Integral and strong solutions and smoothing effect in quasi-autonomous case.

10.1. The general case $u' \in Au+f$

We will continue the study of the initial value problem

$$u'(t) \in Au(t)+f(t), \quad u(o)=x_o \in \overline{D(A)}, \quad o \le t \le T \tag{10.1}$$

which has been considered in Chapter 1, § 4.3.

We have already proved (in a Banach space X)

Theorem 10.1. Let $A \subset X \times X$ be m-dissipative and $f \in L^1(o,T;X)$. Then for every $x_o \in \overline{D(A)}$, there is a unique integral solution u to (10.1), i.e. a continuous function $u:[o,T] \to \overline{D(A)}$, satisfying $u(o)=x_o$ and

$$\|u(\bar{t})-x\| \le \|u(t)-x\| + \int_t^{\bar{t}} \langle y+f(s), u(s)-x \rangle_+ ds \tag{10.2}$$

for all $o \le t \le \bar{t} \le T$, $[x,y] \in A$ and $T > o$. Moreover, if $g \in L^1(o,T,X)$ and $v=v(t;o,y_o)$ is the integral solution to

$$v' \in Av+g(t), \quad v(o)=y_o \in D(A), \quad o \le t \le T \tag{10.3}$$

then

$$\|u(t)-v(t)\| \le \|u(h)-v(h)\| + \int_h^t \|f(s)-g(s)\|ds, \quad o \le h \le t \le T \tag{10.4}$$

Recall that if A is the infinitesimal generator of a C_o-contraction semigroup S (see § 6.2), then the function given by the variation of parameters formula:

$$u(t)=S(t)x_o + \int_o^t S(t-s)f(s)ds, \quad o \le t \le T \tag{10.5}$$

with $x_o \in X=\overline{D(A)}$ and $f \in L^1(o,T;X)$ is said to be the mild solution of the problem

$$u'(t)=Au(t)+f(t),\ u(o)=x_o \in X,\ o \le t \le T \tag{10.6}$$

Now let us point out that in this (linear) case, the notions of "integral" and "mild" solution on [o,T] to (10.6), are equivalent. Indeed, let $f_n \in C^1(o,T;X)$ be such that $f_n \to f$ as $n \to \infty$ (in $L^1(o,T;X)$. Choose $x_n \in D(A)$ with $x_n \to x_o$ an $n \to \infty$. It is known that the problem

$$u'(t)=Au_n(t)+f_n(t),\ u_n(o)=x_n,\ o \le t \le T \tag{10.7}$$

has a unique strong solution u_n given by

$$u_n(t)=S(t)x_n + \int_0^t S(t-s)f_n(s)ds,\ o \le t \le T \tag{10.8}$$

Indeed, in this case $\int_o^t S(t-s)f_n(s)ds=v(t)$ is continuously differentiable $(v'(t)=S(t)f_n(o) + \int_0^t S(t-s)f_n'(s)ds)$

On the other hand it is easy to check that

$$\frac{S(h)v(t)-v(t)}{h} = \frac{v(t+h)-v(t)}{h} - \frac{1}{h}\int_t^{t+h} S(t+h-s)f_n(s)ds \tag{10.9}$$

for all $h > o$ sufficiently small and $o \le t \le T$. Clearly, (10.9) implies $Av(t)=v'(t)-f_n(t)$ with $v(o)=o$, $o \le t < T$. Since $u_n(t)=S(t)x_n+v(t)$ and $\frac{d}{dt}S(t)x_n=AS(t)x_n$ it follows that u_n given by (10.8) satisfies (10.7). But the strong solution u_n is the unique integral solution to (10.7) on [o,T], i.e.

$$\|u_n(\bar{t})-x\| \le \|u_n(t)-x\| + \int_t^{\bar{t}} \langle y+f_n(s),\ u_n(s)-x \rangle_+ ds \tag{10.10}$$

for all $o \le t \le \bar{t} \le T$, $[x,y] \in A$ and $n=1,2,\ldots$

Let us observe that (10.8) or even (10.10) yields

$$\|u_n(\bar{t})-u_m(t)\| \le \|x_n-x_m\| + \int_0^t \|f_n(s)-f_m(s)\|ds,\ o \le t \le T,\quad m,n=1,2\ldots \tag{10.11}$$

hence $u_n(t) \to u(t)$ as $n \to \infty$ uniformly on [o,T].

Passing to the limit in (10.8) and (10.10) it follows that the mild solution u given by (10.5) is the integral solution to (10.2), q.e.d.

It is clear that the mild solution u given by (10.5) is the uniform limit of the strong solutions u_n of (10.7), so u may also be called - a "generalized solution" or "weak solution" to (10.6). In other words, we have proved that in linear m-dissipative case, the notions of "mild", "integral", "DS-limit" and "generalized" (or "weak") solution are equivalent. Now let us discuss in details the notion od "strong" solution to (10.1). In linear case (i.e. in the case of (10.6)), if $x_o \in D(A)$ and $f \in C^1([o,T];X)$, then the function u given by (10.5) is strongly differentiable on $[o,T]$ (with $u'(t)=D^+u(t)|_{t=o}$ and $u'(t)=D^-u(t)|_{t=T}$) and

$$u'(t)=Au(t)+f(t) \text{ for all } t \in [o,T] \tag{10.12}$$

Moreover, an elementary calculus shows that the derivative u' of u (given by (10.5)) is just

$$u'(t)=S(t)Ax_o+S(t)f(o) + \int_o^t S(t-s)f'(s)ds, \tag{10.13}$$

for all $t \in [o,T]$ and $x_o \in D(A)$.

Consequently, if $f \in C^1([o,T]; X)$ and $x_o \in D(A)$, then the function u given by (10.5) is also of class $C^1([o,T];X)$ (therefore u is Lipschitz continuous on $[o,T]$) and satisfies (10.12) for all $t \in [o,T]$. Such a strong solution u is said to be a "classical" solution to (10.6). Clearly, (10.13) yields the estimate

$$\|u'(t)\| \le \|Ax_o\| + \|f(o)\| + \int_o^t \|f'(s)\| \, ds, \quad t \ge o, \qquad x_o \in D(A), \ f \in C^1([o,T];X) \tag{10.14}$$

Actually (10.14) remains valid if $f \in W^{1,1}(o,T;X)$ and A nonlinear. Of course in nonlinear case a classical solution to (10.1) cannot be expected (see e.g. Remark 7.1).

Definition 10.1 A function $u \in C([o,T];X)$ (with $u(t) \in D(A)$ for all $t \in [o,T]$) is said to be a strong solution of (10.1) if it is absolutely continuous on every compact of $]o,T[$, a.e. differentiable on $]o,T[$, $u(t) \in D(A)$ a.e. on $]o,T[$

$$u'(t) \in Au(t)+f(t), \text{ a.e. on }]o,T[\text{ and } u(o)=x_o \tag{10.14)'}$$

Clearly, (10.1) has at most one strong solution. We shall see that in many cases (e.g. $f \in W^{1,1}(o,T;X)$ with X reflexive, the strong solution to (10.1) is absolutely continuous on $[o,T]$. However, in general (e.g. if f is only in $L^2(o,T;X)$) the strong solution u is not absolutely continuous on $[o,T]$. (u is absolutely continuous only on compact subsets of $]o,T[$).

A strong solution u to (10.1) on [o,T] is the unique integral solution on [o,T] (Ch. 1, § 4.3).

In what follows we give criteria in order for an integral solution of (10.1) to be a strong solution. We start with a simple lemma.

Lemma 10.1 Let $A \subset X \times X$ be m-dissipative and $f \in L^1(o,T;X)$. If t_o is a Lebesgue point (at the right) of f and if the integral solution u of (10.1) is (weakly) differentiable at the right of $t_o \in [o,T]$ then $u(t_o) \in D(A)$ and $u'(t_o^+) \in Au(t_o)+f(t_o)$ were $u'(t_o^+)$ denotes the right derivative of u at t_o).

Proof. Take an arbitrary $x \in D(A)$ and $x^* \in F(u(t_o)-x)$. Then

$$x^*(u(t)-u(t_o))=x^*(u(t)-x)-x^*(u(t_o)-x) \leq \|u(t_o)-x\|(\|u(t)-x\| -$$
$$- \|u(t_o)-x\|) \leq \|u(t_o)-x\| \int_{t_o}^{t} \langle y+f(s),u(s)-x \rangle_+ ds \tag{10.15}$$

for all $y \in Ax$, $t_o < t < T$. Recall that

$$\langle y+f(s),u(s)-x \rangle_+ \leq \frac{\|u(s)-x+a(y+f(s))\| - \|u(s)-x\|}{a}, \ \forall\, a > o \tag{10.16}$$

Since t_o is a Lebesgue point (at the right) of f i.e.

$$\lim_{h \downarrow o} \frac{1}{h} \int_{t_o}^{t_o+h} \|f(s)-f(t_o)\| ds = o \tag{10.17}$$

and because u is continuous on [o,T], it follows that t_o is also a Lebesgue point (at the right) of the functions $s \to u(s)-x+\lambda(y+f(s))$ and $s \to u(s)-x$. Combining (10.15) and (10.16), dividing by $t-t_o$ letting $t \downarrow t_o$ and then $\lambda \downarrow o$, we obviously get

$$x^*(u'(t_o+)) \leq \|u(t_o)-x\| \langle y+f(t_o),u(t_o)-x \rangle_+ = \langle y+f(t_o),u(t_o)-x \rangle_s$$

which yields

$$\langle u'(t_o+)-f(t_o)-y,u(t_o)-x \rangle_i \leq o, \forall\, [x,y] \in A \tag{10.18}$$

Inasmuch as A is maximal dissipative, (10.18) implies the conclusion of the lemma.

We are now in a position to prove

Theorem 10.2. Suppose that X is a reflexive space and $A \subset X \times X$ is m-dissipative. If $x_o \in D(A)$ and $f \in W^{1,1}(o,T;X)$, then the initial value problem (10.1) has a(unique) strong solution u on [o,T], with $u' \in L^\infty(o,T;X)$ (so $u \in W^{1,\infty}(o,T;X)$)

$$\|u'(t)\| \le \|Ax_o\| + \|f(o)\| + \int_o^t \|f'(s)\|\, ds, \text{ a.e. on } [o,T] \tag{10.19}$$

Proof. Since X is reflexive, $f \in W^{1,1}(o,T;X)$ means f absolutely continuous on $|o,T|$. According to Theorem 10.1, the problem (10.1) has a unique integral solution u: $[o,T] \to \overline{D(A)}$.

For $x=x_o$, $\bar{t}=h$, t=o and $y \in Ax_o$, by (10.2) we have

$$\|u(h)-x_o\| \le \int_o^h (\|y+f(s)\|)ds, \quad o < h < T \tag{10.20}$$

with $x_o=u(o)$.

On the other hand, $t \to u(t+h)$ is the integral solution of

$$u'(t+h) \in Au(t+h)+f(t+h) \quad o \le t < t+h \le T$$

Consequently, with g(t)=f(t+h) and v(t)=u(t+h), (10.4) yields

$$\|u(t+h)-u(t)\| \le \|u(h)-u(o)\| + \int_o^t \|f(s+h)-f(s)\|ds, \quad o \le t < t+h < T \tag{10.21}$$

Then, for all $o \le t < t+h \le T$, we have

$$\int_o^t \|f(s+h)-f(s)\|ds \le \int_o^{T-h} \|f(s)+h)-f(s)\|ds \le hV_f(o,T) \tag{10.22}$$

where the finite number of $V_f(o,T)$ denotes the variation of f over [o,T] (see Appendix). In view of (10.20)-(10.22) it follows that u is Lipschitz continuous on [o,T] and therefore it is a.e. differetiable on]o,T[(since X is reflexive).

In view of Lemma 10.1, it follows that $u'(t) \in Au(t)+f(t)$ a.e. on]o,T[. Finally, with $y=A_ox_o$ (an element of the least norm of Ax_o) we have $\|A_ox_o\| = \inf\{\|y\|, y \in Ax_o\}$, (10.21) along with (10.20) imply (10.19). The proof is complete.

Remark 10.1 For the conclusion of Theorem 10.2 the condition $x_o \in D(A)$ is essential (indeed, if $x_o \in \overline{D(A)} \setminus D(A)$, then the integral solution u need not be differentiable). We actually have

$$u'(t) \in (Au(t)+f(t))_o, \text{ a.e. on }]o,T[\tag{10.22'}$$

where $(Au(t)+f(t))_o$ are the elements of the least norm of the closed convex set Au(t)+f(t). Indeed, if u is differentiable at t, then $u(t) \in D(A)$, and for x=u(t) (10.2) gives

$$||u(\bar{t})-u(t)|| \leq \int_t^{\bar{t}} ||y+f(s)||ds \tag{10.22)"}$$

which yields $||u'(t)|| \leq ||f(t)+y||$, $\forall y \in Au(t)$.

Other details are given in Section 10.3.

10.2. The case $A=-\partial j$

In the case $A=-\partial j$, the condition $f \in W^{1,1}(o,T;X)$ in Theorem 10.2 can be weakened to $f \in L^2(o,T;H)$ (H-a real Hilbert space of inner product $<,>$ and norm $||.||$). As in the autonomous case (see Theorem 7.2), the solution u of the problem

$$u'(t)+ \partial j(u(t)) \ni f(t) \text{ a.e. on }]o,T[, \; u(o)=x_o \in \overline{D(\partial j)} \tag{10.23}$$

has additional properties. Assume for simplicity that $j:H \to [o,+\infty[$ is a proper lower semicontinuous convex function (l.s.c.) and that there are $z \in H$ such that $j(z)=o$. Recall that

$$D(\partial j) \subset D_e(j) \subset \overline{D_e(j)} = \overline{D(\partial j)} \tag{10.24}$$

where $D_e(j) = \{ x \in H, \; j(x) < \infty \}$ is the effective domain of j.

The condition $R(j) \subset [o,+\infty]$ we are using in Th. 10.3 is not essential. See Remark 10.2.

The following result (important in application to PDE), due to Brezis [3] holds.

Theorem 10.3. For every $x_o \in \overline{D(\partial j)}$ (with $j:H \to \bar{R}_+$, l.s.c.) and $f \in L^2(o,T;H)$, the problem (10.23) has a unique strong solution u (in the sense of Definition 10.1) with the additional properties

$t \to \sqrt{t}\, u'(t)$ is in $L^2(o,T;X)$, $j(u) \in L^1(o,T.H)$

$t \to j(u(t))$ is absolutely continuous on every compact $[d,T]$ and $u \in W^{1,2}(d,T;H)$ for every $d \in]o,T[$. If $x_o \in D_e(j)$, then $u' \in L^2(o,T;H)$ (so $u \in W^{1,2}(o,T;H)$) and $t \to j(u(t))$ is absolutely continuous on $[o,T]$.

Proof. We will prove the following estimates (where u is the strong solution of (10.23), whose existence we have to prove)

$$\int_o^T j(u(t))dt \leq (||u(o)-z|| + \int_o^T ||f(t)||dt)^2 \tag{10.25}$$

$$\int_o^T t||u'(t)||^2 dt \leq \int_o^T t||f(t)||^2 dt + 2(||u(o)-z|| + \int_o^T ||f(t)||dt)^2 \equiv M_f \tag{10.26}$$

with $j(z)=o$ and $M_f < +\infty$. Clearly, for every $d \in]o,T[$, (10.26) implies

$$\int_d^T \|u'(t)\|^2 dt \leq \frac{1}{d} M_f \tag{10.27}$$

Moreover, if $x_o \in D_e(j)$, the solution u satisfies

$$\int_o^T \|u'(t)\|^2 dt \leq j(x_o) + \int_o^T \|f(t)\|^2 dt \ (\text{so } u \in W^{1,2}(o,T;H)) \qquad (10.27)'$$

$$j(u(t)) \leq j(x_o) + \int_o^t \|f(s)\|^2 ds, \ o \leq t \leq T \tag{10.28}$$

In order to prove the existence of the solution u of (10.23), take $x_n \in D(\partial j)$ with $x_n \to x_o$ and $f_n \in W^{1,2}(o,T;H)$ with $f_n \to f$ in $L^2(o,T;H)$, as $n \to \infty$. According to Theorem 10.2, there is a unique strong solution u_n, of the problem

$$u_n'(t) + \partial j(u_n(t)) \ni f_n(t), \text{ a.e. on }]o,T[, \ u_n(o)=x_n \tag{10.29}$$

such that $t \to j(u_n(t))$ is absolutely continuous on $[o,T]$ (Appendix (2.59))
The definition of $\partial j(u_n(t))$, $j(z)= o$ and (10.29) lead us to

$$j(u_n(t)) \leq \langle u_n(t)-z, f_n(t)\rangle - \frac{1}{2}\frac{d}{dt}\|u_n(t)-z\|^2, \ t \in [o,T]$$

Integrating over $[o,T]$ one obtains

$$\int_o^T j(u_n(t))dt \leq \frac{1}{2}\|u_n(o)-z\|^2 + \int_o^T \|f_n(t)\| \ \|u_n(t)-z\| dt \tag{10.30}$$

Since $o \in \partial j(z)$ we easily derive

$$\|u_n(t)-z\| \leq \|u_n(o)-z\| + \int_o^T \|f_n(t)dt$$

so (10.30) implies

$$\int_o^T j(u_n(t))dt \leq (\|u_n(o)-z\| + \int_o^T \|f_n(t)\|dt)^2 \tag{10.31}$$

Multiplying (10.29) by $tu_n'(t)$ we get (see Appendix)

$$t\|u_n'(t)\|^2 + \frac{td}{dt} j(u_n(t)) = \langle f(t), tu_n'(t)\rangle \qquad (10.31)'$$

which gives

$$\int_o^T t\|u_n'(t)\|^2 dt \leq \frac{1}{2}\int_o^T t\|f_n(t)\|^2 dt + \frac{1}{2}\int_o^T t\|u_n'(t)\|^2 dt + \int_o^T j(u_n(t))dt \tag{10.32}$$

By (10.32) and (10.31) it follows

$$\int_0^T t\|u_n'(t)\|^2 dt \le \int_0^T t\|f_n(t)\|^2 dt + 2(\|x_n - z\| +$$

$$+ \int_0^T \|f_n(t)\| dt)^2 \equiv M_{f_n} \tag{10.33}$$

Consequently, for every $d \in]o,T[$

$$\int_d^T \|u_n'(t)\|^2 dt \le \frac{1}{d} M_{f_n} \tag{10.33)'$$

On the other hand (10.4) (or (10.29)) show that

$$\|u_n(t) - u_m(t)\| \le \|x_n - x_m\| + \int_0^t \|f_n(s) - f_m(s)\| ds, \quad o \le t \le T \tag{10.33)"$$

and therefore $u_n(t) \to u(t)$ (strongly in H) as $n \to \infty$) uniformly on $[o,T]$. Clearly, $\lim_{n\to\infty} M_{f_n} = M_f$ (see (10.26)). Consequently, by (10.33)' the sequence u_n' is bounded in $L^2(d,T;H)$ so we may assume $u_n' \rightharpoonup v$ (weakly) in $L^2(d,T;H)$ as $n \to \infty$. Since $u_n \to u$ in $C([o,T];H)$ it follows from

$$\int_d^T \langle u_n(t), \varphi'(t)\rangle dt = - \int_d^T \langle u_n'(t), \varphi(t)\rangle dt, \ \forall \varphi \in C_o^1(]d,T[;H)$$

that

$$\int_d^T \langle u(t), \varphi'(t)\rangle dt = - \int_d^T \langle v(t), \varphi(t)\rangle dt, \ \forall \varphi \in C_o^1(]d,T[;H)$$

that is v is the distributional derivative u' of u. Since $u' = v \in L^2(d,T;H)$, $u \in W^{1,2}(d,T;H)$ so u is absolutely continuous on $[d,T]$, hence u is a.e. differentiable on $]d,T[$. Thus the distributional derivative u' of u is a.e. just the strong derivative u' of u on $]d,T[$. (so $u(t) \in D(A)$ a.e. on $]o,T[$).

Since d is arbitrary in $]o,T[$ it follows from Theorem 10.2 that u is the (unique) strong solution of the problem (10.23). Passing to the limit in (10.31), (10.33) and (10.34) one obtained (10.25), (10.26) and (10.17) respectively.

In (10.30) one observes that $\lim_{n\to\infty} j(u_n(t)) \ge j(u(t))$ a.e. on $]o,T[$ and then one applies Fatou's lemma. In the case of (10.33) one observes that $\sqrt{t}\, u_n' \rightharpoonup tu'$, hence

$$\int_0^T t\|u'(t)\|^2 dt \le \liminf_{n\to\infty} \int_0^T t\|u_n'(t)\|^2 dt \quad \text{and so on.}$$

Finally, since u is absolutely continuous on every compact of $[o,T]$, so is $t \to j(u(t))$ (this is because $u'(t)-f(t) \in \partial j(u(t))$ with $u'-f \in L^2(d;T;H)$, $o < d < T$ (see Appendix). Multiplying (10.23) by $u'(t)$ we derive

$$\|u'(t)\|^2 + \frac{d}{dt} j(u(t)) \leq \|f(t)\|^2 \quad \text{a.e. on }]o,T[\tag{10.34}$$

Obviously, from (10.34) (and $x_o = u(o) \in D_e j$) one obtains both (10.27)' and (10.28).

Remark 10.2. If j is not positively-valued, take $w \in \partial j(z)$ with $z \in D(\partial j)$. Set

$$g(x) = j(x) - \langle w,x \rangle - (j(z) - \langle w,z \rangle), \; x \in H \tag{10.34'}$$

Then $g(x) \geq o$, $\min_{x \in H} g(x) = g(z) = o$. By Theorem 10.3, there is a strong solution u of $u'(t) + \partial g(u(t)) \ni f(t) - w$ a.e. on $]o,T[$. Since $\partial g(u) = \partial j(u) - w$ it follows $u' + \partial j(u) \ni f$. Thereofre the condition $R(j) \subset \bar{R}_+$ in Th. 10.3 is not essential.

10.3. Solutions everywhere differentiable at the right.

We will prove that under natural additional restrictions either on f or on X, the strong solution u of $u' \in Au+f$ (which is a.e. differentiable on $]o,T[$) is everywhere differentiable at the right, on $[o,T[$. A first result in this direction is given by

Theorem 10.4. In addition to the hypotheses of Theorem 10.2, suppose that X is uniformly convex. Then the strong solution u to the problem (10.1) (with $x_o \in D(A)$) is differentiable at the right of every $t \in [o,T[$ (hence $u(t) \in D(A)$, $\forall\, t \in [o,T[)$ and

$$\frac{d^+u}{dt}(t) = (Au(t)+f(t))_o = f(t) - \mathrm{Proj}_{-Au(t)} f(t) \tag{10.35}$$

for all $t \in [o,T[$ (see (10.36)). Recall that if C is a closed convex set of the uniformly convex space X, then for every $x \in X$, there is a unique element $x \in C$ such that $\|x-\bar{x}\| = d[x,C] = \inf_{y \in C} \|x-y\|$ (Appendix)

One denotes $\bar{x} = \mathrm{proj}_C x$. In other words $x - \mathrm{Proj}_C x$ is the element of the least norm of the (closed convex) set $x-C$. Set

$$x - \mathrm{Proj}_C x = (x-C)_o \tag{10.36}$$

Recall also that for every $x,y \in X$

$$\|x\| \langle y,x \rangle_+ = \|x\| \lim (\|x+ty\| - \|x\|)t^{-1} = \langle y,x \rangle_s = \langle y,F(x) \rangle \tag{10.36'}$$

(see (2.1) in Chapter 1) where F is the duality mapping of X and $\langle y,F(x) \rangle$

$\equiv (F(x)(y)$. Clearly, $<y,x>_+ \leq \|y\|$ and in case of the real Hilbert space H, $\|x\| <y,x>_+ = <x,y> = \|x\| <y,x>_-$. Finally, recall

Lemma 10.2. Let X be uniformly convex. If $\{u_n\} \subset X$ is weakly convergent to u and $\limsup \|u_n\| \leq \|u\|$ (i.e. $\|u_n\| \to \|u\|$) then u_n is strongly convergent to u.

(For the proof see e.g. Pavel [15]).

Proof of Theorem 10.4. Fix an arbitrary $t \in [o,T]$. Then for all $h \in [o,T-t[$ we have

$$h^{-1} \| u(t+h)-u(t) \| \leq \frac{1}{h} \int_0^h \|y+f(s)\| ds + \int_0^t h^{-1} \|f(s+h)-f(s)\| ds, \ y \in Ax_0 \tag{10.37}$$

(see (10.21)). Consequently, since f is absolutely continuous

$$\limsup_{h \downarrow o} h^{-1} \|u(t+h)-u(t)\| \leq \|(Ax_0+f(o))_0\| + \int_0^t \|f'(s)\| ds \tag{10.37)'$$

Let h_n be an arbitrary sequence with $h_n \downarrow o$ as $n \to \infty$. Set $u_n = h_n^{-1}(u(t+h_n)-u(t))$. Since u_n is bounded (on the basis of (10.37)'), it contains a weakly convergent subsequence u_{n_k}. Say $u_{n_k} \rightharpoonup v$ as $k \to \infty$. Racall that

$$<u(t+h_n)-u(t), x^*> \leq \|u(t)\| \int_t^{t+h_n} <y+f(s), u(s)-x>_+ ds$$

for all $[x,y] \in A$ where $x^* = F(u(t)-x)$ (cf:(10.15)).

Dividing by h_{n_k} and letting $k \to \infty$ we get

$$<v,F(u(t)-x)> \leq \|u(t)-x\| <y+f(t),u(t)-x>_+ = <y+f(t),F(u(t)-x)>$$

i.e.

$$<v-f(t)-y, F(u(t)-x)> \leq o, \forall [x,y] \in A$$

In view of the maximal dissipativity of A, it follows $u(t) \in D(A)$, and $v \in Au(t)+f(t)$. Going back to (10.2) with $x=u(t), t=t+h_n$ and $y \in Au(t)$ we have

$$\|u_n\| \leq \frac{1}{h_n} \int_t^{t+h_n} \|y+f(s)\| ds, \ \forall y \in Au(t) \tag{10.38}$$

that is

$$\|v\| \leq \limsup_{n \to \infty} \|u_n\| \leq \|y+f(t)\|, \ \forall y \in Au(t) \tag{10.38)'$$

In other words, v is the element of the least norm of the closed convex set $Au(t)+f(t)$ ($Au(t)$ is closed and convex).

Thus $v = (Au(t)+f(t))_0 = f(t) - \mathrm{Proj}_{-Au(t)} f(t)$

On the other hand, by (10.38)' we have $\|u_{n_k}\| \to \|v\|$ and therefore $u_{n_k} \to v$ as $k \to \infty$ (Lemma 10.2). We have proved that every weakly convergent subsequence u_{n_k} (of the bounded sequence u_n) is strongly convergent to $(Au(t)+f(t))_o$.

This means that even $u_n \to (Au(t)+f(t))_o$ as $n \to \infty$. But $\lim_{n\to\infty} u_n = \frac{d^+u}{dt}(t)$ so (10.35) holds, q.e.d.

Remark 10.3. (1) Let X be uniformly convex. If $f \in L^1(o,T;X)$ and o is a Lebesgue point (at the right) of f, then the integral solution u of (10.1) with $x_o \in D(A)$ is differentiable at the right of o and

$$\frac{d^+u}{dt}(o) = (Ax_o + f(o))_o$$

This is the case if $f:[o,T]$ is continuous. To prove this assertion one uses (10.37) with t=o and the rest of the proof of Theorem 10.4.

(2) The Hypothesis $x_o \in D(A)$ has been used only (in (10.37)') to prove the boundedness of $\|u(t+h)-u(t)\|h^{-1}$ with respect to $h > o$.

In other words the following result holds

Theorem 10.4.' Let X be uniformly convex, $A \subset X \times X$ m-dissipative, $f \in L^1(o,T;X)$ and $x_o \in \overline{D(A)}$. If $t \in]o,T[$ is a Lebesgue point (at the right) of f and if the integral solution u satisfies $\|u(t+h)-u(t)\| \leq hK(t)$ for all $h > o$ sufficiently small (with $K(t) > o$ independent of h) then $\lim_{h\downarrow o} \frac{u(t+h)-u(t)}{h} = \frac{d^+u}{dt}(t)$ exists and (10.35) holds.

We also have

Theorem 10.5. Let X be uniformly convex $A \subset X \times X$ m-dissipative and $f \in L^1(o,T;X)$. If either f is continuous on $[o,T[$ or $t \in [o,T[$ is a Lebesgue point (at the right of t, then the following conditions are equivalent

(1) $u(t) \in D(A)$ ($t > o$ if $u(o)=x_o \in \overline{D(A)}$)

(2) $h^{-1}\|u(t+h)-u(t)\| \leq K(t)$, $o < h < T-t$ (where u is the integral solution of (10.1) and $K(t) > o$ is independent of h).

(3) $h^{-1}\|u(t+h)-u(t)\|$ is strongly convergent for $h \downarrow o$ (which implies (10.35)).

Proof. On the basis of the proof of Theorem 10.4, it remains to point out only that (1)$\Rightarrow$(2). As in the case of (10.38), if $u(t) \in D(A)$, then we set $x=u(t)$ in (10.2). It follows

$$\|u(t+h)-u(t)\| \leq \int_t^{t+h} \|y+f(s)\|\, ds, \quad \forall y \in Au(t) \tag{10.39}$$

hence

$$\limsup_{h\downarrow o} h^{-1}\|u(t+h)-u(t)\| \le \|(f(t)+Au(t))_o\| \tag{10.40}$$

that is (1) $\Rightarrow$ (2) q.e.d.

We shall prove that in the case $A=-\partial j$ and $X=H$, the condition (2) in Th. 10.5 is satisfied even if $u(o)=x_o \in \overline{D(A)}$. In other words, in this case, the conclusion of Theorem 10.4 holds (for $t>o$) even if $x_o \in \overline{D(A)}\setminus D(A$ Precisely, we have

<u>Theorem 10.6.</u> Let $f\in W^{1,1}(o,T;H)$ and j as in § 10.2. Then for every $x_o \in \overline{D(\partial j)}$, the integral solution, u of

$$u'(t)+\partial j(ut))\ni f(t),\ u(o)=x_o \tag{10.41}$$

has the properties

(a) $u(t)\in D(\partial j)$ for all $t\in]o,T[$

(b) $\lim_{h\downarrow o} h^{-1}(u(t+h)-u(t)) = \frac{d^+u}{dt}(t)$ exists for all $t\in]o,T[$ and $tu'(t)\in L^\infty(o,T;H)$. Precisely,

$$\|\frac{d^+u}{dt}(t)\| \le \|f(t)\| + \frac{1}{t}\|u(o)-z\| + \frac{1}{t^2}\int_o^t s^2\|\frac{df}{ds}(s)\|\,ds$$

$$+\frac{\sqrt{2}}{t}\left\{\int_o^t s\|\frac{df}{d}(s)\|\right\}^{1/2}(\|u(o)-z\| + \int_o^t \|f(s)\|ds)^{1/2} \equiv K(x_o,t) \tag{10.42}$$

for all $t\in]o,T[$ (where $j(z)=o$ and $x_o=u(o)$)

(c) $\frac{d^+u}{dt}(t) + \mathrm{Proj}_{\partial j(u(t))} f(t)=f(t)\quad \forall\, t\in]o,T[$

Proof. First we take $x_o \in D(\partial j)$. According to Theorems 10.4 and 10.3 there exists a unique strong solution u of (10.41) satisfying: $t\to j(u(t))$ is absolutely continuous on $[o,T]$, and

$$u(t)\in D(\partial j),\ \frac{d^+u}{dt}(t) + \mathrm{Proj}_{\partial ju(t)} f(t)=f(t);\ \text{for all } t\in [o,T[\tag{10.43}$$

The definition of $\partial j(u(t))$ and $j(z)=o$, lead us to

$$j(u(t)) \le \langle u(t)-z,f(t)\rangle - \frac{1}{2}\frac{d}{dt}\|u(t)-z\|^2,\ \text{a.e. on }]o,T[$$

$$\int_o^T j(u(t))dt \le \int_o^T \langle u(t)-z,f(t)\rangle\,dt + \frac{1}{2}\|u(o)-z\|^2 - \frac{1}{2}\|u(T)-z\|^2 \tag{10.43'}$$

On the other hand we have (see (10.31)')

$$t\|u'(t)\|^2 + t\frac{d}{dt}j(u(t)) = \langle tf(t), u'(t)\rangle \quad \text{a.e. on} \quad]o,T[\qquad (10.44)$$

Integrating (10.44) over [o,T] (with u'(t)=(u(t)-z)') one obtains

$$\int_o^T t\|u'(t)\|^2 dt + Tj(u(T)) = \int_o^T j(u(t))dt + T\langle f(T), u(T)-z\rangle$$

$$-\int_o^T \langle u(t)-z, f(t)\rangle\, dt - \int_o^T \langle u(t)-z, tf'(t)\rangle\, dt \qquad (10.44)'$$

Inasmuch as $o \in \partial j(z)$, (10.2) (or (10.41)) gives

$$\|u(t)-z\| \leq \|u(o)-z\| + \int_o^T \|f(s)\|\, ds$$

Combining (10.43)' and (10.44)' we get

$$\int_o^T t\|u'(t)\|^2 dt \leq \frac{T}{2}\|f(T)\|^2 + \frac{1}{2}\|u(o)-z\|^2$$

$$+ (\|u(o)-z\| + \int_o^T \|f(s)\|ds)\int_o^T t\|f'(t)\|dt \qquad (10.45)$$

Clearly, (10.45) is valid for any $\bar{T} \in]o,T[$. Using either (10.41) or (10.4) with v(t)=u(t+h), we see that

$$\|u(\bar{T}+h)-u(\bar{T})\| \leq \|u(t+h)-u(t)\| + \int_t^{\bar{T}} \|f(s+h)-f(s)\|ds$$

Therefore

$$\|\frac{d^+u}{dt}(T)\| \leq \|\frac{d^+u}{dt}(t)\| + \int_t^{\bar{T}} \|\frac{df}{dt}(s)\|ds, \quad o < t < \bar{T} < T \qquad (10.46)$$

Multiplying by t, integrating over $[o,\bar{T}]$ and using Fubini's theorem, we derive

$$\frac{\bar{T}^2}{2}\|\frac{d^+u}{dt}(T)\| \leq \int_o^{\bar{T}} t\|\frac{d^+u}{dt}(t)\|\, dt + \frac{1}{2}\int_o^{\bar{T}} s^2\|\frac{df}{dt}(s)\|\, ds$$

$$\frac{T}{\sqrt{2}}(\int_o^{\bar{T}} t\|\frac{d^+u}{dt}(t)\|^2 dt)^{1/2} + \frac{1}{2}\int_o^{\bar{T}} s^2\|\frac{df}{ds}(s)\|ds \qquad (10.47)$$

for all $\bar{T} \in]o,T[$.

A simple combination of (10.47) and (10.45) (with $\bar{T}$ in place of T) yields (10.42). Now let us consider the general case $x_o \in \overline{D(\partial j)}$. Take $x_n \in D(\partial j)$ such that $x_n \to x_o$ as $n \to \infty$. Then the solution u_n of $u_n'(t) + \partial j(u_n(t)) \ni f(t)$, $u_n(o) = x_n$ is uniformly convergent to u (see (10.11)) on $|o,T|$ and (by the previous case)

$$\| \frac{d^+u_n}{dt}(t) \| \leq K(x_n,t), \ \forall t \in]o,T[\qquad (10.48)$$

where $K(x_n,t)$ is given by (10.42) with x_n in place of x_o. Obviously, $K(x_n,t) \to K(x_o,t)$ as $n \to \infty$. But we have

$$u_n(t+h)-u_n(t) = \int_t^{t+h} u'(s)ds \quad o < t < t+h < T$$

hence

$$\|u_n(t+h)-u_n(t)\| \leq hK(x_n,t), \ o < t < t+h < T, n=1,2,...$$

Letting $n \to \infty$ one has

$$\|u(t+h)-u(t)\| \leq hK(x_o,t), \ o < t < t+h < T \qquad (10.49)$$

Finally, by Theorem 10.5, the inequality (10.49) gives $u(t) \in D(\partial j)$ and $\|\frac{d^+u}{dt}(t)\| \leq K(x_o,t)$ for all $t \in]o,T[$. The proof is complete.

Remark 10.4. If j is not positively-valued with $\min_{x\in H} j(x)=j(z)=o$ (as in Th. 10.6) then (10.42) holds with $f(t)-w$ in place of $f(t)$, with w as in (10.34)'.

10.4. Some results on the asymptotic behaviour of solutions.

We are going to sketch some results on the existence of periodic solutions u of (10.1) and $\lim_{t\to\infty} u(t)$. There exists an extensive literature devoted to this subject (see § 12). Note that in order to prove the existence of $\lim u(t)$ we need appropriate estimates. The proof of the existence of periodic solutions is based on the theory of perturbation of maximal dissipative sets (§ 8).

First of all, the following lemma is needed

Lemma 10.3. Let $A \subset H\times H$ be m-dissipative and $f \in L^1_{loc}(R_+;H)$ ($R_+=[o,+\infty[$) Suppose that $\lim f(t)=f_\infty$ and the integral solution $u \in C(R_+,H)$ of $u' \in Au+f$ has also the property $\lim u(t)=u_\infty$

Then we have $[u_\infty, -f_\infty] \in A$ (i.e. $u_\infty \in D(A)$ and $-f_\infty \in Au_\infty$).

Proof. We know that

$$\langle u(t)-u(t_o), u(t_o)-x\rangle \leq \|u(t_o)-x\| \int_{t_o}^{t} \langle y+f(s),u(s)-x \rangle_+ ds \qquad (10.50)$$

for all $o \le t_o \le t$, $[x,y] \in A$ (see e.g. (10.15))

$$\lim_{t_o \to +\infty} \int_{t_o}^{t_o+1} \langle y+f(s),\ u(s)-x \rangle_+ \, ds \le \langle y+f_\infty,\ u_\infty - x \rangle_+$$

and therefore (10.50) implies $\langle y+f_\infty,\ x-u_\infty \rangle \le o, \forall x \in D(A)$ and $y \in Ax$ so $u_\infty \in D(A)$, and $-f_\infty \in Au_\infty$, q.e.d.

In other words, in the theory of the existence of $\lim_{t\to\infty} u(t)=u_\infty$, the hypothesis $-f_\infty \in R(A)$ is necessary (when $\lim_{t\to\infty} f(t) = f_\infty$ exists)

Remark 10.5. Lemma 10.3 is valid in any real Banach space X (with the same proof, by replacing $\langle \cdot,\cdot \rangle = \langle \cdot,\cdot \rangle_i$) and so on.

Theorem 10.7. (1) Let $A-\omega I \subset X \times X$ be m-dissipative with $\omega < o$ and $f \in L^1_{loc}(R_+;X)$ with $\lim_{t\to\infty} f(t)=f_\infty \in -R(A)$. Then there exists a unique $u_\infty \in D(A)$ such that $-f_\infty \in Au_\infty$ and $\lim_{t\to\infty} u(t)=u_\infty$. Precisely the integral solution u of (10.1) satisfies

$$\|u(t)-u_\infty\| \le e^{\omega t}\|u(o)-u_\infty\| + \int_o^t e^{\omega(t-s)}\|f(s)-f_\infty\| ds \qquad (10.51)$$

for all $t \ge o$.

(2) If in addition $f' \in L^1(R;X)$, $u(o) \in D(A)$ and X is uniformly convex then

$$\left\|\frac{d^+u}{dt}(t)\right\| \le e^{\omega t}\|(Au(o)+f(o))_o\| + \int_o^t e^{\omega(t-s)}\left\|\frac{dt}{ds}(s)\right\| ds \qquad (10.52)$$

for all $t \ge o$ (hence $\frac{d^+u}{dt}(t) \to o$ as $t\to\infty$) and

$$\int_o^\infty \left\|\frac{d^+u}{dt}(t)\right\| dt \le -\frac{1}{\omega}\|(Au(o)+f(o))_o\| - \frac{1}{\omega}\int_o^\infty \left\|\frac{df}{ds}(s)\right\| ds \qquad (10.52)'$$

Proof. (1) The hypothesis $-f_\infty \in R(A)$ means the existence of an element $u_\infty \in D(A)$ such that $o \in Au_\infty + f_\infty$. Therefore $v(t)=u_\infty$ for $t \ge o$ is the strong solution to $v'(t) \in Av(t)+f_\infty$.

Then according to (4.45) in Chapter 1, we have the inequality (10.51). But $\omega < o$, so (10.51) implies $\lim_{t\to\infty}\|u(t)-u_\infty\| = o$ hence the solution u of $-f_\infty \in Au_\infty$ is unique.

(2) In these conditions $\frac{d^+u}{dt}(t)$ exists for all $t \ge o$ (by Thoerem 10.4) Invoking again (4.45) in Ch. 1, we have

$$\|u(t+h)-u(t)\| \le e^{\omega t}\|u(h)-u(o)\| + e^{\omega t}\int_o^t e^{-\omega s}\|f(s+h)-f(s)\| ds \quad (10.53)$$

for $o \le t \le t+h$, with $u(o) \in D(A)$. By (10.35) we have $\frac{d^+u}{dt}(o)=(Au(o)+f(o))_o$

and thus (10.53) implies (10.52). Obviously (10.52)' is a consequence of (10.52). The proof is complete.

OF course, the condition $A-\omega I$ dissipative with $\omega<0$ (it is also called strongly dissipative) is quite restrictive. However, without severe restrictions, $\lim_{t\to\infty} u(t)$ may not exist. Even in the case of $A=-\partial j$ (which is not ω-dissipative with $\omega<0$) we need to assume additional conditions (e.g. the compactness of the level sets $\{x\in H;\ j(x)+\|x\|^2\le r\}$, $r\in R$.

<u>Theorem 10.8.</u> Let $j:H\to]-\infty,+\infty]$ be l.s.c. and $f:[0,+\infty[\to H$ absolutely continuous on every compact in $[0,+\infty[$, with $f'\in L^1(R_+;H)$ and $\lim_{t\to\infty} f(t)=f_\infty\in H$. Suppose that $f_\infty\in R(\partial j)$. Let u be a strong solution of

$$\frac{d^+u}{dt}(t)+\partial j(u(t))\ni f(t),\ \forall t\in[0,+\infty[\tag{10.54}$$

(see Th. 10.6). Then

(i) $\lim_{t\to\infty}\frac{d^+u}{dt}(t)=0$

(ii) If in addition the function $t\to tf'(t)$ is in $L^1(R_+;H)$ then $\|\frac{d^+u}{dt}(t)\|=o(t^{-1})$ as $t\to+\infty$.

Proof. Let $z\in D(\partial j)$ be such that $f_\infty\in\partial j(z)$. Then $\operatorname{Min}_{x\in H}\{j(x)-\langle f_\infty,x\rangle\}=$ $=j(z)-\langle f_\infty,z\rangle$. Set

$$g(x)=j(x)-\langle f_\infty,x\rangle-(j(z)-\langle f_\infty,z\rangle),x\in H \tag{10.55}$$

Then $g(x)\ge 0$, $\forall x\in H$ and $g(z)=0$. According to Theorem 10.6 there is a strong solution u of the problem

$$\frac{d^+u}{dt}(t)+\partial g(u(t))\ni f(t)-f_\infty,\ \forall t>0 \tag{10.55'}$$

This is because, for every $T>0$, $f-f_\infty\in W^{1,1}(0,T;H)$. Obviously, $\partial g=\partial j-f_\infty$ so (10.55) is just (10.54). In the case of (10.55), (10.42) holds with $f(t)-f_\infty$ in place of $f(t)$. In other words we have

$$\begin{aligned}\|\frac{d^+u}{dt}(t)\|&\le\|f(t)-f_\infty\|+\frac{1}{t}\|u(0)-z\|+\int_0^t\frac{s^2}{t^2}\|f'(s)\|ds+\\&\sqrt{2}\,(\int_0^t\frac{s}{t}\|f'(s)\|ds)^{1/2}(\frac{1}{t}\|u(0)-z\|+\int_0^t\|f(s)-f_\infty\|ds)^{1/2}\end{aligned} \tag{10.56}$$

for all $t>0$.

For every $\varepsilon>0$, there exists $N=N(\varepsilon)>0$ such that

$$\int_N^{\infty} \|f'(s)\|ds < \varepsilon \quad , \|f(t)-f_{\infty}\| < \varepsilon, \ \forall t \geq N$$

$$G(t) = \int_o^t \frac{s^2}{t^2} \|f'(s)\|ds \leq \int_o^t \frac{s}{t}\|f'(s)\|ds \leq \frac{1}{t}\int_o^N \|f'(s)\|s\, ds \ +$$

$$+ \int_N^t \|f'(s)\|ds \leq \frac{1}{t}\int_o^N \|f'(s)\|s\, ds + \varepsilon \ , \quad t \geq N \tag{10.57}$$

This shows that $G(t) \to o$ as $t \to \infty$. Similarly,

$$\frac{1}{t}\int_o^t \|f(s)-f_{\infty}\|ds \leq \frac{1}{t}\int_o^N \|f(s)-f_{\infty}\|ds + \frac{1}{t}\int_N^t \|f(s)-f_{\infty}\|ds$$

$$\leq \frac{1}{t}\int_o^N \|f(s)-f_{\infty}\|ds + \varepsilon \ , \ t > N \tag{10.58}$$

On the basis of (10.56)-(10.58) it follows $\frac{d^+u}{dt}(t) \to o$ as $t \to +\infty$. As far as (ii) is concerned, we note that $tf'(t) \in L^1(R_+;H)$ yields

$$\|f(t)-f_{\infty}\| \leq \int_t^{\infty} \|f'(s)\|ds \leq \frac{1}{t}\int_t^{\infty} s\|f'(s)\|ds \leq \frac{M}{t}, \ \forall t > o$$

with

$$M = \int_o^{\infty} s\|f'(s)\|ds$$

Going back to (10.57), $G(t) \leq \frac{1}{t}\int_o^{\infty} s\|f'(s)\|ds$, $\forall t > o$. Finally we have

$$\int_o^t \|f(s)-f_{\infty}\|ds \leq \int_o^t \int_s^{\infty} \|f'(r)\|dr\, ds \leq \int_o^t \Big(\int_s^t \|f'(r)\|dr +$$

$$+ \int_t^{\infty} \|f'(r)\|dr\Big)ds \leq$$

$$t\int_t^{\infty} \|f'(r)\|dr + \int_o^t ds \int_s^t \|f'(r)\|dr \leq \int_t^{\infty} r\|f'(r)\|dr + \int_o^t r\|f'(r)\|dr = M$$

and (ii) follows. For the existence of $\lim_{t\to\infty} u(t) = u_{\infty}$ we need (in general) compactness conditions, namely

For every $r \in R$, the level set

$$C_r = \{x \in D_e(j),\ j(x) \leq r,\ \|x\| \leq r\} \quad \text{(see (6.7))} \tag{10.59}$$

is compact (i.e. $(I+\partial j)^{-1}$ is compact)

<u>Theorem 10.9.</u> Let $j: H \to]-\infty,+\infty]$ be l.s.c. and satisfying Condition (10.59). Let also $f_{\infty} \in R(\partial j)$ and $f: R_+ \to H$ with $f - f_{\infty} \in L^1(R_+;H)$. Then every integral solution u of the equation $u' + \partial j(u) \ni f$ has a (strong) limit $u_{\infty} = \lim_{t\to\infty} u(t)$ and $f_{\infty} \in \partial j(u_{\infty})$.

Proof. We first prove the existence of $\lim_{t\to\infty} v(t) = v_{\infty}$ for the strong solution v of $v' + \partial j(v) \ni f_{\infty}$.

Let $z \in D(\partial j)$ be such that $f_\infty \in \partial j(z)$. Thus, $v_o(t)=z$ satisfies $v_o'+\partial j(v_o) \ni f_\infty$. Consequently

$$\|v(t)-z\| \leq \|v(o)-z\|, \quad \forall\ t \geq o \tag{10.59}$$

Moreover, by Theorem 10.8, $\frac{d^+v}{dt}(t) \to o$ as $t \to \infty$. On the other hand, the definition of $\partial j(v(t)) \ni f_\infty - v'(t)$ yields

$$j(v(t)) \leq j(z) + \|\frac{d^+v}{dt}(t)-f_\infty\| \ \|v(t)-z\|, \ t > o \tag{10.60}$$

It follows that there exists $r > o$ such that $\{v(t),\ t > o\} \subset C_r$, that is $\{v(t),\ t > o\}$ is precompact. Choose t_n such that $\lim_{n\to\infty} v(t_n)=v_\infty$ exists. Since $\frac{d^+v}{dt}(t_n) + \partial j(v(t_n)) \ni f_\infty$ and $\frac{d^+v}{dt}(t_n) \to o$ it follows $f_\infty \in \partial j(v_\infty)$. This implies $\|v(t)-v_\infty\| \leq \|v(s)-v_\infty\|$, $o \leq s \leq t$, hence $\lim_{t\to\infty} v(t)=v_\infty$

Given $\varepsilon > o$, there exists $N=N(\varepsilon) > o$ such that $\int_N^{+\infty} \|f(t)-f_\infty\|dt < \varepsilon$. Let $\tilde{v}$ be the solution to the problem

$$\tilde{v}'(t) + \partial j(\tilde{v}(t)) \ni f_\infty \ , \ \tilde{v}(o)=u(N)$$

where u is the integral solution of $u' + \partial j(u) \ni f$. Then by (10.4) with

$$v(t)=\tilde{v}(t-N),\ t \geq N,\ v'+\partial j(v) \ni f_\infty,\ v(N)=\tilde{v}(o)$$

we have (for all $t \geq N$)

$$\|u(t)-\tilde{v}(t-N)\| \leq \|u(N)-\tilde{v}(o)\| + \int_N^t \|f(s)-f_\infty\|ds$$

it follows that

$$\|u(t)-u(s)\| \leq \|v(t-N)-v(s-N)\| + 2\varepsilon \ , \ \forall t,s \geq N$$

But $\lim_{t,s\to\infty} \|v(t-N)-v(s-N)\| = o$, so $u(t)$ is a Cauchy sequence as $t \to +\infty$. Thus, $\lim_{t\to\infty} u(t)=u_\infty$ exists, and by Lemma 10.3, $f_\infty \in \partial j(u_\infty)$.

<u>Remark 10.6.</u> In the case $f=o$ and $f_\infty = o$ from Theorem 10.9 one obtains Corollary 9.2. An extension of Proposition 9.3 to a general (real) Banach space X is the following one.

<u>Theorem 10.10.</u> Let $A \subset X \times X$ be m-dissipative with $(I-A)^{-1}$ compact. If $f \in L^1(R_+;X)$ then every bounded integral solution $u:R_+ \to \overline{D(A)}$ of $u' \in Au+f$ has a precompact range (i.e. if $R(u)=\{u(t), t > o\}$ is bounded, then $R(u)$ is precompact).

Proof. Let S be the semigroup generated by A. For $\lambda = t$, (3.12) becomes (with $J_\lambda = (I-\lambda A)^{-1}$

$$\|J_\lambda x-x\| \le \frac{4}{\lambda}\int_0^\lambda \|S(s)x-x\|\,ds,\ \forall x\in\overline{D(A)} \tag{10.61}$$

Let u be a bounded integral solution of $u'\in Au-f$. Since $u(t)\in\overline{D(A)}$, $\forall t\ge o$, (10.61) with $x=u(t)$ gives

$$\|J_\lambda u(t)-u(t)\| \le \frac{4}{\lambda}\int_0^\lambda \|S(s)u(t)-u(t)\|\,ds,\ \forall\ t\ge o \tag{10.62}$$

Note that if $f=o$, then $u(t)=S(t)x$ with $u(o)=x\in D(A)$, so given $\varepsilon>o$, there exists $\delta(\varepsilon,x)=\delta(\varepsilon)$ such that

$$\|S(s)u(t)-u(t)\| = \|S(t+s)x-S(t)x\| \le \|S(s)x-x\| < \varepsilon/4$$

for $o < s\le\delta(\varepsilon)$. Consequently, with $\lambda=\delta(\varepsilon)$ (10.62) implies

$$\|J_\lambda u(t)-u(t)\|\le\varepsilon\ ,\ \forall t\ge o,\ \lambda=\delta(\varepsilon) \tag{10.63}$$

Since $J_{\delta(\varepsilon)}$ is compact and R(u) is bounded, $J_{\delta(\varepsilon)}R(u)$ is relatively compact. By (10.63) so is R(u) (see Lemma 5.1 in Ch. 1). Now let us consider the general case $f\in L^1(R_+,H)$.

In this case, the integral solution $v(t+s)$ of $v'(t+s)\in Av(t+s)$, $v(t)=u(t)$, $s\ge o$, is given by $v(t+s)=S(t+s-t)u(t)=S(s)u(t)$. Therefore

$$\|S(s)u(t)-u(t+s)\|\le\int_t^{t+s}\|f(\theta)\|d\theta\ ,\ t\ge o,\ s\ge o \tag{10.64}$$

(Since us is the integral solution of $u'\in Au+f$). But we have also

$$\begin{aligned}\|u(t+s)-u(t)\| &\le \|u(s)-u(o)\| + \int_0^t \|f(s+\theta)-f(\theta)\|d\theta\\ &\le \|u(s)-u(o)\| + \int_0^\infty \|f(s+\theta)-f(\theta)\|d\theta\end{aligned} \tag{10.65}$$

A simple combination of (10.64) and (10.65) lead us to

$$\begin{aligned}\|J_\lambda u(t)-u(t)\| &\le \frac{4}{\lambda}\int_0^\lambda \|u(s)-u(o)\|ds + \frac{4}{\lambda}\int_0^\lambda ds\int_t^{t+s}\|f(\theta)\|d\theta\\ &+\frac{4}{\lambda}\int_0^\lambda ds\int_0^\infty \|f(s+\theta)-f(\theta)\|d\theta\end{aligned} \tag{10.66}$$

for all λ , $t,s>o$.

It is easy to check that (10.66) (in view of the continuity of u and of the hypothesis $f\in L^1(R_+;H)$) implies (10.63). This completes the proof.

<u>Remark 10.7</u>. If X is uniformly convex, then the existence of a bounded integral solution of $u'\in Au+f$ ($f\in L^1(R_+;H)$) is equivalent to $A^{-1}o\ne\emptyset$. Indeed, we have

$$\|u(t)-S(t)x\| \le \|u(o)-x\| + \int_o^t \|f(s)\|ds, \ t>o, \ x\in D(A) \quad (10.67)$$

This shows that $\{S(t)x, \ t\geq o\}$ is bounded, and hence every trajectory $\{S(t)y, \ t\geq o\}$; $y\in D(A)$ is bounded. Therefore there exists $x_o\in\overline{D(A)}$ with $S(t)x_o=x_o$, $\forall t\geq o$ (i.e. $o\in Ax_o$) (see § 9).

Conversely, if $S(t)x_o=x_o$, $\forall \ t\geq o$, then

$$\|u(t)-x_o\| \le \|u(o)-x_o\| + \int_o^\infty \|f(s)\| \ ds, \ \forall \ t\geq o \quad (10.68)$$

that is every integral solution u is bounded on $[o,+\infty[\ = R_+$.

In other words, a reasonable statement of Theorem 10.10 in uniformly convex spaces is the following one.

Theorem 10.10'. Suppose that X is uniformly convex, $A\subset X\times X$ is m-dissipative, $(I-A)^{-1}$ is compact and $f\in L^1(R_+;X)$. Then $A^{-1}o\neq\emptyset$ iff all integral solutions u of $u'\in Au+f$ have a precompact range.

Remark 10.8. Let $o\in Az$. Then (for $x\in\overline{D(A)}$)

$$\|S(t)x-z\| \le \|S(s)x-z\|, \quad \forall o<s<t \quad (10.69)$$

Therefore $\lim \|S(t)x-z\| = v_\infty$ exists. If $\{S(t)x, t\geq o\}$ is precompact, it follows that for every convergent subsequence $S(t_n)x$, we have $v_\infty = \lim_{n\to\infty} \|S(t_n)x-z\|$ i.e.

$$v_\infty = \|y-z\|, \ \forall y\in\omega(x), \ x\in\overline{D(A)} \quad (10.70)$$

(see (9.4)). In the case $\omega(x)\subset A^{-1}o$

$$\|S(t)x-y\| \le \|S(s)x-y\|, \ \forall \ t>s>o \quad (10.71)$$

with $o\in Ay$, $y=\lim S(t_n)x$ for some $t_n\to+\infty$.

Thus $\omega(x)\subset A^{-1}o$ implies the existence of $\lim_{t\to\infty} S(t)x=y$.

§ 11. The correspondence $A\to S_A(t)$. The existence of accretive sets.

Let $A\subset X\times X$ be m-dissipative and let $S_A(t)$ be the semigroup generated by A via the exponential formula (3.10). We will discuss the dependence of S_A on A (the continuity of the integral solution to (1.1) with respect to A). To this goal we need

Definition 11.1. Let A_n be nonempty subsets of $X \times X$ (X - a general Banach space), n=1,2,... We say that $A \subset \lim_{n\to\infty} A_n$, if for every $[x,y] \in A$ there exist $[x_n, y_n] \in A_n$ such that $x_n \to x$ and $y_n \to y$ as $n \to \infty$.

We now state without proof the following results

Theorem 11.1. Suppose that A_n are m-dissipative subsets of $X \times X$ (n=1,2,...) and $A \subset \lim_{n\to\infty} A_n$. Then

$$\lim_{n\to\infty} S_{A_n}(t)x_n = S_A(t)x, \quad t \geq 0 \tag{11.1}$$

for every $x \in \overline{D(A)}$ and every $x_n \in \overline{D(A_n)}$ with $x_n \to x$ as $n \to \infty$. Moreover, the convergence in (11.1) holds uniformly with respect to t on bounded intervals of R_+.

Note that the conclusion of this Theorem holds in more general conditions on A and A_n (see Miyadera and Kobayashi [1]).

Theorem 11.2. (Brezis [4]). Let H be a real Hilbert space H and let A and $\{A_n\}_{n\geq 1}$ be maximal monotone sets of $H \times H$. Then the following conditions are equivalent

(i) $\forall x \in \overline{D(A)}$, $\forall \lambda > 0$, $(I+\lambda A_n)^{-1}x \to (I+\lambda A)^{-1}x$, as $n \to \infty$

(ii) For every $x \in D(A)$ there exists $x_n \in D(A_n)$, n=1,2,... such that $x_n \to x$ and $A_n^o x_n \to A_o x$ (A_n^o, A_o are the minimal sections of A_n and A respectively.

(iii) For every $x \in \overline{D(A)}$ there exist $x_n \in \overline{D(A_n)}$ n=1,2,... such that $x_n \to x$ and $S_{A_n}(t)x_n \to S_A(t)x$, for every $t \geq 0$.

In addition, the convergence in (i) and (iii) is uniform with respect to and t on bounded subsets.

The proof of the above results can also be found in Pavel [12, Ch. 2, § 12].

Now let A_n and A be dissipative sets in the real Banach space X, satisfying Range condition

$$R(I-\lambda A_n) \supset \overline{D(A_n)}, \; n=1,\ldots \quad R(I-\lambda A) \supset \overline{D(A)}, \lambda > 0 \tag{11.2}$$

and

$$\overline{D(A)} \subset \overline{D(A_n)}, \; n=1,\ldots \tag{11.3}$$

Let us now consider the conditions

(1) $\lim_{n\to\infty} (I-\lambda A_n)^{-1}x = (I-\lambda A)^{-1}x,\ \forall \lambda > 0,\ x \in \overline{D(A)}$

(2) $\lim_{n\to\infty} S_{A_n}(t)x = S(t)x,\quad \forall x \in \overline{D(A)}$, uniformly on bounded t-intervals.

It was shown by Brezis and Pazy [1] that (1) implies (2). If in addition to (11.2)-(11.3) one assumes that $\overline{D(A)}$ is convex and X* is uniformly convex then (2) implies (1). In particular, if A is maximal dissipative and X,X* are uniformly convex, that $\overline{D(A)}$ is convex so (1) and (2) are equivalent (see Y. Kobayashi [2] for (2) $\Rightarrow$ (1)).

Now let C be a closed convex subset of the Banach space X, and let $S(t):C \to C$ be a semigroup of type ω (see § 2). Recall that in the case $\omega = 0$ (i.e. $\|S(t)x-S(t)y\| \le \|x-y\|$, $\forall x,y \in C$), S(t) is said to be <u>nonexpansive or a semigroup of contractions</u>. The infinitesimal generator A_o of S(t) is defined by

$$A_o x = \lim_{t\downarrow 0} \frac{S(t)x-x}{t}\quad,\ x \in D(A_o) \tag{11.4}$$

where

$$D(A_o) = \left\{ x \in C,\ \lim_{t\downarrow 0} \frac{S(t)x-x}{t} \text{ exists} \right\} \tag{11.5}$$

Set also

$$\tilde{D} = \left\{ x \in C,\ \lim_{t\downarrow 0} \frac{\|S(t)x-x\|}{t} < \infty \right\} \tag{11.6}$$

In linear case (and C=X) it is well known that $\overline{D(A_o)}=X$ (see§ 6) and A_o is m-dissipative (if S(t) is nonexpansive). Conversely, the semigroup generated by A_o via the exponential formula, is just S(t). (This is Hille-Yoshida theorem). Even in nonlinear case, on Hilbert spaces this theory is essentially complete (i.e. there is also bijective correspondence between the nonexpansive semigroups S(t) on C and their infinitesimal generators A_o). Precisely, the set of infinitesimal generators of contractions on $C \subset H$ is given by $\{$the minimal sections A_o of maximal dissipative sets $A \subset H \times H\}$. On these results the reader is referred to Komura [1], Crandall and Pazy [2], and Dorroh [1]. Recently, these results have been extended by Baillon |1|, from the Hilbert spaces H to the Banach spaces X with X* uniformly convex. In other words, Baillon has obtained the following result:

<u>If</u> X* <u>is uniformly convex, then the infinitesimal generator</u> A_o <u>of</u> S(t) <u>has a dense domain in</u> C (<u>i.e.</u> $\overline{D(A_o)}=C$), <u>and there is a set</u> $A \subset X \times X$ <u>such</u>

<u>that</u> (a) $A_o \subset A$, $\overline{D(A)}=C$, (b) A <u>is dissipative,</u> $R(I-\lambda A) \supset C$, $\forall \lambda > 0$, <u>and</u> (c) $S(t)x= \lim_{n \to \infty} (I - \frac{t}{n} A)^{-n}x$, $\forall x \in C$.

Let us recall that on a general Banach space X, the infinitesimal generator A_o is not (necessarily) densely defined any more. This aspect has been pointed out by Webb (see Remark 4.3). It is also known that there are semigroups of contractions on X without infinitesimal generators (Cf. Crandall and Liggett |1|). However, an open problem is whether or not $\tilde{D}$ given by (11.6) is nonempty whenever S(t) is a nonexpansive semigroup on C.

Chapter 3

Applications to partial differential equations

In this chapter one uses evolution operators and semigroup approach in order to unify the treatment of some partial differential equations of parabolic, hyperbolic and Schrödinger type.

§ 1. A class of (nonlinear) evolution equations associated with time-dependent domain multivalued operators.

1.1. A general existence theorem

In this section we shall consider the abstract evolution equation (inclusion)

$$u'(t) \in A(t)u(t)+F(t)u(t), \quad u(s)=x_o \in \overline{D(A(s))}, \quad o \leq s \leq t \leq T \tag{1.1}$$

where $\{A(t), o \leq t \leq T\}$ is a family of dissipative (possible multivalued) sets $A(t) \subset X \times X$, satisfying Hypotheses H(2.1) with $\omega = o$ and H(2.2) in Chapter 1, § 2. As far as F is concerned, suppose that $D(F(t))=\overline{D(A(t))}$ and that $(t,x) \to F(t)x$ is continuous (i.e. if $t_n \to t$ and $x_n \in D(F(t_n))$ is convergent to x, then $x \in D(F(t))$ and $F(t_n)x_n \to F(t)x$.
On the basis of (4.11)", by an integral solution on [s,T] to the problem (1.1) we mean a continuous function $u:[s,T] \to X$ which satisfies: $u(s)=x_o$, $u(t) \in \overline{D(A(t))}$,

$$\|u(t)-x\| \leq \|u(\tau)-x\| + \int_\tau^t (\langle y+F(\tau)u(\tau),u(\tau)-x\rangle_+ + C\|f(\tau)-f(r)\|)d\tau \tag{1.2}$$

for all $s \leq \tau \leq t \leq T$, $r \in [s,T]$ and $[x,y] \in A(r)$, with $C=\max(L(\|u\|),L(\|x\|))$ $\|u\|=\sup\{\|u(t)\|, s \leq t \leq T\}$ and f appearing in H(2.1).
Throughout this section X is a real Banach space.

We are now in a position to state and prove the main result of this section, namely

Theorem 1.1. Let $A(t) \subset X \times X$ for every $t \in [o,T]$ satisfy H(2.1) with $\omega = o$ and H(2.2) in Ch. 1, §2. Suppose that the evolution operator U(t,)(generated by $\{A(t), o \leq t \leq T\}$ via Theorem 3.6) is compact for $t \in]s,T]$ in the sense of Definition 5.1 (in Ch. 1) with $D(s)=\overline{D(A(s))}$. If $D(F(t))=\overline{D(A(t))}$ for $t \in [o,T]$ and the function $(t,x) \to F(t)x$ is continuous, then the problem (1.1) has at least one integral solution u on $[s,T_1]$, with $T_1 = T_1(s,x_o) \in]s,T]$.

Remark 1.1. In the case $A(t)=A$ independent of t one obtains a result of Vrabie [1]. Finally, the above result of Vrabie contains the result of Pazy [2] in the case A-a linear (unbounded) m-dissipative operator which generates a compact-semigroup $S(t)$, $t>0$. For proving Theorem 1.1 we cannot make use of standard fixed point theorems since $\overline{D(A(t))}$ is not convex in general (unless X is uniformly convex). We shall use the method of steps in the theory of differential equations with delay.

Proof of Theorem 1.1. First we prove that for each $\lambda\in]0,T]$, with $T_1=s+T_0\leq T$ the delay equation

$$u_\lambda'(t)\in A(t)u_\lambda(t)+F(t-\lambda)u_\lambda(t-\lambda),\ u_\lambda(\theta)=x_0,\ \text{if } s-\lambda\leq\theta\leq s, \tag{1.3}$$

has an integral solution $u_\lambda:[s-\lambda,T_1]\to X$, i.e. u_λ is continuous $u_\lambda(t)\in\overline{D(A(t))}$, $u_\lambda(\theta)=x_0$ for $\theta\in[s-\lambda,s]$ and

$$\|u_\lambda(t)-x\|\leq\|u_\lambda(t_0)-x\|+\int_{t_0}^{t}(\langle y+F(\tau-\lambda,u_\lambda(\tau-\lambda)),u_\lambda(\tau)-x\rangle_+ + C\|f(\tau)-f(r)\|)d\tau \tag{1.4}$$

for all $0\leq s\leq t_0\leq t\leq s+T_0$, $r\in[s,s+T_0]$, $[x,y]\in A(r)$.

where $C_{max}\ L(\|x\|),L(\rho+\|x_0\|)$ with ρ and T_0 defined as below.

Since F is continuous at (s,x_0), it is locally bounded at (s,x_0), that is there are $\rho,M,T_0>0$ such that

$$\|F(t)x\|\leq M,\ \text{for } |t-s|\leq T_0,\ \|x-x_0\|\leq\rho,\ x\in\overline{D(A(t))} \tag{1.5}$$

and

$$MT_0+\sup_{|t-s|\leq T_0}\|U(t,s)x_0-x_0\|\leq\rho \tag{1.6}$$

Clearly T_0 depends on s and x_0 (and $T_0\leq T-s$). Set $T_1=s+T_0$.

Let u_λ^1 be the integral solution of the problem

$$(u_\lambda^1)'(t)\in A(t)u_\lambda^1(t)+F(t-\lambda)x_0,\ u_\lambda(s)=x_0,\ s\leq t\leq\lambda+s \tag{1.7}$$

Such a continuous function $u_\lambda^1:[s,s+\lambda]\to X$ satisfying $u_\lambda^1(s)=x_0$, $u_\lambda^1(t)\in\overline{D(A(t))}$ and (1.4) on $[s,s+\lambda]$ exists on the basis of Theorem 4.1' in Ch. 1, with $g(t)=F(t-\lambda)x_0$. Defining $u_\lambda^1(\theta)=x_0$ for $s-\lambda\leq\theta\leq s$ we have (1.4) on $[s-\lambda,S+\lambda]$. Set

$$v(t)=U(t,s)x_0,\ s\leq t\leq T \tag{1.8}$$

Since v is the integral solution of

$$v'\in A(t)v,\ \ v(s)=x_0,\ s\leq t\leq T$$

by Theorem 4.1 we have

$$\|u_\lambda^1(t)-v(t)\| \le \int_s^t \|F(\tau-\lambda)x_o\|\, d\tau \le M\lambda \le MT_o \qquad (1.8)'$$

for $s \le t \le s+\lambda \le s+T_o=T_1$. Therefore

$$\|u_\lambda^1(t)-x_o\| \le \|u_\lambda^1(t)-v(t)\|+\|v(t)-x_o\| \le MT_o+\sup_{s\le t\le s+T_o}\|U(t,s)x_o-x_o\| \le \rho$$

Set $u_\lambda(t)=u_\lambda^1(t)$ for $s-\lambda \le t \le s+\lambda$. Similarly we extend u on $[s+\lambda, S+2\lambda]$. Indeed, set $g(t)=F(t-\lambda, u_\lambda^1(t-\lambda))$ for $s+\lambda \le t \le s+2\lambda$ and denote by u_λ^2 the integral solution of

$$(u_\lambda^2)'(t) \in A(t)u_\lambda^2(t)+F(t-\lambda, u_\lambda^1(t-\lambda)),\; u_\lambda^2(s+\lambda)=u_\lambda^1(s+\lambda),\; s+\lambda \le t \le s+2\lambda$$

Set $u_\lambda(t)=u_\lambda^2(t)$ for $t\in[s+\lambda, s+2\lambda]$. Then u_λ is an integral solution of (1.3) on $[s,s+2\lambda]$. Arguing as for (1.8)' we have

$$\|u_\lambda(t)-v(t)\| \le \int_s^t \|F(\tau-\lambda)u_\lambda(\tau-\lambda)\|d\tau \le M(t-s) \le 2\lambda M \le MT_o$$

for $s \le t \le 2\lambda+s \le s+T_o$ so (1.8)' holds also with u_λ in place of u_λ^1 and $t \in [s,s+2\lambda]$ i.e.

$$\|u_\lambda(t)-x_o\| \le \rho \quad \text{for all } t\in[s,s+2\lambda]$$

In such a manner we extend u_λ to $[s,s+T_o]$ as an integral solution to (1.3). Therefore there exists $u_\lambda:[s-\lambda, s+T_o]\to X$ satisfying (1.4) and $\|u_\lambda(t)-x_o\| \le \rho$, $\forall t \in [s,s+T_o]$, for $\lambda > o$ small enough.

Next we prove that the sequence u_λ is precompact in $C([s,T);X)$. To this goal, let $t\in]s,T]$. Take an $\varepsilon>o$ such that $s < t-\varepsilon < t$. Denote $v_\lambda(t)=U(t,t-\varepsilon)u_\lambda(t-\varepsilon)$.

By one of the hypotheses $U(t,t-\varepsilon)$ is compact so it maps the bounded subset $\{u_\lambda(t-\varepsilon), \lambda>o\}$ of $\overline{D(A(t-\varepsilon))}$ into a precompact subset. In other words $K_\varepsilon(t)=\{v_\lambda(t), \lambda>o\}$ is precompact. On the other hand, v_λ is the integral solution of the problem

$$v_\lambda'(\tau) \in A(\tau)v_\lambda(\tau),\; v_\lambda(t-\varepsilon)=u_\lambda(t-\varepsilon),\; t-\varepsilon \le \tau$$

so the following estimate holds

$$\|u_\lambda(t)-v_\lambda(t)\| \le \int_{t-\varepsilon}^t \|F(\tau-\lambda)u_\lambda(\tau-\lambda)\|d\tau \le \varepsilon M$$

This implies (in view of Lemma 5.1 in Ch. 1) that the set $\{u_\lambda(t), \lambda > o\}$ is also precompact. It remains to prove the equicontinuity of the family $\{u_\lambda, \lambda > o\}$ (for sufficiently small λ). Take an arbitrary $t_o\in[s,T_1]$,

$(T_1=s+T_o)$. The case $t_o=s$ is obvious, since we have

$$\|u_\lambda(t)-u_\lambda(s)\| \leq \|u_\lambda(t)-v(t)\|+\|v(t)-x_o\|$$

$$\leq \int_s^t \|F(\tau-\lambda)u_\lambda(\tau-\lambda)\|d\tau + \|v(t)-x_o\| \leq M(t-s)+\|v(t)-v(s)\|$$

for all $\lambda > o$ and $t \in [s,T_1]$ (so u_λ are equicontinuous at $t_o=s$). Therefore take $s < t_o < T_1$. Choose $o < d \leq \min\{\frac{\varepsilon}{6M}, t_o-s, T_1-t_o\}$ and consider

$$V_d^\lambda(t) \equiv v_d(t)=U(t,t_o-d)u_\lambda(t_o-d), \quad t_o-d \leq t.$$

Then

$$\|u_\lambda(t)-v_d(t)\| \leq \int_{t-d}^t \|F(\tau-\lambda)u_\lambda(\tau-\lambda)\|d\tau \leq (|t-t_o|+d)M \leq 2dM \leq \varepsilon/3 \tag{1.9}$$

for all $|t-t_o| \leq d$. The main point now is to observe that $\{v_d^\lambda, \lambda > o\}$ are equicontinuous at t_o. Indeed

$$\|v_d^\lambda(t)-v_d^\lambda(t_o)\| = \|U(t,t_o-d)u_\lambda(t_o-d)-U(t_o,t_o-d)u_\lambda(t_o-d)\| \tag{1.10}$$

But the compactness of U implies the equicontinuity of $t \to U(t,t_o-d)z$ at $t=t_o$ on the bounded subset $Y = \{u_\lambda(t_o-d), \lambda > o\}$ i.e. for every $\varepsilon > o$, there exists $\eta = \eta(\varepsilon, t_o, d) < d$ such that $\|v_d^\lambda(t)-v_d^\lambda(t_o)\| < \varepsilon$, $|t-t_o| < \eta$, for all $z \in Y$. Finally,

$$\|u_\lambda(t)-u_\lambda(t_o)\| \leq \|u_n(t)-v_d(t)\|+\|v_d(t_o)\|+\|v_d(t_o)-u_\lambda(t_o)\| < \varepsilon$$

for all $|t-t_o| \leq \eta$, and $\lambda > o$. Selecting a convergent subsequence $u_{\lambda_k} \to u$ as $\lambda_k \downarrow o$ (for $k \to \infty$) in $C([s,T];X)$ and passing to the limit in (1.4) one obtains (1.2) as follows. Clearly, $u_{\lambda_k}(t) \to u(t)$ and $F(t-\lambda)u_{\lambda_k}(t-\lambda) \to F(t)u(t)$ as $k \to \infty$ uniformly on $[s,T_1]$. This is because a continuous function maps a uniformly convergent sequence into a uniformly convergent sequence. Suppose that even $u_\lambda \to u$ (for simplicity of writing).

Using an inequality of the form (10.16) we have

$$\overline{\lim_{\lambda \downarrow o}} \int_{t_o}^t < y+F(\tau-\lambda)u_\lambda(\tau-\lambda), u_\lambda(\tau)-x>_+ d\tau \equiv I \leq$$

$$\lim_{\lambda \downarrow o} \int_{t_o}^t (\|u_\lambda(\tau)-x+a(y+F(\tau-\lambda)u_\lambda(\tau-\lambda))\|-\|u_\lambda(\tau)-x\|)a^{-1}\, d\tau =$$

$$\int_{t_o}^{t} (\|u(\tau)-x+a(y+F(\tau)u(\tau))\|-\|u(\tau)-x\|a^{-1}d\tau \ , \ \forall\ a>o \qquad (1.11)$$

Letting $a \downarrow o$, it follows (by (2.7) in Ch. 1 and Lebesgue dominating convergence theorem)

$$I \leq \int_{t_o}^{t} <y+F(\tau)u(\tau),u(\tau)-a>_+ \, d\tau$$

It is now clear that (1.4) implies (for $\lambda \downarrow o$) (1.2). The proof is complete.

1.2. The simplest proof of Peano's existence theorem.

Let us consider the particular case of (1.1), namely $A(t)=o$ and $X=R^n$. Let us write (1.1) in classical form

$$x'=F(t,x), \quad x(t_o)=x_o \qquad (1.12)$$

with F continuous on the rectangle $D: |t-t_o| \leq a, \ \|x-x_o\| \leq b$

Theorem 1.2. The problem (1.12) admits at least one solution defined on the interval $|t-t_o| \leq \rho = \min\{a, \frac{b}{M}\}$ where $M=\sup\{\|F(t,x)\|, (t,x)\in D\}$

Proof. Let u_λ be the solution of the delay equation

$$u_\lambda'(t)=F(t,u_\lambda(t-\lambda)), \ u_\lambda(\theta)=x_o \text{ for } t_o-\lambda \leq \theta \leq t_o, \ t\in[t_o-\lambda, t_o+\rho] \qquad (1.13)$$

with $\lambda \in\,]o,\rho[$. Clearly

$$u(t)=x_o + \int_{t_o}^{t} F(s,u(s-\ )) \, ds, \quad |t-t_o| \leq \rho \qquad (1.14)$$

which implies $\|u_\lambda(t)-x_o\| \leq b$ and $\|u_\lambda(t)-u_\lambda(s)\| \leq M|t-s|$ for all $t \in [t_o,t_o+\rho]$ and $\lambda \in\,]o,\rho[$. By Arzela lemma, we may assume that $u_\lambda(t) \to u(t)$ as $\lambda \downarrow o$, uniformly on $[t_o,t_o+\rho]$. Letting $\lambda \downarrow o$ in (1.14), the conclusion of the theorem follows.

As far as we know this is the simplest proof of Peano's existence theorem (i.e. Theorem 1.2).

1.3. The behaviour of maximal (integral) solution as $t \uparrow t_{max}$.

We shall prove that the classical result on the behaviour of $u(t)$ as $t \uparrow t_{max}$ remains valid and in the case of the integral solution of (1.1).

Theorem 1.3. Let $A(t) \subset X\times X$ for $t \in [o,T]$ satisfy H(2.1) with $\omega=o$ and H(2.2) in Ch. 1 for every $T>o$. Suppose that $D(F(t))=\overline{D(A(t))}$ for $t\in[o,T]$

$(t,x) \to F(t)x$ is continuous and maps bounded sets into bounded sets. Moreover, assume that for every $s > 0, x_0 \in \overline{D(A(s))}$, the problem (1.1) has at least one (local) integral solution. Then there exists a maximal integral solution $u: [s,t_{max}[\to X$ of (1.1) and we have either 1°) $t_{max} = +\infty$; or 2°) $t_{max} < \infty$ and $\lim_{t \uparrow t_{max}} \|u(t)\| = +\infty$

Proof. The existence of maximal solutions follows from Zorn's lemma. We have to prove that if $t_{max} \equiv t_m < +\infty$ then $\lim_{t \uparrow t_m} \|u(t)\| = +\infty$. First we prove that in this case (i.e. $t_m < +\infty$) u is unbounded on $[s,t_m[$. The proof of this fact is carried out by reductio ad absurdum, as follows: suppose that there exists $M > 0$ such that $\|u(t)\| \leq M$, $\forall t \ [s,t_m[$ $(t_m < +\infty)$. Then there exists $M_1 > 0$ such that $\|F(t)u(t)\| \leq M_1$, for all $t \in [s,t_m[$ (since F is bounded on the bounded set $\{(t,u(t)),\ s \leq t < t_m\}$ Let $\varepsilon > 0$. Choose $d = d(\varepsilon) > 0$ such that $dM_1 < \varepsilon/3$. Denote $v(t) = U(t,t_m-d)u(t_m-d)$, for $t \leq t_m-d$. Then

$$\|u(t)-v(t)\| \leq \int_{t_m-d}^{t} \|F(s)u(s)\|ds \leq dM < \varepsilon/3 \quad , \ t_m-d \leq t < t_m \tag{1.15}$$

On the other hand v is continuous on $[s,t_m]$ so we may assume d small enough in order to have

$$\|v(t)-v(\tau)\| < \varepsilon/3 \text{ for all } t,\tau \in [t_m-d,t_m]$$

It follows

$$\|u(t)-u(\tau)\| < \varepsilon \quad , \ \forall t,\tau \in [t_m-d,t_m[$$

hence $\lim_{t \uparrow t_m} u(t)$ exists. This implies that u can be extended to the right of t_m, which is a contradiction. Consequently, $\limsup_{t \uparrow t_m} \|u(t)\| = +\infty$ (provided that $t_m < +\infty$). We now prove that we have also

$$\liminf_{t \uparrow t_m} \|u(t)\| = +\infty \tag{1.16}$$

Suppose for contradiction, that $\liminf_{t \uparrow t_m} \|u(t)\| < +\infty$. Then there exists $r_0 > 0$ and a sequence $t_k \uparrow t_m$ with the properties

$$\|u(t_k)\| < r_0, s < t_k < t_{k+1}, \ k=1,2,\ldots \tag{1.17}$$

Given $b > 0$, there exists $c > 0$ such that

$$\|U(t_k+h,t_k)u(t_k)\| \leq c, \ k=1,2,\ldots;\ h \in [0,b],\ r_0 \leq c \tag{1.18}$$

This is because $(t,s,x) \to U(t,s)x$ is bounded on bounded sets (where

U is the evolution operator generated by S(t)). Precisely

$$\|U(t,t_k)u(t_k)\| \le \|U(t,t_k)u(t_k)-U(t,t_k)U(t_k,s)x_0\| + \|U(t,s)x_0\|$$

$$\le \|u(t_k)-U(t_k,s)x\| + \|U(t,s)x_0\|, \ \forall t \ge t_k \tag{1.19}$$

which yields (1.18) for $t=t_k+h$. Denote

$$B=\sup \{ \|F(t)y\|, \ t \in [s,t_m], \ y \in \overline{D(A(t))}, \ \|y\| \le 3c\} \tag{1.20}$$

We may assume $bB < c$. Let k_0 be with the property $t_{k_0} > t_m-b$ (of course $o < b < t_m$). Then it follows

$$\|u(t)\| \le 3c, \ \forall t \in [t_{k_0}, t_m[\tag{1.20'}$$

The proof of (1.20)' shall also be carried out by reductio ad absurdum Therefore if (1.20) were not true, there would exist $h_1 > o$ satisfying

$$h_1 < b, \ \|u(t_{k_0}+t)\| < 3c, \ \forall t \in [o,h_1[, \ \|u(t_{k_0}+h_1)\| = 3c \tag{1.21}$$

$$t_{k_0}+h_1 < t_m$$

Let us consider $v(t)=U(t,t_{k_0})u(t_{k_0})$, with $t \ge t_{k_0}$. We know how to estimate the difference of the integral solutions u and v. Namely

$$\|u(t_{k_0}+h_1)-v(t_{k_0}+h_1)\| \le \int_{t_{k_0}}^{t_{k_0}+h_1} \|F(\tau)u(\tau)\|d\tau$$

$$\le h_1B \le bB < c \tag{1.22}$$

We have reached the contradiction below (see also (1.18))

$$3c=\|u(t_{k_0}+h_1)\| \le \|u(t_{k_0}+h_1)-v(t_{k_0}+h_1)\|+\|v(t_{k_0}+h_1)\|<$$

$$c + \|U(t_{k_0}+h_1,t_{k_0})u(t_{k_0})\| \le \quad 2c$$

and thus (1.20)' holds. But (1.20)' contradicts the unboundedness of u on $[s,t_m[$ and therefore (1.16) is necessarily true. This completes the proof.

§ 2. A class of partial differential equations with time-dependent domain, modelling long water waves of small amplitude. Evolution operators approach.

2.1. Model equations for long waves in nonlinear dispersive systems

In this section we are concerned with the initial-boundary value problem (IBVP) for the nonlinear dispersive equation

$$
\text{(IBVP)}\quad\begin{cases} u_t+(F(u))_x-u_{xxt}=0, \quad 0<t<T,\ 0<x<1 & (2.1)\\ u(0,x)=u_0(x) \quad ,\ 0\leq x\leq 1, & (2.2)\\ u(t,0)=h_0(t),\ u(t,1)=h_1(t), \quad 0\leq t\leq T & (2.3)\end{cases}
$$

where T is an arbitrary positive number, h_0, $h_1 \in C^1[0,T]$; $F:R\to R$ is a nonlinear function of class $C^1(R)$ with $F(0)=0$ satisfying also the growth condition

$$|F(y)|\leq a|y|^2+b|y|+c,\ \forall y\in R \tag{2.4}$$

with $a,b,c\in R_+ = [0,+\infty[$

The initial condition u_0 is a function in the Sobolev space $H^1(0,1)$ (which is identified with the absolutely continuous functions from $[0,1]$ into R). Of course u_0 has to satisfy the compatibility conditions (consistency)

$$u_0(0)=h_0(0),\ u_0(1)=h_1(0) \qquad (2.2)'$$

In the particular case

$$F(y)=y+y^2/2,\ y\in R \tag{2.5}$$

the equation (2.1) becomes (with $y=u(t,x)$)

$$u_t+u_x+uu_x-u_{xxt}=0 \tag{2.6}$$

Equation (2.6) is a model for long water waves of small amplitude, generated in a uniform open channel by a wavemaker at one end. This equation (i.e. (2.6)) was proposed by Benjamin, Bona and Mahony [1] as an alternative to the Kortweg - de Vries equation (KdV)

$$u_t+u_x+uu_x+u_{xxx}=0 \tag{2.7}$$

which is also an approximate model for long waves of small amplitude in a class of non-linear dispersive systems.

The presence of u_{xxx} in (2.7) (i.e. an extra x-derivative) in place of u_{xxt} in (2.6), imposes constraints on the initial-boundary data (which appear unnatural to the physical problem). For example, in the case of (2.7) we have to choose $u_o \in H^3(R)$ to get only a local solution (see Pazy [4,p.251]), while in the case of (2.6) we shall prove that for $u_o \in H^1(o,1)$ we get a weak solution $u=u(t)$ on R_+.

If u_o is of class $C^2[o,1]$, then the weak solution u of (2.6)+(2.2) +(2.3) is just a classical solution.

Moreover, we shall prove that the problem (2.1)-(2.3) is well-posed in Hadamard's sense, i.e. the weak or classical solution exists, is unique and depends continuously on the specified data. Actually, the solution u ot (2.1)-(2.3) will be obtained via nonlinear evolution operators (as given in Chapter 1, § 3).

Note that in the case of a surface water waves (in a uniform open channel) $u=u(t,x)$ represents the location of the free surface with respect to its undisturbed position while the independent variables x and t are proportional to distance along the channel and to elapsed time, respectively. The functions h_o and h_1 are boundary forcing functions. The two-point boundary-value problem (2.1)+(2.3) is natural for many reasons.

For example, any numerical algorithm for (2.1) requires a bounded domain (e.g. $o \leq x \leq 1$) even though for the existence of solution we can work on R_+ (or with periodic condition $u(t,o)=u(t,1), o \leq t \leq T$). Moreover, the initial-boundary conditions (2.2)+(2.3) are more convenient for direct comparisons with laboratory experiments on surface water waves (Bona and Dougalis).

Now let us proceed to convert the problem (2.1)-(2.3) (following Oharu and Takahashi [1]) into a Cauchy problem for an abstract evolution equation

$$u'(t)=A(t)u'(t) \quad o < t < T \tag{2.8}$$

$$u(o)=u_o \in \overline{D(A(o))} \tag{2.9}$$

(with $\overline{D(A(t))}$-dependent of t) in the Sobolev space $H^1(o,1) \equiv H^1 \equiv W^{1,2}(o,1)=W^{1,2}(I), I=]o,1[$.

Denote by $(u,v)_k$ and $||u||_k$ the inner product and norm of H^k respectively (k=1,o). Clearly, $H^o = L^2(o,1) \equiv L^2$. Therefore

$$(u,v)_1 = \int_o^1 u(x)v(x)dx + \int_o^1 u'(x)v'(x)ds = (u,v)_o (u',v')_o \tag{2.10}$$

for $u,v \in H^1$, and

$$||u||_1 = \left(\int_o^1 |u(x)|^2 dx + \int_o^1 |u'(x)|^2 dx\right)^{1/2} = (||u||_o^2 + ||u'||_o^2)^{1/2} \tag{2.11}$$

It is easy to see that

$$\sup_{o \leq x \leq 1} |v(x)| \equiv ||v||_{C([o,1])} \leq 2||v||_1, v \in H^1 \tag{2.12}$$

i.e.

$$H^1(o,1) \subsetneq C([o,1]) \tag{2.13}$$

and the injection is compact.

For $u \in H^1$, set $D_x u = \frac{du}{dx}$ in the sense of L^2. Thus, if $u'(x)$ is the classical derivative of u at x, then $u'_x = D_x u$ a.e. on $]o,1[$. The operator D is skew-symetric, i.e.

$$(D_x u,w)_o = -(u,D_x w)_o, \ \forall\, u \in H^1, \forall w \in H_o^1 \tag{2.14}$$

where

$$H_o^1 = \{u \in H^1(o,1),\ u(o)=u(1)=o\} \tag{2.14'}$$

Clearly, a classical solution u of the problem (2.1)-(2.3) is a function $u=u(t,x)$, $o \leq t \leq T$, $o \leq x \leq 1$ satisfying (2.1)-(2.3), where the derivatives involved are taken in classical sense and are continuous functions.

A crucial remark is the following one. If u is a classical solution of (IBVP) with $u_{xxt} = u_{txx}$, then $y = u_t$ satisfies the following two-point boundary-value problem in $C([o,1])$

$$y''-y=(F(u))_x;\ y(o)=h_o'(t),\ y(1)=h_1'(t),\ o \leq x \leq 1,\ o \leq t \leq T \tag{2.15}$$

where $x \to u(t,x)$ is of class $C^2([o,1])$ so in particular it belongs to D(t) given by

$$D(t) = \{v \in H^1(o,1);\ v(o)=h_o(t),\ v(1)=h_1(t)\} \tag{2.16}$$

for $t \in [o,T]$. With the convention $(u(t))(x)=u(t,x)$ we can write $u(t) \in D(t)$. This suggests the introduction of the operator $A(t):D(t) \to H^2(o,1)$ as follows. For each $v \in D(t)$, denote by $A(t)v=z$ the (unique) solution z of the problem

$$z''-z=(Fv))_x;\ z(o)=h_o'(t),\ z(1)=h_1'(t),\ o\le x\le 1 \tag{2.17}$$

where $z''=z_{xx}$. Or, integrating this elementary boundary value problem we get

$$(A(t)v)(x)\equiv z(x)=h_o'(t)p(x)+h_1'(t)q(x)+\int_o^1 K(x,s)F(v(s))ds \tag{2.18}$$

for $t\in[o,T]$, $x\in[o,1]$, $v\in D(t)$, where

$$p(x)=(e^{2-x}-e^{x})/(e^2-1),\ q(x)=(e^{1+x}-e^{1-x})/(e^2-1) \tag{2.19}$$

$$K(x,s)=[e^{2-(x+s)}-e^{x+s}+\mathrm{sign}(x-s)e^{2-|x-s|}-e^{|x-s|}]/2(e^2-1) \tag{2.20}$$

$$x\in[o,1],\ s\in[o,1]\ (\text{so } K(o,s)=K(1,s)=o).$$

In other words,

$$D(A(t))\equiv D(t)=\overline{D(A(t))},\ o\le t\le T \tag{2.21}$$

since it is clear that $D(t)$ is a closed convex subset of $H^1(o,1)$.

Let us sketch the proof of (2.18). We start from the fact that the general solution of the second order differential equation

$$w''(x)-w(x)=a(x),\ o\le x\le 1,\ a\in C[o,1]$$

is given by

$$w(x)=C_1e^x+C_2e^{-x}+\int_o^1 (-\tfrac{1}{2})e^{-|x-s|}a(s)ds$$

where C_1,C_2 are arbitrary constants in R.

Consequently, if $v\in D(t)$, the solution $z=A(t)v$ of the problem (2.17) is given by

$$z(x)=C_1e^x+C_2e^{-x}+\int_o^1(-\tfrac{1}{2})e^{-|x-s|}(F(v(s))_s ds \tag{2.22}$$

Writing $\int_o^1=\int_o^x+\int_x^1$ and then integrating by parts we get

$$z(x)=C_1e^x+C_2e^{-x}+\frac{1}{2}F(h_o(t))e^{-x}-\frac{1}{2e}F(h_1(t))e^x+\int_o^1\bar{K}(x,s)F(v(s))ds,\ v\in D(t) \tag{2.23}$$

with

$$\bar{K}(x,s)=\frac{1}{2}\,\mathrm{sign}\,(x-s)e^{-|x-s|} \tag{2.24}$$

The constants C_1 and C_2 have to satisfy the system $z(o)=h_o'(t)$, $z(1)=h_1'(t)$. Solving this system one obtains

$$C_1e^x = \frac{1}{e^2-1}(eh_1'(t)-h_0'(t))e^x + \frac{1}{2e}Fh_1(t))e^x - \frac{1}{2(e^2-1)}\int_0^1 (e^{x+s}+e^{x-s})F(v(s))ds$$

and

$$C_2e^{-x} = \frac{e}{e^2-1}(eh_0'(t)-h_1'(t))e^{-x} - \frac{1}{2}F(h_0(t))e^{-x} + \frac{e^2}{2(e^2-1)}\int_0^1 (e^{s-2-x}+e^{-s-x})F(v(s))ds$$

Replacing C_1e^x and C_2e^{-x} in (2.23) we obtain (2.18) with

$$K(x,s) = \frac{1}{2}\text{ sign }(x-s)e^{-|x-s|} + \frac{e^{s-x}-e^{x-s}+e^{2-s-x}-e^{s+x}}{2(e^2-1)} \tag{2.25}$$

Inasmuch as

$$e^{s-x}-e^{x-s} + (\text{sign}(x-s))(e^{|x-s|}-e^{-|x-s|})=0$$

We see that $K(x,s)$ given by (2.25) can be written in the form given by (2.20). Thus, the proof of (2.18) is complete.

Therefore, the problem (2.1)-(2.3) is reduced to the abstract Cauchy problem (2.8)+(2.9) in H^1, with $u_0 \in \overline{D(o)} \subset H^1$, and $A(t)$ given by the integral representation (2.18). Since $\overline{D(A(t))}$ depends on t, the existence of the solution to (2.8)+(2.9) cannot be proved in the classical framework. However, we shall see in next section that (2.8)+(2.9) falls just in the theory of evolution equation with time dependent domain as developed in Ch. 1.

To this goal some other preliminaries are needed. Let us introduce the function

$$(g(t))(x)=h_0(t)p(x)+h_1(t)q(x),\quad t\in[o,T],\ x\in[o,1] \tag{2.26}$$

Hence $g(.)\in C^1([o,T];\ H^1)$, $D^2g'(t)=g'(t), g(t)\in D(t)$

$$(g'(t))(x)=h_0'(t)p(x)+h_1'(t)q(x),\quad t\in[o,T],\ x\in[o,1] \tag{2.26)'$$

For simplicity one writes sometimes $g(t,x)$ and $g'(t,x)$ instead of $(g(t))(x)$ and $(g'(t))(x)$ respectively (or $g'(t,x)=g_t(t,x)$).

Therefore (2.18) can be written in the form

$$(A(t)v)(x)=(g'(t))(x) + \int_0^1 K(x,s)F(v(s))ds, \quad (t,x)\in [0,T]\times[0,1] \tag{2.27}$$

for $v\in D(t)$, and $A(t)v\in D'(t)$, where

$$D'(t)=\{ z\in H^1(o,1);\ z(o)=h_0'(t), z(1)=h_1'(t)\} \tag{2.28}$$

We also have

$$A(t)v-g'(t)\in H_0^1(o,1) \text{ (given by (2.14)')} \tag{2.29}$$

$$D_x^2A(t)v-A(t)v=D_xF(v),\ A(t)v\in D'(t), o\le t\le T, v\in D(t) \tag{2.30}$$

where

$$(F(v))(x)=F(v(x)),\ x\in[o,1] \tag{2.31}$$

Obviously, since $F\in C^1(R)$, it follows that the operator $v\to F(v)$ defined by (2.31) maps H^1 into itself (and H_0^1 into H_0^1 since $F(o)=o$ for $o\in R$). Moreover, on the basis of (2.12) and of Lagrange formula it follows that F is locally Lipschitz continuous on H^1 with respect to the norm of L^2. Precisely we have

$$\|F(v_1)-F(v_2)\|_o\le M_r\|v_1-v_2\|_o,\ r>o \tag{2.32}$$

for all $v_1,v_2\in H^1$ with $\|v_i\|_1 \le r$, $i=1,2$ and

$$M_r = \sup\{ |F'(y)|;\ |y|\le 2r,\ y\in R\} \tag{2.33}$$

It is straightforward that (2.32) implies Locally Lipschitz continuity of A(t) given by (2.27). In fact we can prove much more about A(t), namely

<u>Lemma 2.1.</u> (1) Let $F\in C^1(R)$, $h_0,h_1\in C^1[o,T]$. Then

$$\|A(t)v-A(s)w\|_1\le B(\|F(v)-F(w)\|_o+g'(t)-g'(s)\|_1) \tag{2.34}$$

for all $t,s\in[o,T]$, $v\in D(t)$ and $w\in D(s)$ where B is a positive constant (given by (2.42)). (2) The tangential condition holds

$$\lim_{\lambda\downarrow o}\frac{1}{\lambda} d[v+\lambda A(t)v;\ D(t+\lambda)] = o,\quad \forall t\in[o,T[,\ \forall v\in D(t) \tag{2.35}$$

Precisely, for each $t\in[o,T[$, $\lambda\in]o,T-t[$ and $v\in D(t)$ we have

$$v+\lambda A(t)v+g(t+\lambda)-g(t)-\lambda g'(t)\in D(t+\lambda) \tag{2.35'}$$

(3) If in addition F satisfies the growth condition (2.4), then there

are continuous functions $\alpha, \beta, \gamma : [o,T] \to R_+$ such that

$$(A(t)v-g'(t), v-g(t))_1 \le \alpha(t)\|v-g(t)\|_1^2 + \beta(t)\|v-g(t)\|_1 + \gamma(t) \quad (2.36)$$

for all $t \in [o,T]$ and $v \in D(t)$.

Remark 2.1 Obviously, the inequality (2.34) and (2.32) show that:

$$\|A(t)v-A(s)w\|_1 \le B(M_r\|v-w\|_1 + \|g'(t)-g'(s)\|_1) \quad (2.37)$$

for $t,s \in [o,T]$, $v \in D(t)$, $w \in D(s)$, $\|v\|_1 \le r$, $\|w\|_1 \le r$.

In particular (2.37) implies the continuity of $(t,v) \to A(t)v$ in H^1 (i.e. if $t_n \to s$, $v_n \in D(t_n)$ and $v_n \to w$ in H^1, then $w \in D(s)$ and $A(t_n)v_n \to A(s)w$ in H^1. Moreover, (2.37) implies locally (2.11) in Chapter 1 with $\omega = BM_r$, $L \equiv 1$ and $g'(t)$ in place of f. Finally, we see from (2.37) that $A(t)$ is locally Lipschitz continuous, namely

$$\|A(t)v_1-A(t)v_2\|_1 \le BM_r\|v_1-v_2\|_1, \; v_i \in D(t), \; \|v_i\| \le r, i=1,2 \quad (3.38)$$

According to Proposition 7.1 in Chapter 1, (2.35) implies

$$\lim_{\lambda \downarrow o} \frac{1}{\lambda} d[v; R(I-\lambda A(t+\lambda))| = o, \; \forall t \in [o,T[, \; v \in D(t) \quad (2.39)$$

It is interesting to note that any of the usual range condition (see (1.10) in Ch. 1) is not satisfied in this case. Precisely, let us point out that $R(I-\lambda A(t+\lambda)) \supset \overline{D(A(t))} = D(t)$ is not valid in this case. Indeed, if we assume that for $v \in D(t)$, there exists $v_\lambda \in D(t+\lambda)$ such that $v_\lambda - \lambda A(t+\lambda)v_\lambda = v$, then $h_o(t+\lambda) - \lambda h_o'(t+\lambda) = h_o(t)$, $\lambda > o$, $o \le t \le T$ (which is not valid in general). Note also that (via contracting mapping principle) Oharu and Takahashi have shown more than (2.39), namely: for every $r > o$ there exists $\lambda_o = \lambda_o(r) > o$ such that for $t \in [o,T[$, $\lambda \in]o, \lambda_o[$, $\lambda < T-t$ and $v \in D(t)$ with $\|v\|_1 \le r$ one can find $v \in D(t+\lambda)$ with the property

$$v_\lambda - \lambda A(t+\lambda)v_\lambda = v+g(t+\lambda)-g(t)-\lambda g'(t) \quad (2.39)'$$

Proof of Lemma 2.1. Combining (2.10), (2.14), (2.29), (2.30) and $D^2g'(t) = g'(t)$, we easily derive

$$\|A(t)v-A(s)w-g'(t)+g'(s)\|_1^2 = (D_xF(v)-D_xF(w), A(t)v-A(s)w-g'(t)+g'(s))_o$$

$$= -(F(v)-F(w), D_xA(t)v-g'(t))-D_x(A(s)w-g'(s)))_o$$

$$\le \|F(v)-F(w)\|_o \|D_x(A(t)v-g'(t))-D_x(A(s)w-g'(s))\|_o \quad (2.40)$$

for all $s,t \in [o,T]$, $v \in D(t)$ and $w \in D(s)$.

On the other hand, by (2.27) we have

$$\|D_x(A(t)v-g'(t))-D_x(A(s)w-g'(s))\|_o \leq B^2\|F(v)-F(w)\|_o \tag{2.41}$$

where $B^2=B_1$ with

$$B_1 = \left\{\int_o^1 \int_o^1 |D_xK(x,s)|^2 \, ds \, dx\right\}^{1/2} \tag{2.42}$$

Actually, taking into account

$$\frac{d}{dx}((\text{sign } \bar{x})e^{|\bar{x}|})=e^{|\bar{x}|}, \quad \frac{d}{dx}(\text{sign } \bar{x})e^{-|\bar{x}|})=-e^{-|\bar{x}|}, \bar{x} \in R$$

We see that $x \to K(x,s))$ given by (2.20) is of class C^1 and we have (with $a_1=2(e^2-1)$

$$D_xK(x,s)=(K(x,s))_x = (-e^{2-(x+s)}-e^{x+s}-e^{2-|x-s|}-e^{|x-s|})/a_1 \tag{2.43}$$

for all $s,x \in [o,1]$. Consequently

$$B_1 \leq B_2 = \sup\{|K(x,s))_x|, \; x,s \in [o,1]\} \tag{2.44}$$

so we can take $B = \sqrt{B_2}$.

It is now clear that (2.40) along with (2.41) yields (2.34). The relation (2.35)' is automatically satisfied since $(A(t)v)(o)=(g'(t))(o)= =h_o'(t)$, $v(o)=h_o(t)=(g(t)(o)$.

The relation (2.35)' is true since $v-g(t)$ and $A(t)v-g'(t)$ belong to H_o^1 and $g(t+\lambda) \in D(t+\lambda)$. Obviously, (2.35)' implies (2.35) (this is because $\lambda^{-1}(g(t+\lambda)-g(t))-g'(t) \to o$ in H^1 as $\lambda \downarrow o$. It remains to prove (2.36). To this end, we frist recall that $v-g(t) \in H_o^1$ for $v \in D(t)$, and arguing as before, we have

$$\begin{aligned}(A(t)v-g'(t),v-g(t))_1 &= (F(v), D_x(v-g(t)))_o = \\ &= \int_{h_o(t)}^{h_1(t)} F(y)dy - (F(v), D_x(g(t)))_o\end{aligned} \tag{2.45}$$

Or, by (2.4) we have

$$|F(v(x))| \leq a|v(x)|^2+b|v(x)|+c \tag{2.46}$$

Inasmuch as $(D_xg(t)(x)=h_o(t)p_o'(x)+h_1(t)q'(x)$, a simple combination of (2.45), (2.46) and $v(x)=(v(x)-g(t))+g(t)$ leads us to (2.36). This completes the proof.

2.2 Weak solutions and classical solutions

Recall that $u:[o,T]\times[o,1]\to R$ is said to be a classical solution to the initial-boundary value problem (2.1)-(2.3) if all derivatives involved in (2.1) exist in classical sense ($u_{xxt}=u_{txx}$) they are continuous and (2.1)-(2.3) are fulfilled.

In view of the remarkable properties of A(t) (as indicated in Lemma 2.1) we shall prove that (2.8)+(2.9) has a unique strong solution in $H^1(o,1)$ for every $u_o\in D(o)\equiv D(A(o))$.

We have already proved (in the previous section) that if u is a classical solution to (2.1)-(2.3), then $y=u_t$ satisfies (2.15), that is (by 2.18))

$$u_t = A(t)u(t), \text{ for all } t\in[o,T],\ u(o)=u_o \tag{2.47}$$

Integrating (2.47) over [o,t] and taking into account (2.18) we get $u(t)=u(o)+\int_o^t A(\tau)u(\tau)d\tau$, i.e.

$$u(t,x)=u_o(x)+(h_o(t)-h_o(o))p(x)+(h_1(t)-h_1(o))q(x)+$$

$$+\int_o^t\int_o^1 K(x,s)F(u(\tau,s))ds\,d\tau\quad t\in[o,T],\ x\in[o,1],\ u_o\in D(o) \tag{2.48}$$

where $u(\tau,s)\equiv(u(\tau))(s)$. Therefore, a classical solution satisfies necessarily the integral equation (2.48). The converse assertion is also true, provided that $u_o\in C^2(o,1)$ and $h_o,h_1\in C^1[o,T]$. These remarks suggest the following definition

<u>Definition 2.1.</u> A function $u:[o,T]\times[o,1]\ \ R$ is said to be a weak or generalized, solution to (2.1)-(2.3) if the function $t\to u(t)\equiv u(t,.)$ is continuously differentiable from [o,T] into H^1 and satisfies (in the strong sense of H^1)

$$u'(t)=A(t)u(t),\ u(t)\ \ D(t),\ o\le t_o\le t\le T,\ u(t_o)=u_o \tag{2.49}$$

with A(t) given by (2.18) $u_o\in D(t_o)$, $t_o\in[o,T]$ and $D(A(t))=D(t)=D(A(t))$.

A more natural way to define the notion of weak solution of (IBVP) is the following standard device.

If u is a classical solution of (2.1), and if we multiply (2.1) by

$w\in H_o^1$ and integrate over $o\leq x\leq 1$ we obtain easily

$$(u_t,w)_1 + (D_x F(u(t)),w)_o=o,\quad o < t < T,\quad w\in H_o^1 \tag{2.50}$$

This suggests

Definition 2.2. A function u: $[o,T]\times[o,1]\to R$ is said to be a weak solution of (2.1)-(2.3) if $t\to u(t)\equiv u(t,.)$ is in $C^1([o,T];\,H^1)$, $u(o)=$ $=u_o$, $u(t)\in D(t)$ for $t\in[o,T]$ and

$$(u'(t),w)_1 + (D_x F(u(t)),\,w)_o=o, \tag{2.50)'}$$

for all $t\in[o,T]$ and $w\in H_o^1$ where $u'\equiv u_t$ denotes the strong derivative in H^1.

In view of the property

$$D_x^2 A(t)u(t)-A(t)u(t)=D_x F(u(t)) \tag{2.51}$$

it follows at once that Definition 2.1 implies Definition 2.2. Conversely, if u is a weak solution of (IBVP) in the sense of Definition 1.1 and in addition $u'(t)\in D'(t)\cap H^2$ for $t\in[o,T]$, then u is a weak solution in the sense of Definition 2.1. Let us prove this latter fact. By definition of $(.,.)_1$, (2.50)' means

$$\int_o^1 u_t(t,x)w(x)dx +\int_o^1 u_{tx}w_x dx + \int_o^1 (F(u(t,x)))_x w(x)dx=o \tag{2.52}$$

Since $u_t\in H^2(o,1)$ (by hypothesis), (2.52) means that the distributional derivative u_{txx} of u_{tx} is just

$$u_t(t,x)+(F(u(t,x)))_x=u_{txx} \tag{2.53}$$

hence u_{txx} is classical derivative. We have supposed also that

$$u_t(t,o)=h_o'(t),\quad u_t(t,1)=h_1'(t) \tag{2.53)'}$$

On the basis of (2.15)+(2.18), it follows from (2.53)+(2.53)' that $u_t(t,x)=(A(t)u(t)(x)$ which implies $u_t=A(t)u(t)$ in the sense of H^1. We conclude that the following result holds.

Proposition 2.1. Definition 2.1 implies Definition 2.2. COnversely if u is a weak solution of (IBVP) in the sense of Definition 2.2 and $u'(t)$ $\in D'(t)\cap H^2$ for $t\in[o,T]$, then u is a weak solution of (IBVP) in the sense of Definition 2.1.

It is also clear that u is a weak solution in the sense of Definition 2.1 (with $u_0 \in D(o)$ iff us satisfies the integral equation (2.48) and $u:[o,T]\times[o,1]\to R$ is continuous.

If u is a classical solution of (IBVP) on $\tilde{D}=[o,T]\times[o,1]$, then u is a weak solution in each of the above senses.

Conversely, a weak solution in the sense of Definition 2.1 with $u_0 \in D(o)\cap C^2[o,1]$ is a strong solution of (IBVP) on $\tilde{D}$. This fact follows directly from (2.49). Indeed, $u(.)\in C^1([o,T];H^1)$ implies the continuity of $(t,x)\to u(t,x)$ from $\tilde{D}$ into R. Then by (2.49) and (2.48) it follows that u is a classical solution of (IBVP). Therefore it remains to prove the existence uniqueness and continuous dependence on the initial data of solution to the problem (2.49). To this goal we can proceed in two ways.

A first one consists in considering the Banach space $C([o,T_1]\times[o,1])$ with supremum $\|u\|=\sup\{|u(t,x)|,\ o\le t\le T_1,\ o\le x\le 1\}$ for $T_1 < T$ sufficiently small and applying Banach fixed point principle to the operator $(Gu)(t,x)$ defined by the right hand side of (2.48). Roughly speaking, this method has been used by Bona and Bryant [1]. A second way consists in proving the existence of strong solution (in H^1) of the evolution equation (2.49). The advantage of this way over the first one consists of

1°) It provides approximate solutions in the form of poligonal lines which are more convenient for numerical schemes.

2°) It provides a unifying method to study the properties of solution. This is because the solution is expressed in terms of evolution operators, whose properties are known in Ch. 1, § 3.3).

This second way was followed by Oharu and Takahashi [1]. We will follow the same way but in a simplified manner in the sense that in the construction of approximate solutions we proceed directly, without using fixed point principle of Banach. Oharu and Takahashi [1] have used this principle to prove in a simple manner Tangential condition (2.39) or a surjectivity property in Lemma 3.3 of their paper [1].

<u>Theorem 2.1.</u> Let $F\in C^1(R)$, $h_0, h_1 \in C^1[o,T]$ and $A(t)$ given by (2.27). Let also $u_0 \in D(t_0)$ and $M, T_1, r > o$ with the properties

$$\|A(t)v\|_1 \le M,\ \forall |t| \le T_1, u\in D(t) \text{ with } \|u-u_o\|_1 \le r \tag{2.54}$$

$T_1 - t_o = \min\{T-t_o, r/_{(M+1)}\}$. Let $o < T_2 < T_1$ (so $T_2 = T_2(t_o, x_o)$)

1°) Then for every positive integer n there exist a positive integer N_n, $\{t_i^n\}$ with $t_o^n = t_o < t_1^n < \ldots < t_i^n < t_{i+1}^n < \ldots < t_{N_n-1}^n < t_{N_n}^n = T_2$ and $u_i^n \in D(t_i^n)$, $i=o,1,\ldots,N_n$ with the properties

$$u_o^n = u_o,\ u_{i+1}^n = u_i^n (t_{i+1}^n - t_i^n)\ (A(t_i^n)u_i^n + p_i^n) \tag{2.55}$$

$$p_i^n = ((g(t_i^n + d_i^n) - g(t_i^n))/d_i^n) - g'(t_i), d_i = t_{i+1} - t_i \tag{2.56}$$

where $d_i^n = d_i$ is the largest number in $]o, \frac{1}{n}]$ satisfying

$$\|A(t)u - A(t_i)u_i\|_1 \le \frac{1}{n},\ \text{for } |t-t_i| \le d_i,\ \|u-u_i\|_1 \le d_i(M+1) \tag{2.57}$$
$$u\in D(t)$$

$$\|p_i\|_1 \le \frac{1}{n},\ i=o,1,\ldots,N_n,\ \text{with } p_i \equiv p_i^n,\ t_i \equiv t_i^n,\ u_i \equiv u_i^n \tag{2.58}$$

2°) The sequence of poligonal lines

$$u_n(t) = u_i + (t-t_i)(A(t_i)u_i + p_i),\ t_i \le t \le t_{i+1},\ i=\overline{o,N_n} \tag{2.59}$$

is uniformly convergent in H^1 to a function $u\in H^1$ which is the unique solution of (2.49). Write it $u=u(t;t_o,u_o)$, $t_o \le t \le T_2$.

3°) The operator $U(t,t_o): D(t_o) \to D(t)$ defined by $U(t,t_o)x_o = u(t;t_o,x_o)$, $x_o \in D(t_o)$, $t_o \le t \le T_2$ ($T_2 = T_2(t_o,x_o)$) is an evolution operator in the sense of Definition 3.3 in Ch. 1.

4°) If in addition F satisfies the growth condition (2.4) then the local solution $u(t;t_o,u_o)$ can be extended to the whole $[t_o,T]$ and $U(t,t_o)$ is an evolution operator on $o \le t_o \le t \le T$.

5°) If $u_o \in C^2[o,1]$ then the generalized solution u defined at 4°) is a classical solution of (IBVP) and depends continuously on the initial boundary data u_o, h_o and h_1 in the following sense:

If $u_{o,n} \to u_o$ in $H^1(o,1)$, $h_{o,n} \to h_o$ and $h_{1,n} \to h_1$ in $C^1[o,T]$ then the solution $u_n = u_n(t,o;u_{o,n})$ of (2.49) corresponding to $u_{o,n}$, $h_{o,n}$ and

$h_{1,n}$ tends to $u=u(t;o,u_o)$ in $C([o,T];H^1)$, where $u=u(t;o,u_o)$ is the solution of (2.49) corresponding to u_o,h_o,h_1.

Remark 2.2. The convergence of u_n to u in $C([o,T];H^1)$ (i.e. $\|u_n(t)-u(t)\|_1$ goes to zero as $n\to\infty$ uniformly with respect to $t\in[o,T]$) implies, according to (2.12):

$$|u_n(t,x)-u(t,x)| \le \|u_n(t)-u(t)\|_{C[o,1]} \le 2\|u_n(t)-u(t)\|_1 \qquad (2.60)$$

for all $x\in[o,1]$, $t\in[o,T]$ and $n=1,2,\dots$

Consequently, $u_n(t,x)\to u(t,x)$ as $n\to\infty$, uniformly with respect to $(t,x)\in[o,T]\times[o,1]$ (in the topology of R).

Proof of Theorem 2.1. In view of the continuity of $(t,v)\to A(t)v$ in the norm of H^1 (see Remark 2.1), there are positive constants r,T_1 and M, as indicated in (2.54). 1°) Let d_o be the largest number on $]o,\frac{1}{n}]$ with the properties

$$\|A(t)u-A(t_o)u_o\|_1 \le \frac{1}{n},\ |t-t_o|<d_o, u\in D(t);\ \|u-u_o\|_1 \le d_o(M+1)$$

$$\|p_o\| \le \frac{1}{n} \text{ where } p_o=d_o^{-1}(g(t_o+d_o)-g(t_o))-g'(t_o).$$

Set $t_1=t_o+d_o$ and

$$u_1=u_o+(t_1-t_o)(A(t_o)u_o+p_o$$

Since $u_o\in D(t_o)$, by (2.35)' with $\lambda=d_o$ we see that $u_1\in D(t_1)$. Moreover, $\|u_1-u_o\|_1\le d_o(M+\frac{1}{n}) \le (T_2-t_o)M+1) < r$. Consequently

$$\|A(t_1)u_1-A(t_o)u_o\|_1 \le \frac{1}{n},\ \|A(t_1)u_1\|_1 \le M$$

In such a manner, we define t_i, u_i and p_i as indicated by (2.55)-(2.58). Since $t_{i+1}-t_i=d_i$ and $\|u_{i+1}-u_i\| \le d_i(M+\frac{1}{n})$, $u_{i+1}\in D(t_{i+1})$, it follows

$$\|A(t_{i+1})u_{i+1}-A(t_i)u_i\|_1 \le \frac{1}{n},\ i=o,1,\dots \qquad (2.61)$$

We will prove that $\lim_{i\to\infty} t_i=T_1$ (which implies the existence of $N_n=i_n$ with the property $t_{N_n-1} < T_2 \le t_{N_n}$. Then we take t_{N_n} just T_2). The proof of $t_i\to T_1$ shall be carried out by reductio ad absurdum. Hence, let us suppose that $\lim_{i\to\infty} t_i=t^* < T_1$. then it follows: $\lim_{i\to\infty} d_i=o$ and (from $\|u_{i+1}-u_i\| \le d_i(M+1)$) $\lim_{i\to\infty} u_i=u^*$ exists. Moreover u^* $D(t^*)$. Choose $d^*\in]o,\frac{1}{n}]$ with the properties:

$$\|(d^*)^{-1}(g(t^*+d^*)-g(t^*))-g'(t^*)\|_1 \leq \frac{1}{2n} \qquad (2.62)$$

$$\|A(t)u-A(t^*)u^*\|_1 \leq \frac{1}{2n},\ |t-t^*| \leq 2d^*,\ \|u-u^*\|_1 \leq 2d^*(M+1) \qquad (2.63)$$

with $u \in D(t)$, $t^*+2d^* < T_1$. It is now easy to check that

$$\|A(t)u-A(t_i)u_i\|_1 \leq \frac{1}{n},\ |t-t_i| \leq d^*,\ \|u-u_i\|_1 \leq d^*(M+1) \qquad (2.64)$$

with $u \in D(t)$, for all $i \geq i_o$, where i_o is such that $d_i < d^*, t^*-t_i \leq d^*$ and $\|u_i-u^*\|_1 \leq d^*$, $\forall i \leq i_o$.

On the basis of (2.64) and of the maximality of d_i satisfying (2.57) and (2.58), it follows that

$$\|(d^*)^{-1}(g(t_i+d^*)-g(t_i))-g'(t_i)\|_1 > \frac{1}{n},\ \forall \geq i\ \ i_o \qquad (2.65)$$

Letting $i \to +\infty$ in (2.65) we get an inequality which contradicts (2.62). Consequently $\lim_{i\to\infty} t_i^n = T_1$ for all $n=1,2,\ldots$

2°) Let us observe that

$$\frac{u_i-u_{i-1}}{t_i-t_{i-1}} - \tilde{p}_i = A(t_i)u_i,\ i=1,2,\ldots,N_n \qquad (2.66)$$

with $\tilde{p}_i = p_{i-1}+A(t_{i-1})u_{i-1}-A(t_i)u_i$, hence (by (2.61)) $\|\tilde{p}_i\| \leq \frac{2}{n}$. Inasmuch as $\|u_i-u_o\|_1 \leq r$, $i=o,\ldots,N_n$ and that $A(t)$ satisfies locally H(2.1) in Ch. 1 (see Remark 2.1) it follows from Theorem 3.1 in Chapter 1, that the sequence y_n of step functions

$$y_n(t)=u_o \text{ for } t=t_o \text{ and } y_n(t)=u_i^n \text{ for } t_{i-1}^n < t \leq t_i^n \qquad (2.67)$$

$i=1,\ldots,N_n$ is uniformly convergent on $[o,T_2]$ to a function $u=u(t;t_o,u_o)$. On the other hand, it is easy to check that the difference of y_n and u_n given by (2.67) and (2.59) is estimated by $\frac{2}{n}(M+1)$. Indeed

$$\|y_n(t)-u_n(t)\|_1 \leq \|u_i^n-u_{i-1}^n\|_1 + (t-t_{i-1}^n)(M+\frac{1}{n}) \leq \frac{2}{n}(M+1) \qquad (2.68)$$

for $t \in]t_{i-1},t_i]$, $i=1,\ldots,N_n$ (since $t-t_{i-1} \leq d_{i-1} \leq \frac{1}{n}$).

Thus, (2.68) holds for all $t \in [t_o,T_2]$ and therefore $\lim_{n\to\infty} y_n(t) = \lim_{n\to\infty} u_n(t)=u(t;t_o,x_o)$ in the sense of H^1, uniformly on $[t_o,T_2]$. In this case the DS-limit (integral)solution $u=u\ t;t_o,u_o)$ of (2.49) is even a strong solution of in H^1. This can be proved as follows. First, we know that $u(t;t_o,u_o)\ \overline{D(t)}=D(t)$ in our case here. Similarly to (5.25), y_n

can be written in the form

$$y_n(t)=u_o + \int_{t_o}^{t} A(b_n(s))y_n(b_n(s))ds+g_n(t),\ t\in[t_o,T_2] \quad (2.69)$$

with $g_n(t)$ given by (5.25)' with p_i in place of p_i $(i=o,\dots,N_n$, hence $\|g_n(t)\|\le\frac{T_2}{n}$, $t\in[t_o,T_2]$.

Since $b_n(s)\to b(s)$ in R and $y_n(b_n(s))\to u(s)=u(s;t_o,u_o)$ as $n\to\infty$ in H^1, it follows from (2.69) that

$$u(t;t_o,u_o)=u_o + \int_{t_o}^{t} A(s)u(s);t_o,u_o)ds,\ t_o\le t\le T_2 \quad (2.70)$$

Taking into account H^1-continuity of $s\to A(s)u(s;t_o,u_o)$, (2.70) implies that u is the (unique) strong solution of (2.49) (in H^1) on$[t_o,T_2]$. Part 3°) is a consequence of the uniqueness.

4°) Let $u:[t_o,t_m[\to H^1$ be the maximal solution of (2.49), where $t_m\equiv t_{max}\in[t_o,T[$. Of course if $t_{max}=T$ we have to prove that $\lim_{t\uparrow T} u(t)=y$ exists and then defining u(T)=y it follows u'(T)=A(T)u(T) (where u'(T) denotes the left derivative of u at T in H^1). This aspect will be derived from the proof of the fact that t_m T is a contradiction. Indeed let us suppose that $t_m<T$.

Then by (2.36) we derive (with $v=u(t)=u(t;t_o,u_o)$)

$$(u'(t)-g'(t),u(t)-g(t))_1\le\alpha(t)\|u(t)-g(t)\|_1^2 + \beta(t)\|u(t)-g(t)\|+ \gamma(t) \quad (2.71)$$

for all $t\in[t_o,t_m[$, where $\alpha\ \beta\ \gamma:[t_o,R]\to R_+$ are continuous in the norm of H^1. This is because $\alpha\ \beta\ \gamma$ are expressed in terms of polynomials (of at most second degree) in g(t) (see the proof of (2.36)). Set $h(t)=\|u(t)-g(t)\|_1$, $a_1\ge\max\{\sup\alpha(t), \sup\beta(t), \sup\gamma(t), t_o\le t\le T$. By (2.71) we have

$$\frac{d}{dt}h^2(t)\le 2a_1(h^2(t)+h(t)+1),\ t_o\le t<t_m \quad (2.72)$$

which implies the boundedness of h(t) on $[t_o,t_m]$. Indeed if $h(t)\le 1$ for all $t\in[t_o,t_m[$ we have nothing to prove. Therefore, let us suppose that $h(t)\ge 1$ on $[t_o,t_m[$. Then (2.72) yields

$$\frac{d}{dt}h^2(t)\le 6a_1h^2(t),\ \text{i.e. } h^2(t)\le\|u_o-g(o)\|_1^2\ e^{6a_1(t_m-t_o)}$$

Thus h and therefore u are bounded in H^1 on $[t_o,t_m[$. Since $(s,v)\to A(s)v$ is bounded on bounded sets (see (2.37)) we can use (2.70) for

$t\in[t_o,t_m[$ in order to prove that $\lim_{t\uparrow t} u(t)\equiv y$ exists (in H^1). By using once again (2.70), we conclude that

$$\lim_{t\uparrow t_m} \frac{u(t)-u(t_m)}{t-t_m} = A(t_m)u(t_m)=u'(t_m) \text{ (where } u(t_m)\equiv y)$$

In other words, u is a solution on $[t_o,t_m[$, which contradicts (according to local existence) the significance of t_m. Therefore, the hypothesis $t_m<T$ is false, so $t_m=T$.

5°) If $u_o\in C^2[o,1]$, it follows from (2.48) that u_t, u_x and u_{xxt} exists (classically) and are continuous. Moreover $u(t)\in D(t)$ and

$$u_t(t,x)=h_o'(t)p(x)+h_1'(t)q(x) + \int_o^1 K(x,s)F(u(t,s)ds =$$

$$= (A(t)u(t))(x),\ t\in[o,T],\ x\in[o,1]$$

This implies (using also (2.30))

$$D^2u_t=D^2A(t)u(t)=A(t)u(t)+DF(u(t)), \text{ i.e.}$$

$$u_{txx}(t,x)=u_t(t,x)+(F(u(t,x)))_x$$

Since $u_{txx}=u_{xxt}$, (2.1) follows. It remains to prove the continuous dependence of the solution on the speoified data u_o, h_o and h_1 as indicated at 5°). To this aim set

$$(g_n'(t)v)(x)=h_{o,n}'(t)p(x)+h_{1,n}'(t)q(x)\equiv g_n(t,x)$$

$$(A_n(t)v)(x)=(g_n'(t))(x)+\int_o^1 K(x,s)F(v(s))ds \tag{2.73}$$

for $v\ D_n(t)\equiv\{v\ H^1, v(o)=h_{o,n}(t), v(1)=h_{1,n}(t)\}$

Let u_n be the solution of the problem

$$u_n'(t)=A_n(t),\ u_n(o)=u_{o,n}, u_n(t)\ D_n(t),\ o\leq t\leq T \tag{2.74}$$

Similarly to the proof of boundedness of $u=u(t;t_o,o)$ on $[t_o,t_{max}[$ one proves that u_n is bounded on $[o,T]$. Say $\|u_n(t)\|_1\leq r$, $\|u(t)\|_1\leq r$ for all $t\in[o,T]$ and $|K(x,s)|\leq K_1$, $|DK(x,s)|\leq K_2$ for all $x,s\in[o,1]$.

Then the definition of $\|.\|_1$, along with (2.73), (2.32) and (2.27) lead us to

$$||A_n(t)u_n(t)-A(t)u(t)||_1 \leq ||A_n(t)u_n(t)-A(t)u(t)||_o +$$

$$+ ||D_x(A_n(t)u_n(t)-A(t)u(t)||_o \leq$$

$$||g_n'(t)-g'(t)||_o + K_1M_r||u_n(t)-u(t)||_o +$$

$$||D_x(g_n'(t)-g'(t))||_o + K_2M_r||u_n(t)-u(t)||_o, \quad o \leq t \leq T \tag{2.74'}$$

But $h_{o,n} \to h_o$ and $h_{1,n} \to h_1$ in $C^1[o,T]$, implies $g_n' \to g'$ in $C[o,T]$ as $n \to \infty$. Thus, given $\varepsilon > o$ there exists n_ε such that

$$||g_n'(t)-g'(t)||_o+||D_x(g_n'(t)-g'(t))||_o < \varepsilon, \; n \geq n_\varepsilon, \; t \in [o,T] \tag{2.75}$$

On the other hand, by (2.49) and (2.74),

$$||u_n(t)-u(t)||_1 \leq ||u_{o,n}-u_o||_1 + \int_o^t ||A_n(s)u_n(s)-A(s)u(s)||_1 \, ds \tag{2.76}$$

It is now clear that (2.74)-(2.76) in conjunction with $||u_n(t)-u(t)||_o \leq ||u_n(t)-u(t)||_1$, yield

$$||u_n(t)-u(t)||_1 \leq ||u_{o,n}-u_o||_1 \, e^{rT}, \forall \; n \geq n_\varepsilon \;, \; t \in [o,T] \tag{2.77}$$

with $r=(K_1+K_2)M_r$, which completes the proof.

Remark 2.3. Obviously (e.g. from (2.48) it follows that if the grater regularity of u_o, h_o and h_1 is assumed, the solution u of IBVP acquires correspondingly greater regularity (e.g. if $u_o \in C\ [o,1]$ then $x \to u(t,x)$ is also of class $C^\infty[o,1]$).

Remark 2.4. For a practical numerical scheme the definition of d_i as in (2.57)-(2.58) in not convenient. This definition is necessary for the general case in which $(t,v) \to A(t)v$ in only continuous (see Proposition 2.2 below). However in our case here we can choose $d_i=d(n)$, for $i=o,1,\ldots,N_n$, where $d(n) \in \,]o,\frac{1}{n}]$ is defined by the uniform continuity of h_o and h_1, i.e.

$$|f'(t)-f'(s)| \leq \frac{1}{n}Q^{-1}, \text{ for } |t-s| \leq d(n), \; t,s \in [o,T] \tag{2.78}$$

with $f=h_o$ and $f=h_1$. Set $t_1=t_o+1\ d(n)$. Then

$$||d^{-1}(n)(g(t_i+d(n))-g(t_i))-g'(t_i)||_1 \leq \frac{1}{n} \;, \; i=o,1,\ldots,N_n$$

where $Q=\max \sup(p(x)), \sup(q(x), \; o \leq x \leq 1$ and n_n is the greatest integer satisfying $t_o+N_n d(n) \geq T_1$. Thus, we can work on $[o,T_1]$. Since we have $\|u_i\|_1 \leq \|u_i-u_o\|_1 + \|u_o\|_1 \leq r+\|u_o\|_1 \equiv r_1$ by (2.37) it follows

$$\|A(t_{i+1})u_{i+1}-A(t_i)u_i\|_1 \leq B(M_{r_1}\|u_{i+1}-u_i\|_1 + \|g'(t_{i+1})-g'(t_i)\|_1) \tag{2.79}$$

We may assume that $\|g'(t)-g'(s)\|_1 \leq \frac{1}{n}$ for $|t-s| \leq d(n)$. Then (2.79) yields (2.61) with $B(M+1)M_{r_1}+1)/n$ in place of $1/n$, which doesn't change (2.68) and (2.70).

A general case which guarantees the existence of approximate solutions of the form (2.59) is pointed out in the next proposition below

<u>Proposition 2.2.</u> Suppose that X is a Banach space (of norm $\|.\|$) and the following conditions hold

1°) For each $t \in [o,T[$, $A(t):D(A(t)) \to X$ satisfies

$$\lim_{\lambda \downarrow o} \frac{1}{\lambda} d(v+\lambda A(t)v; D(A(t+\lambda)))=o, \ \forall\ v \in D(A(t)) \tag{2.80}$$

2°) $D(A(t))=\overline{D(A(t))}=D(t)$; $t \to D(A(t))$ is closed (in the sense of H(2.2) in Ch. 1, §2)

3°) $(t,v) \to A(t)v$ is continuous. Let $u_o \in D(A(t_o))$, M,T_1,r,T_2 be as indicated by (2.54) with $\|.\|$ in place of $\|.\|_1$. Then for every positive integer n, there exists u_n of the form (2.59) with d_i and p_i defined as below: $d_i \in]o, \frac{1}{n}]$ is the greatest number $(\leq \frac{1}{n})$ satisfying (2.57) (with $\|.\|_1=\|.\|$) and

$$d[u_i+d_iA(t_i)u_i;D(t_i+d_i)] \leq \frac{d_i}{2n} \tag{2.81}$$

while

$$p_i=(u_{i+1}-u_i-d_iA(t_i)u_i)d_i^{-1} \tag{2.82}$$

where $u_{i+1} \in D(t_i+d_i)$ is chosen such that

$$\|u_i+d_iA(t_i)u_i-u_{i+1}\| \leq \frac{d_i}{n} \text{ (hence } \|p_i\| \leq \frac{1}{n}) \tag{2.82)'}$$

$i=o,1,\ldots,N_n$.

Proof. The only fact that remains to check is that $\lim_{i\to\infty} t_i=t^* < T_1$ leads to a contradiction.

Essentially, the proof follows the same idea as that of (2.64), namely; Choose $d^* \in]o, \frac{1}{2a}]$ such that: $t^*+d^* < T_1$

$$d[u^*+d^*A(t^*)u^*;\ D(t^*+d^*)] \leq \frac{d^*}{4n} \tag{2.83}$$

and

$$\|A(t)n-A(t^*)u^*\| \le \frac{1}{2n}, \quad |t-t^*| \le 3d^*, \quad \|u-u^*\| \le 3d^*(M+1) \quad u \in D(t) \tag{2.84}$$

Set $\delta_i = t^*+d^*-t_i$ (so $\delta_i \quad d_i$ and $\delta_i \downarrow d^*$ as $i \to \infty$). Let i_o with the properties: $t^*-t_i \le d^*$, $\|u_i-u^*\| \le d^*$ for $i \ge i_o$. Then $d_i = t_{i+1}-t_i \le t^*-t_i \le d^*$, and $\delta_i \le 2d^*$ for $i \le i_o$. Consequently (2.84) implies

$$\|A(t)u-A(t_i)u_i\| \le \frac{1}{n}, \text{ for } |t-t_i| \le \delta_i, \|u-u_i\| \le \delta_i(M+1) \; u \in D(t), i \ge i_o \tag{2.85}$$

that is (2.57) holds with $\delta_i \le 2d^* \le \frac{1}{n}$ in place of d_i. Therefore, (2.81) is not satisfied with δ_i in place of d_i, i.e.

$$d[u_i + \delta_i A(t_i)u_i; D(t_i + \delta_i)] > \frac{\delta_i}{2n}, \quad \forall i \ge i_o \tag{2.86}$$

Inasmuch as $t_i + \delta_i = t^*+d^*$, $\delta_i \downarrow d^*$, $t_i \uparrow t^*$ and $u_i \to u^*$, letting $i \to \infty$ in (2.86) we derive that the left hand side of (2.83) is $\ge d^*/2n$ which is a contradiction, q.e.d.

For other precise conditions on $D(A(t))$ see author's book [15].

§3. The semigroup approach to some partial differential equations in L^1

3.1. Phenomena that lead to Porous Medium Equation (PME)

In this subsection we present briefly (following Vasques [1] and Hirsch [1]) some phenomena that imply a process of nonlinear diffusion whose (mathematical) study lead to PME (as a convenient mathematical model). In § 3.2 we will (sketch) show how these PME can be treated via nonlinear semigroups.

3.1.1. Thermal conduction in ionized gases at high temperature

Let Ω be a bounded domain in R^3 with sufficiently smooth boundary Γ (for simplicity say Γ is a C^∞-manifold). Denote by Ω_a an arbitrary subdomain of Ω and suppose that in Ω we have a gas. Set:

$u=u(x,t)$ the temperature at $x=(x_1,x_2,x_3) \in \Omega$ at time t.

$Q^1(t) = \int_{\Omega_a} c(t,x,u)dx$ (with $c(t,x,u) > 0$ for $u > 0$) - the amount of heat in Ω_a at time t.

$Q^2(t) = -\int_{\Omega_a} \vec{G}(t,x,u,\nabla u).\vec{n}\, ds$ - amount of heat outflowing from Ω_a per unit of time where $n=n(x)$ is the outward normal to Γ_a at $x \in \Gamma_a$ and Γ_a is the (sufficiently smooth) boundary of Ω_a.

Clearly ds is the element of area of Ω_a.

$Q^3(t)= \int_{\Omega_a} f(t,x,u)dx$ - amount of heat from sources (or absorption) f in Ω_a per unit of time.

The balance of heat in Ω_a is described by

$$\frac{d}{dt}Q^1(t)=-Q^2(t)+Q^3(t) \tag{3.1}$$

Inasmuch as (formally) we have (supposing we are in conditions of Gauss-Ostrogradski formula)

$$\int_{\Gamma_a} \vec{G}.\vec{n}\, ds= \int_{\Omega_a} \text{div}\ \vec{G}\, dx \tag{3.2}$$

and that Ω_a is arbitrary in Ω, from (3.1) one obtains

$$\frac{\partial}{\partial t} c(t,x,u)=\text{div}\ \vec{G}(t,u,\nabla u)+f(t,x,u) \tag{3.3}$$

where $\nabla u=\text{grad}\ u$, i.e.

$$\nabla u= (\frac{\partial}{\partial x_1} u,\ \frac{\partial}{\partial x_2} u,\ \frac{\partial}{\partial x_3} u) \tag{3.4}$$

and with $(G_1,G_2,G_3)=\vec{G})$

$$\text{div}\ \vec{G}= \frac{\partial}{\partial x_1} G_1 + \frac{\partial}{\partial x_2} G_2 + \frac{\partial}{\partial x_3} G_3 \tag{3.5}$$

As in the case of classical heat-equation model, we may assume $c(t,x,u) \simeq c.u$, where c is specific heat constant.

In the case of high temperature phenomenon, the radiation of heat yields

$\vec{G} = ku^{\alpha}\ \nabla u$, $\alpha \in]4,5;5,5[$, $k > o$ -thermal conduction constant

and therefore

$$\text{div}\ \vec{G}=k.\text{div}\ u^{\alpha}\ \nabla u= \frac{k}{\alpha+1}\text{div}\ (\nabla u^{\alpha+1})= \frac{k}{\alpha+1}\Delta u^{\alpha+1} \tag{3.6}$$

where

$$\Delta. = \frac{\partial^2 .}{\partial x_1^2} + \frac{\partial^2 .}{\partial x_2^2} + \frac{\partial^2 .}{\partial x_3^2} = \text{div}(\text{grad}.) \tag{3.7}$$

is the Laplace operator.

Consequently, neglecting the source term f, (3.3) and (3.6) lead us to the following equation

$$\frac{\partial u}{\partial t} = k_1 \, \Delta u^m \quad . \tag{3.8}$$

with $k_1 = \frac{k}{c(\alpha+1)}$, $m=1+\alpha > 1$ (cf. Zeldovich and Raizer [1]).

3.1.2. The flow of a gas in porous media

Suppose that a gas flows through a porous medium which occupies the domain $\Omega \subset R^3$.

Denote by $u=u(t,x)$ the density of this gas and by $p=p(t,x)$ its pressure ($x \in \Omega, t > o$). Then

$$p=p_0 u^\alpha \, , \; \alpha \geq 1 \; (p_0 \text{ is a constant}) \tag{3.9}$$

If $v=v(t,x,u)$ is the velocity of the gas then Darcy's law gives

$$\mu \vec{v} = -k \; \nabla p=-k \; \nabla \; p_0 u^\alpha \tag{3.10}$$

where $\mu > o$ and $k > o$ denote the viscosity and permeability of the medium (respectively). The dynamic of gas is described by the following conservation law

$$b \; \frac{\partial u}{\partial t} + \operatorname{div}(u\vec{v})=o \tag{3.11}$$

where b is the porosity of the medium ($o < b < 1$)

The laws (3.9)-(3.11) in conjunction with

$$\operatorname{div}(u \nabla u^\alpha) = \frac{\alpha}{\alpha+1} \operatorname{div}\nabla u^{\alpha+1} = \frac{\alpha}{\alpha+1} \Delta u^{\alpha+1}$$

yield

$$\frac{\partial u}{\partial t} = \gamma\Delta(u^m) \text{ with } m= \alpha+1 \geq 2 \tag{3.12}$$

where $\gamma = \alpha k p_0/\mu(\alpha+1)$ (cf. Muskat [1] for details).

The equation (3.8) (and (3.12)) is called "porous medium equation" (PME). Se also (3.50-(3.52).

3.1.3. PME as a mathematical model for population dynamic

Now let us suppose that the bounded domain (habitat) $\Omega \subset R^3$ is occupied by a single numerous "population" (e.g. bacteria). Denote by $u=u(t,x)$ the density (spatial concentration) of this population in $x \in \Omega$ at the time $t > o$.

Arguing as in § 3.1.1 we get

$$\frac{\partial u}{\partial t} = \operatorname{div}(c(u)\nabla u)+F(u) \tag{3.13}$$

where c(u) is the "diffusion coefficient" and F(u) denote the competition (survival, growth rate).

Note that in nonlinear dynamics of population one considers (cf. Gustin, Mac Camy)

$$c(u)=ku^{\alpha}\ ,\ \alpha \geq 1,\ k > o,\ \alpha\ =m-1 \tag{3.14}$$

and therefore (3.13) becomes of type (PME) (3.13).
(In the Pearl-Verhulst logistic law one takes F(u)=au(b-u), with some a,b > o).

Remark 3.1. Equation (3.8) (i.e. PME) has recently been proposed as a mathematical model for many other diffusion processes e.g.: spatial spread (interstellar diffusion) of galactic civilizations, boundary layer theory, the Stefan problem, the finite propagation of signals, and so on. A common feature of all these cases is $m > 1$. The case $o < m < 1$ occurs in the theory of plasmas (cf. Berryman and Holland [1]). In this case, by the virtue of (3.14) we have

$$c(u)=k\ u^{m-1} \to \infty \quad \text{as } u \to o\ (o < m < 1)$$

For this reason, (3.13) with c(u) given by (3.14) with $o < m < 1$ is said to be the "fast diffusion" model, while for $m > 1$ "finite velocity" model.

Let us derive (3.13) in the case of R instead of R^3. Therefore suppose that a single numerous species ("population") occupies a long narrow tube T. Such a tube can be modelled by the interval $[x_1, x_2] \equiv T \subset R$. As in (3.13) denote by u=u(t,x) the population density at time t (at $x \in T$). We first suppose that the population (e.g. bacteria) neither procreate nor die but wander randomly over T (without crossing the endpoints x_1 and x_2). Let x_o be an interior point of T. Then for $h \in]o, x_2 - x_o[$ we have $[x_o, x_o+h] \subset]x_1, x_2[$. The "amount" of bacteria (at time t) in $[x_o, x_o+h]$ is $\int_{x_o}^{x_o+h} u(t,s)ds$. A typical plausible hypothesis is to assume that the rate of migration out of $[x_o, x_o+h]$ across x_o is proportional to $-u_x(t,x_o)$ (while that of migration into $[x_o, x_o+h]$ across x_o+h is proportional to $u_x(t, x_o+h)$. Accordingly, the time-rate change of the amount of bacteria in $[x_o, x_o+h]$ is given by

$$\frac{d}{dt}\int_{x_o}^{x_o+h} u(t,s)ds = \int_{x_o}^{x_o+h} \frac{\partial}{\partial t}u(T,s)ds = c(u_x(t,x_o+h)-u_x(t,x_o))$$

where $c=c(x_o)$ is a proportionality factor.

Dividing by h and then letting $h \to o$ one obtains

$$u_t = c(x)u_{xx} \tag{3.15}$$

with Neumann boundary condition

$$u_x(x_1) = u_x(x_2) = o \tag{3.16}$$

Now, if at each $x \in T$ we have a growth rate (owed to procreation or death) $g=g(x,u)$ (depending only on x and $u=u(t,x)$) then to the time-rate changes caused by migration we have to add $g(x,u)$ (i.e. the change owed to the growth). In this case we are lead to the equation

$$u_t = c(x)u_{xx} + g(x,u), \quad x \in T, \quad t \geq o \tag{3.17}$$

with the Neumann (no-flux) boundary conditions (3.16).

If the population does not live in the tube $T \subset R$ but in the habitat $\Omega \subset R^3$, then by a similar reasoning we are led to the "reaction-diffusion" equation

$$u_t = c(x)\,\Delta u + g(x,u), \tag{3.18}$$

with the Neumann condition

$$\left.\frac{\partial u}{\partial n}\right|_{\Gamma} = o \tag{3.19}$$

Celarly, (3.18) is similar to (3.13) (namely, for $c(u)=c(x)$ = const., and $F(u)=g(x,u)$, (3.13) is just (3.18)). However, (3.13) is more complicated than (3.18).

One could model similarly several interacting propulation (species) occupying the same habitat Ω (cf. Capasso and Maddalena [1], Aronsson and Mellander [1], Morris W. Hirsch [1] and the corresponding bibliography). A semigroup approach to population dynamics can be found in the recent book of G.F. Webb [2].

§ 3.2. Some results on PME

In some problems (e.g. the water infiltration) we are led to replace u^m in (3.8) by a more general nonlinearity of the form $g(u)$. Thus we get a more general model

$$u_t = \Delta g(u) \tag{3.20}$$

Of course we need suitable initial and boundary conditions in order to have a "well-posed" problem (this notion has been introduced by Hadamard). Recall that a problem for (3.20) is said to be "well-posed" on an interval I, if the solution exists, is unique on I and (among other properties) depends continuosuly on the initial data. This definition comes from the general case

$$u' \in Au, \quad o \leq t \tag{3.21}$$

with the initial condition (see Ch. 2, § 1, Ch. 1, § 1)

$$u(t_o)=u_o \in \overline{D(A)}, \; t_o > o \tag{3.21)'}$$

As we will see, for concrete PDE , the boundary conditions are included in the domain of D(A) (e.g. for $A=\Delta$ with $D(A)=H_o^1(\Omega)\cap H^2(\Omega)$, $u_o \in D(A)$ means (among other conditions) a Dirichlet boundary condition

$$u_o(x)=o, \text{ a.e. on } \Gamma \tag{3.22}$$

or, in short,

$$u_o\big|_\Gamma = o \tag{3.22)'}$$

More general boundary conditions have the form

$$\frac{\partial u}{\partial n}(x) \in -b(u(x)), \text{ a.e. on } \Gamma \tag{3.23}$$

where $\frac{\partial u}{\partial n}(y)$ denotes the directional derivative of u in the outward normal direction at $y \in \Gamma$ (in short - the normal derivative) and b is a maximal monotone (accretive) set of $R \times R$. Therefore, there is a proper lower semicontinuous convex function $j: R \to]-\infty,+\infty]$ such that $b = \partial f$ (Appendix, Corollary 1.1). Thus (3.23) is equivalent to

$$\frac{\partial u}{\partial n}(x) \in -\partial f(u(x)), \text{ a.e. on } \Gamma \tag{3.24}$$

Condition (3.23) contains many important boundary conditions, e.g.

For $b(r)=o$, $\forall r \in R$, (3.23) becomes $\frac{\partial u}{\partial n}(x)=o$ a.e. on Γ

(a Neumann or "no-flux" boundary, condition).

If $b(r)=\emptyset$ for $r\neq o$, and $b(r)=R$ for $r=o$ then (3.23) means just (3.22) (a Dirichlet condition)

Newton's law

$$\frac{\partial u}{\partial n} = -k(u-u_o)\Big|_\Gamma \quad , k > o \tag{3.25}$$

corresponds to $b(r)=k(r-u_o)$, $k > o$ (where u_o is the surrounding temperature) In connection with (3.20) we give the following result

Theorem 3.1. Let $T > o$ and $g\in C(R) \cap C^1(R\setminus\{o\})$ with $g(o)=o$ and $g'(s) > o$ for $s\neq o$. Then for every $f \in L^1(o,T;L^1(\Omega))$ and $u_o \in L^1(\Omega)$, the problem

$$u_t = \Delta g(u)+f, \text{ a.e. in }]o,T[\times \Omega \tag{3.26}$$

$$u(t,x)=o, \text{ a.e. on }]o,T[\times \Gamma \tag{3.27}$$

$$u(o,x)=u_o(x), \text{ a.e. in } \Omega \tag{3.28}$$

has a unique integral solution $u \in C([o,T];L^1(\Omega))$

For the proof of this theorem we need

Lemma 3.1. Let g be as in Theorem 3.1. Then the operator A defined by

$$D(A)= \{u\in L^1(\Omega);\ g(u)\in W_o^{1,1}(\Omega), \Delta g(u)\ L^1\in(\Omega)\} \tag{3.29}$$

$$Au = \Delta g(u), \text{ for } u \in D(A) \tag{3.30}$$

is m-dissipative (in $L^1(\Omega)$) with $\overline{D(A)}=L^1(\Omega)$.

Proof. Note first that $\Delta g(u)$ in (3.30) means the Laplace operator of $g(u)$, in the sense of distributions.

Let us prove the dissipativity of A in $L^1(\Omega)$. Denote by $\|.\|$ the norm of $L^1(\Omega)$. Then we have to check that

$$\|u-v\|_1 \leq \|u-v-t(Au-Av)\|_1, \ \forall\ t > o, u,v \in D(A) \tag{3.31}$$

(see Proposition 2.1. in Appendix).

To this goal, choose $f_n \in C^1(R)$ with the properties

$$f_n(o)=o,\ |f_n(s)| \leq 1,\ f_n'(s) \geq o,\ \lim_{n\to\infty} f_n(s)=\text{sign } s,\ \forall s \in R \tag{3.32}$$

where sign o=o (In general one sets sign o=[-1,1]).
Such a sequence is e.g.

$$f_n(s) = \frac{ns}{n|s|+1}\ ,\ s\in R \left(\text{i.e. } f_n'(s) = \frac{n}{(n|s|+1)^2}\ ,\ s\in R\right) \tag{3.33}$$

We now take $u,v\in D(A)$. Then

$$\operatorname{grad} f_n(g(n)-g(v))=f_n'(g(n)-g(v)) \operatorname{grad} (g(u)-g(v))$$

and so, by virtue of Green formula, we have

$$\int_\Omega (Au-Av)f_n(g(u)-g(v))dx = -\int_\Omega f_n'(g(u)-g(v))|\operatorname{grad}(g(u)-g(v))|^2 dx \leq o$$

Consequently, for every $t > o$

$$\int_\Omega (u-v)f_n(g(u)-g(v))dx \leq \int_\Omega (u-v-t(Au-Av))f_n(g(u)-g(v))dx$$

$$\leq \int_\Omega |u-v-t(Au-Av)|dx= \|u-v-t(Au-Av)\|_1 \tag{3.34}$$

Inasmuch as $g'(s) > o$ for $s\neq o$ it follows

$$\operatorname{sign} (g(u(x))-g(v(x)))=\operatorname{sign}(u(x)-v(x))$$

and therefore, letting $n\to\infty$ in (3.34) one obtains (3.31). It remains to prove that for each $f\in L^1(\Omega)$, there exists a (unique) $u\in D(A)$, such that

$$u-\Delta g(u)=f \tag{3.35}$$

To this goal set $b(s)=g^{-1}(s)$. Thus, with $v=g(u)$ (3.35) is equivalent to

$$b(v)-\Delta v=f \tag{3.36}$$

According to the previous step, the Laplace operator with

$$D(\Delta) =\{v\in W_o^{1,1}(\Omega),\quad \Delta v\in L^1(\Omega)\} \tag{3.37}$$

is dissipative in $L^1(\Omega)$. Since g is maximal monotone in R, so is $b=g^{-1}$. On the basis of a result of Brézis-Strauss [1], (3.36) has a unique solution $v\in D(\Delta)$. Then $u=g^{-1}(v)$ is the solution of (3.35). Clearly, $D(A)\supset\{u\in L^1(\Omega);\ g(u)\in D(\Delta)\}$ (which is dense in $L^1(\Omega)$) and therefore $\overline{D(A)}=L^1(\Omega)$. This completes the proof.

Proof of Theorem 3.1. One applies Theorem 10.1. in Ch. 2 with $X=L^1(\Omega)$ and A as in Lemma 3.1.

Remark 3.1. In (3.36) the operator b is actually the (natural) realization B of b on $L^1(\Omega)$. In order to be more precise, let $b\subset R\times R$ be maximal monotone with $b(o)=o$. Set

$D(B)=\{u\in L^1(\Omega);\ u(x)\in D(b)$ a.e. on Ω, $\exists\, v\in L^1(\Omega)$ such that

$$v(x)\in b(u(x)) \text{ a.e. on } \Omega\} \tag{3.38}$$

and for $u\in D(B)$, define

$$Bu=\{v\in L^1(\Omega),\ v(x)\in b(u(x)),\ \text{a.e. on } \Omega\} \tag{3.39}$$

It is a simple exercice to check that the realization B of b in $L^1(\Omega)$ is also m-accretive (like b in R). In the case g=I (the identity on R) Lemma 3.1. yields

Corollary 3.1. The realization of Δ in $L^1(\Omega)$ with the domain given by (3.37) is m-dissipative in $L^1(\Omega)$.

Let us comment a generalization of Corollary 3.1. To this goal, set

$$Lu = -\sum_{i,j=1}^{3} (a_{ij}u_{x_i})_{x_j} + \sum_{i=1}^{3} (a_i u)_{x_i} + au \tag{3.40}$$

where $(u)_{x_i}$ denotes the partial derivative of u with respect to x_i, i=1,2,3. In what follows one can replace R^3 by R^N for $N\geq 1$ (N a positive integer). Suppose that

$$a_i,\ a_{ij}\in C^1(\bar{\Omega}),\ a\in L^\infty(\Omega);\ a+\sum_i (a_i)_{x_i}\geq o,\ a\geq o,\ \text{a.e. on } \Omega \tag{3.41}$$

There is a positive number $k>o$ such that

$$\sum_{i,j} a_{ij}y_iy_j\geq k|y|^2,\ \text{a.e. on } \Omega,\ y=(y_1,y_2,y_3)\in R^3 \tag{3.42}$$

where $|y|$ denotes the euclidean norm of y. Set

$$D(A)=\{u\in W_o^{1,1}(\Omega),\ Lu\in L^1(\Omega);\ Au=Lu,\ u\in D(A)\} \tag{3.43}$$

This operator A is the realization of L in $L^1(\Omega)$.

Note that the derivatives involved in Lu are understood in the sense of distributions. In other words, the following two assertions are equivalent

(I) $u\in D(A)$ and $Lu=f$

(II) $u\in W_o^{1,1}(\Omega)$ and for all $v\in W_o^{1,\infty}(\Omega)$

$$\sum_{i,j} (a_{ij}\frac{\partial u}{\partial x_i}, \frac{\partial v}{\partial x_j}) - \sum_j (a_j u, \frac{\partial v}{\partial x_j}) + (au,v)=(f,v)$$

where $(f,v)=\int_\Omega f(x)v(dx)$ and so on.

Finally, denote by A_p the realization of L on L^p with $1<p<\infty$, i.e.

$$D(A_p)=\{W^{2,p}(\Omega)\cap W_o^{1,p}(\Omega);\ A_pu=Lu \text{ for } u\in D(A_p),\ 1<p<\infty \qquad (3.44)$$

The following result holds

Theorem 3.2. (1) $1<p<\infty$, $-A_p$ is the infinitesimal generator of a contraction semigroup in $L^p(\Omega)$.

(2) The operator $-A$ defined by (3.43) is also the infinitesimal generator of a contraction semigroup in $L^1(\Omega)$

(3) $\sup(I+\lambda A)^{-1}f \leq \max\{o, \sup_\Omega f\}$, $\forall\ \lambda>o$, $f\in L^1(\Omega)$

(4) For $1<q<N/(N-1)$, we have $D(A)\subset W_o^{1,q}(\Omega)$ and

$d\|u\|_{1,q} \leq \|Au\|_1$, for $u\in D(A)$, where $d=d(q)$ is a positive constant

(5) A is the closure in $L^1(\Omega)$ of A_2.

The proof of Part (1) can be found in Agmon, Douglis and Nirenberg [1]. parts (2)-(5) are proved in Brézis and Strauss [1] .

Let us go back to the operator $A=\Delta g$ defined by (3.30). The following results hold, with $\Omega\subset R^N$ and

$$\|u\|_p=\int_\Omega |u(x)|^p dx^{\ 1/p},\quad u\in L^p(\Omega),\ p\geq 1 \qquad (3.44)'$$

Theorem 3.3. In addition to the hypotheses of Theorem 3.1, suppose that g satisfies

$$g'(s)>M_1|s|^{b-1},\quad s\neq o, \qquad (3.45)$$

where M_1 is a positive constant, $b>o$ if $N\leq 2$ and $b>(N-2)/N$ if $N\geq 3$. Then (1) The semigroup $S(t):L^1(\Omega)\to L^1(\Omega)$ generated by Δg in $L^1(\Omega)$ is compaxt (for $t>o$). (2) $S(t):L^1(\Omega)\to L^\infty(\Omega)$. (3) There exists a constant $M>o$ such that for each $p\geq 1$,

$$u\in L^p(\Omega) \text{ and } t>o$$

$$\|S(t)u\|_1 \leq M^{-N(2p+N(b-1))}\|u\|_p^{\ 2p/(2p+N(b-1))}$$

A detailed proof of Part (1) can be found in the book of Vrabie [3]. The inequality from (3) is proved by Baras. In the proof of the compactness of S(t), one proves that for each bounded subset B of $L^1(\Omega)$, there exist a constant $C=C(B)$ o such that

$$\|S(t+h)u-S(t)u\|_1 \leq C\left(\frac{h}{t}\right)^{1/2},\quad \forall\ h\geq o,\ t>o \qquad (3.46)$$

for all $u\in B$. This inequality implies the equicontinuity of $t\to S(t)u$ at $t>o$ on bounded subset of $L^1(\Omega)$.

In more general boundary conditions, the equation (3.26) was studied systematically by Benilan (in his thesis [2]). Namely, let us replace Dirichlet condition (3.27) by a general condition of type (3.23), i.e.

$$-\frac{\partial}{\partial n}g(u)\in \beta(u) \text{ on }]0,T[\times \Gamma \tag{3.47}$$

where β is a maximal monotone graph of $R\times R$.

In this case D(A) given by (3.29) has to be replaced by
$D(A_{\beta})=\{u\in L^1(\Omega);\ g(u)\in W^{1,1}(\Omega),\ \Delta g(u)\in L^1(\Omega)$, and

$$\frac{\partial g(u)}{\partial n}(x)\in \beta(u(x)),\text{ a.e. on } \Gamma\} \tag{3.48}$$

Set

$$A_{\beta}u=\Delta g(u),\text{ for } u\in D(A_{\beta}) \tag{3.49}$$

Note that the trace of u on Γ makes sense if $u\in D(A_{\beta})$. Indeed, since $g(u)\in W^{1,1}(\Omega)$, $g(u)|_{\Gamma}$ is defined so we may consider $u|_{\Gamma} = g^{-1}(g(u)|_{\Gamma})$

Roughly speaking, Benilan [2] is concerned with (3.26) with A_{β} in place of A (i.e. with (3.27) replaced by (3.47)).

A typical exmaple of g satisfying Hypotheses of Theorem 3.1 is the following one (see e.g. Ames [1])

$$g(r)=r|r|^{m-1},\ r\in R,\ m\geq 1 \tag{3.50}$$

Indeed, in this case

$$g'(r)=m|r|^{m-1},\ r\in R,\ m\geq 1 \tag{3.51}$$

with $r\neq 0$ if $0<m<1$ (hence (3.45) is satisfied).

The case m=2 (i.e. $g(r)=r|r|$) has been studied by Lions. Therefore, in general (when u(x) is not a.e. positive in Ω) we have to replace (3.12) by

$$\frac{\partial u}{\partial t} = \gamma\Delta u|u|^{m-1} \tag{3.52}$$

4. Other examples of m-dissipative operators and compact semigroups

In the previous section we have seen that the operator g is defined by (3.30), is densely defined and m-dissipative in $L^1(\Omega)$ (Lemma 3.1). Moreover, under additional restriction (3.45), the semigroup S(t) generated by Δg is compact.

It is the purpose of th is section to give other examples of m-dissipative (accretive, or maximal monotone) operators in $L^p(\Omega)$, with $p \geq 1$ and Ω - a bounded set of R^N with smooth enoufh boundary Γ . For some conditions on Γ we refer to Adam's book [1].

Example 4.1. Let b be a maximal monotone graph of $R \times R$. The there exists a l.s.c. function $j: R \to]-\infty, +\infty]$ such that $b = \partial j$ (Corollary 1.1 in Appendix). Set

$$D(A) = \{u \in H^2(\Omega): -\frac{\partial u}{\partial n}(x)\ b(u(x)) \text{ a.e. on } \Gamma\} ; \tag{4.1}$$

$$Au = -\Delta u, u \in D(A)$$

First we have to emphasize that $u \in H^2(\Omega)$ implies $u|_\Gamma \in H^{3/2}(\Omega)$ and therefore $\frac{\partial u}{\partial n} \in H^{1/2}(\Gamma) \subset L^2(\Gamma)$ where $\frac{\partial u}{\partial n}$ is defined by (3.23), and $u \to (u|_\Gamma, \frac{\partial u}{\partial n})$ is continuous from $H^2(\Omega)$ into $H^{3/2}(\Omega) \times H^{1/2}(\Gamma)$ (cf. Lions and Magenes [1], vol. 1). Thus, the boundary condition on Γ in (4.1) makes sense. We now define

$$f(u) = \begin{cases} \frac{1}{2}\int_\Omega |\text{grad } u|^2\, dx + \int_\Gamma j(u) d\sigma & , \text{ if } u \in H^1(\Omega) \\ & \text{ and } j(u) \in L^1(\Gamma) \\ +\infty & , \text{ otherwise} \end{cases} \tag{4.2}$$

When there is no danger of confusion we will write

$$\int_\Omega g(x)dx = \int_\Omega g dx,\ W^{k,p}(\Omega) = W^{k,p} \text{ and so on} \tag{4.2)'}$$

The relationship between f and $\Delta = A$ is given by

Theorem 4.1. (1) The functional $f: L^2(\Omega) \to]-\infty, +\infty]$ is convex and lower semicontinuous (l.s.c.) and its subdifferetial ∂f is just the operator A defined by (4.1) (i.e. $\partial f = -\Delta$ on D(A)), hence A is maximal monotone in $L^2(\Omega)$.

(2) There is a constant $c > o$ such that

$$\|u\|_{H^2(\Omega)} \leq c(1+ \|u- \Delta u\|_{L^2(\Omega)}), \ \forall u \in D(A) \qquad (4.3)$$

(3) If in addition j is positively valued (i.e. $j:R \to [0,\infty]$) then the semigroup $S(t)$ generated by A is compact. Finally if $j(0)=0$, then $D(A)=L^2(\Omega)$ (in words, A is densely defined in $L^2(\Omega)$.

Sketch of the proof: (1) The convexity of f is obvious (since j is convex). The lower-semicontinuity of f in $L^2(\Omega)$, is equivalent to the fact that for each $r > 0$, the level set

$$C_r = \{ u \in L^2(\Omega) : f(u) \leq r \} \qquad (4.4)$$

is closed in $L^2(\Omega)$. Let us prove this latter property. Take $\{u_n\} \subset C_r$ such that $u_n \to u$ in $L^2(\Omega)$ and prove that $u \in C_r$. Indeed, for such a sequence we may assume (relabeling if necessary) that $u_n \to u$ in $H^1(\Omega)$ (this because we may assume that j is positively valued. Recall that j is bounded from below by an affine function - Appendix §1, so we may replace $j(x)$ by $j(x)-ax-a_1 \geq 0$, $\forall x \in R$ for some a, $a_1 \in R$). Moreover, we may assume that $\lim_{n\to\infty} u_n(x)=u(x)$, a.e. in Ω. Thus we have

$$\|u\|_{H^1(\Omega)} \leq \liminf_{n\to\infty} \|u_n\|_{H^1(\Omega)} \qquad (4.5)$$

and by Fatou's lemma along with lower semicontinuity of j,

$$\int_\Gamma j(u(x))d\sigma \leq \liminf_{n\to\infty} \int_\Gamma j(u_n(x))d\sigma \leq r \qquad (4.6)$$

i.e. $u \in H^1(\Omega)$ and $j(u) \in L^1(\Gamma)$. Combining (4.5), (4.6) with the hypotheses $u_n \to u$ in $L^2(\Omega)$ and $f(u_n) \leq r$ we get $f(u) \leq \liminf_{n\to\infty} f(u_n) \leq r$, so $u \in C_r$. Furthermore, for $u,v \in D(A)$

$$f(v)-f(u) = \frac{1}{2}\int_\Omega (|\text{grad } v|^2 - |\text{grad } u|^2)dx +$$

$$+\int_\Gamma (j(v)-j(u))d\sigma \geq \int_\Omega \text{grad } u(\text{grad } v-\text{grad } u)\, dx -$$

$$-\int_\Gamma \frac{\partial u}{\partial n}(v-u)d\sigma = -\int_\Omega \Delta u(v-u)dx \qquad (4.7)$$

where the last inequality is due to Green's formula. Therefore, (4.7) holds for all $v \in L^2(\Omega)$, so $A \subset \partial f$. In order to prove that $A= \partial f$ it suffices to prove that A is maximal monotone in $L^2(\Omega)$, i.e. that $R(I+A) =L^2(\Omega)$.This means to show that for every $f \in L^2(\Omega)$, there exists $u \in H^2(\Omega)$

such that

$$u - \Delta u = f, \text{ a.e. in } \Omega, \ -\frac{\partial u}{\partial n} \in b(u), \text{ a.e. on } \Gamma \tag{4.8}$$

This is a delicate problem, which is solved by Brézis in his thesis [2, pp. 49, 43]. The idea is to replace b by the Lipschitz continuous operator b_λ (Yosida approximation of b, i.e. $b_\lambda = (1-(1+\lambda b)^{-1})/\lambda$) and to find an estimate for the solution u_λ of the problem

$$u_\lambda - \Delta u_\lambda = f, \text{ in } \Omega : -\frac{\partial u}{\partial n} = b_\lambda(u_\lambda) \text{ on } \Gamma \tag{4.9}$$

The main part of the proof is the estimate below (see also Barbu [2, p. 65,66]

$$\|u_\lambda\|_{H^2(\Omega)} \leq c(1+\|f\|_{L^2(\Omega)}) \tag{4.10}$$

with $c > 0$ independent of λ. Since $H^2(\Omega)$ is compactly imbedded into $L^2(\Omega)$, we may assume that $u_\lambda \to u$ in $L^2(\Omega)$, $u_\lambda \to u$ in $H^2(\Omega)$, $\frac{\partial u_\lambda}{\partial n} \to \frac{\partial u}{\partial n}$ in $L^2(\Gamma)$ and $(1+\lambda b)^{-1}u_\lambda = J_\lambda u_\lambda \to u$ in $L^2(\Gamma)$. Since $b_\lambda(u_\lambda) \in b(J_\lambda u_\lambda)$ it follows $-\frac{\partial u}{\partial n} \in b(u)$ on Γ. Thus, letting $\lambda \downarrow 0$ in (4.9) one obtains (4.8). Moreover, $\|u\|_{H^2(\Omega)} \leq \liminf_{\lambda\downarrow 0} \|u_\lambda\|_{H^2(\Omega)}$ in conjunction with (4.8) and (4.10) yield (4.3).

(3) If j is positively valued, then the level set (see Remark 6.1 in Ch. 2)

$$C_r = \{u \in L^2(\Omega);\ \|u\|_2 + f(u) \leq r\} \tag{4.11}$$

is relatively compact in $L^2(\Omega)$. Indeed, C_r is a bounded subset of $H^1(\Omega)$ (which is compactly imbedded in $L^2(\Omega)$). Consequently, the semigroup S(t) generated by $\partial f = A$ is compact (Theorem 6.2 in Ch. 2). Finally, if in addition $j(0)=0$, then $0 \in \partial j(0) = b(0)$ and hence $D(A) \supset C_o^\infty(\Omega)$ (which is dense in $L^2(\Omega)$).

<u>Example 4.2.</u> In the case

$$j(r) = \begin{cases} 0 & , \text{ if } r=0 \\ +\infty, & \text{if } r \neq 0 \end{cases} \tag{4.12}$$

we have

$$D(\partial j) = \{0\}, \partial j(0) =]-\infty, +\infty[\equiv R \tag{4.13}$$

(i.e. $\partial j(r) = \emptyset$ if $r \neq 0$). Inasmuch as $b = \partial j$, (4.1) and (4.2) become respectively

$$D(A) = H^2(\Omega)\cap H_0^1(\Omega),\ Au = -\Delta u, u\in D(A) \tag{4.14}$$

$$f(u) = \begin{cases} \frac{1}{2}\int_\Omega |\text{grad } u|^2 dx, & \text{if } u\in H_0^1(\Omega) \\ +\infty & \text{, otherwise} \end{cases} \tag{4.15}$$

From Theorem 4.1 one derives

<u>Corollary 4.1.</u> The linear operator (4.14) is a densely defined maximal monotone operator in $L^2(\Omega)$ and $A = \partial f$ with f given by (4.15). Moreover, the semigroup S(t) of class C_o generated by -A is compact, and (4.3) holds.

Of course, the direct proof of Corollary 4.1 is much more simple than that of Theorem 4.1. This is left to the reader. Note that S(t) is an analytic semigroup (Pazy [4, Th. 7.2.7]).

In what follows we shall use the notations

$$Q =]0,+\infty[\times\Omega,\ F_\ell =]0,+\infty[\times\Gamma \tag{4.16}$$

Clearly, F is the lateral boundary of the cylinder Q(Γ - the boundary of Ω). In view of the maximum principle (cf. Brézis [5, p. 211])

$$\text{Min}\,\{0, \inf_\Omega u_o\} \le (S(t)u_o)(x) \le \text{Max}\,\{0, \sup_\Omega u_o\},\ (t,x)\in Q \tag{4.17}$$

for every $u_o\in L^2(\Omega)$ (where S(t) is the C_o semigroup generated by (4.14)) it follows

<u>Corollary 4.2</u>. Let $u_o\in L^2(\Omega)$ and let S(t) be the C_o-semigroup generated by $-\Delta$ on $H^2(\Omega)\cap H_0^1(\Omega)$. Then the following properties hold

(1) $u_o \ge 0$ a.e. on Ω implies $S(t)u_o \ge 0$ in Q

(2) $\|S(t)u_o\|_\infty \le \|u_o\|_\infty$, $\forall u_o\in L^\infty(\Omega)$ (i.e. $\|S(t)\|_\infty \le 1$

In other words, Property (2) asserts that the restriction of S(t) to $L^\infty(\Omega)$ is also a contraction (nonexpansive) semigroup. Property (1) implies the fact that

$$S(t) : D_+ \to D_+,\ \forall\, t \ge 0 \tag{4.17)'$$

i.e. S(t) has the "positivity preserving" property (see J.A. van

Casteren [1], for the theory of such semigroups, which are called - positivity preserving semigroups). Here

$$D_+ = \{ u_o \in L^2(\Omega), \ u_o \geq 0, \text{ a.e. in } \Omega \} \tag{4.18}$$

Other properties of S(t) are given in the next section. Note that in the special case $\Omega = R^N$ we have an explicit representation of S(t), namely

$$(S(t)u_o)(x) = \frac{1}{(4\pi t)^{N/2}} \int_{R^N} e^{-\|x-y\|^2/4t} u_o(y)dy, \ t \geq 0, \ x \in R^N \tag{4.18'}$$

for every $u_o \in L^2(R^N)$. The proof is given in the next section.

Example 4.3. With b as in Example 4.1, define

$$D(B_p) = \{u \in H^2(\Omega) \cap W^{1,p^*}(\Omega) \colon u \in L^p(\Omega), \ -\frac{\partial u}{\partial n} \in b(u) \text{ a.e. on } \Gamma\} \tag{4.19}$$

and $B_p = -\Delta u$ for $u \in D(B_p)$ where $\frac{1}{p^*} = \frac{1}{p} - \frac{1}{N}$ if $1 \leq p < N$, and $p^* = +\infty$ if $p \geq N$. Then B_p is m-accretive in $L^p(\Omega)$ (cf. Brézis [2, p.60]).

Example 4.4. Pseudo-Laplace operator. For $p \geq 2$ define the operator $A \colon W_o^{1,p}(\Omega) \to W^{-1,q}(\Omega)(\frac{1}{p} + \frac{1}{q} = 1)$ by

$$< Au, v > = \sum_{i=1}^{N} \int_{\Omega} \left|\frac{\partial u}{\partial x_i}\right|^{p-2} \frac{\partial u}{\partial x_i} \frac{\partial v}{\partial x_i} dx, \ u, v \in W_o^{1,p}(\Omega) \tag{4.20}$$

It is easy ot check that A is the subdifferential (even the Gâteaux derivative) of the functional $f \colon W_o^{1,p}(\Omega) \to R$,

$$f(u) = \frac{1}{p} \sum_{1}^{N} \int_{\Omega} \left|\frac{\partial u}{\partial x_i}\right|^p dx, \ u \in W_o^{1,p}(\Omega) \tag{4.21}$$

Now let us denote

$$\Delta_p u = \sum_{i=1}^{N} \frac{\partial}{\partial x_i} \left(\left|\frac{\partial u}{\partial x_i}\right|^{p-2} \frac{\partial u}{\partial x_i}\right) \tag{4.22}$$

and

$$D(\Delta_p) = \{u \in W^{1,p}(\Omega); \ \Delta_p u \in L^2(\Omega), \ -\frac{\partial u}{\partial n_p}(x) \in b(u(x)), \text{ a.e. on } \Gamma\} \tag{4.23}$$

where $b = \partial j$ and

$$\frac{\partial u}{\partial n_p} = \sum_{i=1}^{N} \left|\frac{\partial u}{\partial x_i}\right|^{p-2} \frac{\partial u}{\partial x_i} \cos(\vec{n}, e_i) \tag{4.24}$$

Here $\vec{n} = n(x)$ is the outward normal to Γ at $x \in \Gamma$, while $\{\vec{e}_i, i=1,\dots,N\}$

is the canonical base in R^N.

Finally define $f_p : L^2(\Omega) \to]-\infty, +\infty]$ by

$$f_p(u) = \begin{cases} \frac{1}{p} \sum_{i=1}^{N} \int_\Omega |\frac{\partial u}{\partial x_i}|^p \, dx + \int_\Gamma j(u(x))d\sigma, & \text{if } u \in W^{1,p}(\Omega) \text{ and } j(u) \in L^1(\Gamma) \\ + \infty & \text{, otherwise} \end{cases} \tag{4.25}$$

Following the proof of Theorem 4.1 (1) we can prove that the "Pseudo-Laplace" operator Δ_p given by (4.22) (with $D(\Delta_p)$ as in (4.23)) is maximal dissipative in $L^2(\Omega)$ and we actually have

$$- \Delta_p u = \partial f_p(u), \; u \in D(\Delta_p) \tag{4.26}$$

In the case p=2 one obtains Example 4.1.

Note that we can consider

$$\tilde{L}_p u = \Delta_p u - u|u|^{p-2} \tag{4.27}$$

in place of Δ_p (see Vrabie [3]). In this case we have to add to f_p the term

$$g_p(u) = \frac{1}{p}||u||_p^p = \frac{1}{p} \int_\Omega |u(x)|^p dx, \; u \in L^p(\Omega) \tag{4.28}$$

Indeed, the subdifferential (which in this case is just Gâteaux differential) of g_p is given by

$$\partial g_p(u) = u|u|^{p-2}, \; u \in L^p(\Omega) \tag{4.29}$$

(see also Proposition 1.5 in Appendix). Note that (4.29) can be proved in two ways. The first one: for $u,v \in L^p(\Omega)$ $(p \geq 2)$ we have

$$\frac{d}{dt} g_p(u+tv)\Big|_{t=0} = \int_\Omega |u(x)|^{p-1} \text{ sign } u(x)v(x)dx \tag{4.30}$$

and (4.29) follows (see Pavel [15, pp. 12-14] for details. The second one: for $u \in L^p(\Omega)$ one proves that

$$\frac{1}{p} \int_\Omega |v(x)|^p dx - \frac{1}{p}\int_\Omega |u(x)|^p dx \geq \int_\Omega u(x)|u(x|^{p-2}(v(x) - \\ - u(x))dx, \;\; \forall \; v \in L^p(\Omega) \tag{4.31}$$

This is a consequence of some elementary remarks along with Young's inequality (which yields)

$$\int_\Omega |v|\,|u|^{p-1}dx \le \frac{1}{p}\int_\Omega |v|^p + \frac{1}{q}\int_\Omega |u|^{(p-1)q}\,dx \quad (\frac{1}{p}+\frac{1}{q}=1) \tag{4.32}$$

Or, (4.31) shows that $u|u|^{p-2}$ (which belongs to $L^q(\Omega)$) is in $\partial g_p(u)$. Since ∂g_p is single valued, (4.29) is established.

Example 4.5. Let $b\subset R\times R$ be a maximal monotone set with $0\in b0$ and let B^2 be the realization of b in $L^2(\Omega)\equiv L^2$ i.e.

$$B^2 = \{[u,v]:\ u,v\in L^2(\Omega),\ v(x)\in b(u(x)) \text{ a.e. on } \Omega\} \tag{4.33}$$

It is easy to check that B^2 is maximal monotone in L^2, $0\in B0$ and $(B^2_\lambda u)(x)=b_\lambda(u(x))$ a.e. in Ω ($b_\lambda 0 = B^2_\lambda 0 = 0$). Set

$$A_2u = -\Delta u + B^2u, \text{ for } u\in D(A_2)= H^2(\Omega)\cap H^1_0(\Omega)\cap D(B^2) \tag{4.34}$$

Equivalently

$$D(A_2)= \{ u\in H^2(\Omega)\cap H^1_0(\Omega);\ \Delta u+B^2u\in L^2(\Omega)\tag{4.35}$$

Theorem 4.2. (1) The operator A_2 defined by (4.34) is maximal monotone (m-accretive) in $L^2(\Omega)$. (2) The operator A_1 below (with b single-valued - for simplicity of writing)

$$A_1u = -\Delta u+Bu, \text{ for } u\in D(A_1) = \{ u\in W^{1,1}_0(\Omega);\ -\Delta u+Bu\in L^1(\Omega)\} \tag{4.36}$$

is m-accretive in $L^1(\Omega)$ (where $(Bu)(x)=b(u(x))$, a.e. in Ω).

Proof. (1) We will make use of Theorem 8.6 in Ch. 2 with $A= -\Delta$, $D(A)= =H^1_0(\Omega)\cap H^2(\Omega)$ and $H = L^2(\Omega)$. Therefore, let us consider the (maximal monotone) operator $-\Delta + B^2$. For $f\in L^2(\Omega)$ and $\lambda>0$, denote $u_\lambda = u_\lambda(f)\in H^2(\Omega)\cap H^1_0(\Omega)$ the unique solution (in $L^2(\Omega)$) of the equation

$$u_\lambda - \Delta u_\lambda + B^2_\lambda u_\lambda = f \text{ in}, \ u_\lambda = 0 \text{ on } \Gamma \tag{4.37}$$

and prove that $B^2_\lambda u_\lambda$ is bounded, namely

$$\|B^2_\lambda u_\lambda\|_2 \le \|f\|_2, \quad \forall\ \lambda>0 \tag{4.38}$$

To this aim we first observe that by Green formula,

$$\int_\Omega B^2_\lambda u_\lambda \Delta u_\lambda\, dx = -\int_\Omega \nabla B^2_\lambda u_\lambda \nabla u_\lambda\, dx = -\int_\Omega |\nabla u_\lambda|^2 b'_\lambda(u_\lambda(x))dx \le 0 \tag{4.38'}$$

Multiplying (4.37) by $B^2_\lambda u_\lambda$ and using (4.38)' along with $u_\lambda B^2_\lambda u_\lambda \ge 0$ a.e. in Ω, one obtains (4.38). Passing to the limit as $\lambda\downarrow 0$ in (4.37) (on the basis of Theorem 8.4 in Ch. 2) we get $u_\lambda \to u$ in $L^2(\Omega)$ with

$$u\in D(A_2),\ f\in u - \Delta u + B^2u \tag{4.39}$$

that is $R(I+A_2) = L^2$.

(2) For proving the m-accretivity of A_1 in $L^1(\Omega)$ we proceed as in the proof of Lemma 3.1. Let f_n be as in (3.32). First we take an arbitrary $g_i \in L^2(\Omega)$. Then, according to the previous step, there is $u_i \in D(A_2)$ such that

$$u_i - \lambda\Delta u_i + \lambda B u_i = g_i, \; i = 1,2 \tag{4.40}$$

By Green's formula,

$$-\int_\Omega f_n(u_1-u_2)\,\Delta(u_1-u_2)dx = \int_\Omega |\nabla(u_1-u_2)|^2 f_n(u_1-u_2)dx \geq 0 \tag{4.41}$$

A simple combination of (4.40) and (4.41) along with the properties of f_n lead us to

$$\int_\Omega (u_1-u_2)f_n(u_1-u_2)dx + \lambda\int_\Omega (Bu_1-Bu_2)f(u_1-u_2)dx \tag{4.42}$$

$$\leq \int_\Omega (g_1-g_2)f_n(u_1-u_2)dx \leq \int_\Omega |g_1(x)-g_2(x)|dx = \|g_1-g_2\|_1$$

where $f(u_1-u_2) \equiv f(u_1(x)-u_2(x))$ and $(Bu_i)(x) \equiv b(u_i(x))$ a.e. in Ω, $i=1,2$. Letting $n \to \infty$ we get

$$\|u_1-u_2\|_1 + \lambda\|Bu_1-Bu_2\|_1 \leq \|g_1-g_2\|_1 \tag{4.43}$$

Now let h be arbitrary in $L^1(\Omega)$. Then there is a sequence $h_n \in L^2(\Omega)$ with $h_n \to h$ in $L^1(\Omega)$ as $n \to \infty$. Let u_n be the corresponding solution of (4.40) with h_n in place of g_i, i.e.

$$u_n - \lambda\Delta u_n + Bu_n = g_n, \; u_n \in D(A_2), \; n=1,2,\ldots \tag{4.44}$$

According to (4.43), we have

$$\|u_n-u_m\|_1 + \lambda\|Bu_n - Bu_m\|_1 \leq \|g_n-g_m\|_1 \tag{4.45}$$

which shows that u_n and Bu_n are convergent in L^1 as $n \to \infty$. Say $u_n \to u$ in L^1. Then it follows $Bu_n \to Bu$ in L^1 and by (4.44) we have $\Delta u_n \to w$ in L^1 as $n \to \infty$. Since Δ with $D(\Delta)$ given by (3.37) is m-dissipativity in L^1 (hence it is closed) it follows that u $W_o^{1,1}(\Omega)$ and $w = \Delta u$. We conclude that $u \in D(A_1)$ and

$$u - \lambda\Delta u + \lambda Bu = h, \text{ i.e. } R(I+\lambda A_1) = L^1, \; \forall \lambda > 0$$

It remains to show the accretivity of A_1 in L^1. Let $u_1, u_2 \in D(A_1)$. Set $g_i = u_i + \lambda A_1 u_i$, $i=1,2$. In this case (4.43) still holds (by a similar proof) so

$$\|u_1-u_2\|_1 \leq \|u_1-u_2 + \lambda(A_1u_1-A_1u_2)\|_1 \tag{4.46}$$

which means just the accretivity of A in L^1 (see Appendix, Proposition 2.7).

Example 4.6. Let C= C([0,1]; R) be the sapce of all continuous functions u: [0,1] → R endowed with supremum norm $\| \|_C$. Define the linear unbounded operator

$$Au=-u' \text{ for } u\in D(A) = \{u\in C;\ u'\in C;\ u(0) = 0\} \tag{4.47}$$

where $u'(x) = u_x$ is the derivative of u at $x \in [0,1]$.

It is easy to check that A is the infinitesimal generator of the translation semigroup $S(t) : C \to C$ given by

$$(S(t)u(x)= \begin{cases} u(x-t), & \text{if } x-t \in [0,1] \\ 0 & , \text{if } x-t \in [0,1] \end{cases} \tag{4.48}$$

where $u\in C$, $t \geq 0, x\in[0,1]$.

Clearly, this C_o-semigroup is not compact, e.g. $S(\frac{1}{2})$ does not map bounded subsets of C into relatively compact subsets of C. However, $(\lambda I-A)^{-1}=J_\lambda$ is a compact operator. This is easily seen from the representation of J_λ, namely

$$(J_\lambda f)(x)= \frac{1}{\lambda}\int_0^x e^{-\frac{1}{\lambda}(x-s)}f(s)ds,\ f\in C, \lambda>0,\ x\in[0,1] \tag{4.49}$$

The integral representation (4.49) of J_λ follows from the fact that $J_\lambda f= u$ is the solution of the equation

$$u(x)+ \lambda u'(x)=f(x),\ u(0)=0 \tag{4.50}$$

Finally, the compactness of J_λ is a consequence of Arzela's lemma. Indeed, let M be a bounded subset of C (say $\|f\|_C \leq q,\ \forall\ f\in M$). Then $J_\lambda M$ is bounded ($\|J_\lambda f\|_C \leq q$, $\forall f\in M$, since J_λ is nonexpansive). Moreover, the functions $J_\lambda M$ are equicontinuous. This is because, from (4.50) we have

$$(J_\lambda f)'(x)\leq \frac{1}{\lambda}|f(x)-(J_\lambda f)(x)| \leq \frac{2}{\lambda}\|f\|_C \leq \frac{2}{\lambda}q,\ \forall\ \ f\in M$$

This example shows that Conditions (I) and (II) in Theorem 6.1 of Brézis (Ch. 2) are independent even in the linear case (Theorem 6.3 of Pazy, Ch. 2). Indeed, in this case (I) holds while (II) does not. Conversely, in the case of subdifferentials (II) holds while (I) may not hold.

The evolution equation associated with A is

$$u_t = Au = -u_x \text{ which yields } u_{tt} = u_{xx} \tag{4.51}$$

so we are in "Hyperbolic" case.

In the case of $A = -\Delta$ given by (4.14) $-A = \Delta$ is the infinitesimal generator of a compact C_o-semigroup, so the corresponding evolution equation is $u_t = \Delta u$ (which is of Parabolic type, see § 5). The conclusion is

Proposition 4.1. Let A be defined by (4.47). Then a) $(\lambda I - A)^{-1} \equiv J_\lambda$ (with $\lambda > 0$) is compact. b) The semigroup $S(t)$ (as defined by (4.48)) generated by A is not compact.

Example 4.7. In the study of waves equation (next section) the following unbounded operator plays a crucial role

$$AU = \begin{pmatrix} -v \\ -\Delta u \end{pmatrix} \equiv \begin{pmatrix} 0 & -I \\ -\Delta & 0 \end{pmatrix}\begin{pmatrix} u \\ v \end{pmatrix}, \quad U = \begin{pmatrix} u \\ v \end{pmatrix} \in H_o^1(\Omega) \times L^2(\Omega) \equiv H \tag{4.52}$$

with

$$D(A) = (H^2(\Omega) \cap H_o^1(\Omega)) \times H_o^1(\Omega) \tag{4.53}$$

We will discuss the following three cases
(1) Ω - a bounded domain (of R^N) with smooth boundary Γ. (2) Ω - a domain of R^N (not necessarily bounded) with smooth Γ (e.g. $\Omega = R_+^N$).
(3) $\Omega = R^N$.

Case (1). In view of Poincaré inequality

$$\|u\|_p \le k\|\nabla u\|_p, \ \forall\, u \in W_o^{1,p}(\Omega), \ 1 \le p < \infty \tag{4.54}$$

it follows that $\|u\|_{1,p} = \|\nabla u\|_p$ is a norm on $W_o^{1,p}(\Omega)$ and this norm is equivalent to the norm of $W^{1,p}(\Omega)$. Therefore, we consider on $H = H_o^1 \times L^2$ the inner product

$$\langle U_1, U_2 \rangle = \int_\Omega \nabla u_1 \nabla u_2 dx + \int_\Omega v_1 v_2 dx, \text{ where } U_1 = \begin{pmatrix} u_1 \\ v_1 \end{pmatrix}, \tag{4.55}$$

$U_2 = \begin{pmatrix} u_2 \\ v_2 \end{pmatrix} \in H$ and norm: $\|U_1\|^2 = \int_\Omega |\nabla u_1|^2 dx + \int_\Omega |v_1|^2 dx$

In view of Green formula, (4.52) and (4.55) yield

$$\langle AU, U \rangle = -\int_\Omega \nabla v \, \nabla u \, dx - \int_\Omega (\Delta u) v \, dx = 0, \ \forall\, U = \begin{pmatrix} u \\ v \end{pmatrix} \in D(A) \tag{4.56}$$

so A and -A are monotone. It is easy to prove that

$$R(I-A) = R(I+A) = H(H=R(A+tI),\ \forall\ t\in R) \tag{4.57}$$

so A and -A are maximal monotone in H.

Let us show for example that R(I-A)=H. This means to prove that for $\binom{f}{g}\in H$, there are $\binom{u}{v}\in D(A)$ such that

$$u+v = f,\ v+\Delta u = g \tag{4.58}$$

This system is equivalent to

$$v = f - u,\ u - \Delta u = f - g \tag{4.58)'$$

Since $f - g\in L^2(\Omega)$ and $R(I-\Delta) = L^2$ with $D(\Delta) = H^2\cap H_o^1$, there is $u\in H^2\cap H_o^1$ satisfying $u - \Delta u = f-g$ (Corollary 4.1). Then $v = f-u\in H_o^1(\Omega)$ and (4.58) holds. Similarly one proves that R(I+A)=H. Let $S_A(t)$, $t\geq 0$ be the semigroup generated by A. We can extend S_A to $]-\infty, 0]$ as follows

$$S_A(-t) = S_{-A}(t) \text{ for } t\geq 0 \tag{4.59}$$

In other words we have defined

$$S(t)\begin{cases} S_A(t) & ,\text{ if } t\geq 0 \\ S_{-A}(t) & ,\text{ if } t\leq 0 \end{cases} \tag{4.60}$$

It is easy to prove that S is a group of unitary operators on H, i.e. $S(0) = I$, $S(t+s) = S(t)S(s)$, $\forall t,s\in R$

$$\|S(t)U\| = \|U\|,\ \forall U\in H,\ \forall\ t\in R \tag{4.61}$$

Say for example that $t > 0$ and $U\in D(A)$. Then $v(t) = S(t)U = S_A(t)U$ is the strong solution of

$$v'(t)=Av(t),\ v(0) = U,\ t\geq 0 \tag{4.62}$$

Since $\langle Av(t),v(t)\rangle = 0$, $t\geq 0$, by (4.62) we get

$\frac{d}{dt}\|v(t)\|^2 = 0$, $\forall\ t>0$ and therefore $\|v(t\| = \|v(0)\|$, $\forall t\geq 0$, i.e. $\|S(t)U\| = \|U\|$, for all $t\geq 0$ and $U\in D(A)$. Since $D(A)=H$ we actually have $\|S(t)U\| = \|U\|$ for all $U\in H$ and thus the last property in (4.61) is proved for $t\geq 0$ (the case $t\leq 0$ proceeds similarly). Set

$$F(t) = S_A(t)S_{-A}(t)x,\ t\geq 0,\ x\in D(A) \tag{4.63}$$

One verifies that $F'(t)=AF(t)-AF(t)=0$, $\forall t\geq 0$. Since $F(0)=x$, it follows $F(t)=x$, $\forall\ t\geq 0$. In such a manner one proves that

$$S_A(t)S_{-A}(t)x = S_{-A}(t)S_A(t)x = x, \ \forall\, t \geq 0, \ x \in H \tag{4.64}$$

which yields $S(t+s) = S(t)S(s)$, $\forall$ $t,s \in R$.

It is also easy to check that

$$< AU, V > = < U, -AV > , \ \forall\, U,V \in D(A) \tag{4.65}$$

which implies $A^* = -A$, i.e. A is skew-adjoint.

It is actually known that A is skew-adjoint if and only if A and -A are maximal monotone in H. Since $A^*=-A$ we have $S_A^*(t)=S_{A^*}(t)=S_{-A}(t)$, and by (4.63), $S_A^{-1}(t)=S_{-A}(t)=S_A^*(t)$ (i.e. $S_A(t)$ is unitary). In other words we have proved

Corollary 4.3. The operator A defined by (4.52) (with Ω bounded) is a maximal monotone skew-adjoint operator and (consequently) the semigroup $S_A(t)$, $t \geq 0$ generated by A can be extended to a group $S(t)$ (as indicated by (4.60)) of unitary operators.

Remark 4.1. The property $\|S_A(t)U\| = \|U\|$, $\forall t \geq 0$, $U \in H$ in (4.61) can be interpreted as a conservation of energy (see § 6).

Case (2). If Ω is not bounded, then we have to use the usual inner product (and the corresponding norm) on H, i.e. in place of (4.55) we set

$$(U_1,U_2) = \int_\Omega \nabla u_1 \nabla u_2 \, dx + \int_\Omega u_1 u_2 \, dx + \int_\Omega v_1 v_2 \, dx \tag{4.66}$$

where $U_i = \begin{pmatrix} u_i \\ v_i \end{pmatrix} \in H$, $i=1,2$. Consequently, in this case

$$\|U_1\|^2 = \int_\Omega |\nabla u_1|^2 dx + \int_\Omega u_1^2 \, dx + \int_\Omega v_1^2 \, dx = \|u_1\|_{1,2}^2 + \|v_1\|_2^2 \tag{4.67}$$

where $\|u_1\|_{1,2}$ ($\|v_1\|_2$) denote the norm of $H_o^1(\Omega)$ and $L^2(\Omega)$ respectively.

Corollary 4.4. The operator A given by (4.52) has the properties (a) $A+tI$ is surjective for every $t \in R$ (i.e. $R(A+tI)=H$) (b) $A+I$ is maximal monotone (in H).

Proof. The proof of Property (a) is similar to that of (4.57) and relies on $R(\lambda I-\Delta)=L^2(\Omega)$ with $D(\Delta) = H^2(\Omega) \cap H_o^1(\Omega)$. It remains to prove that $A+I$ is monotone (accretive). Indeed for every $U= \binom{u}{v} \in D(A)$ we have

$$(AU,U)+\|U\|^2 = -\int_\Omega \nabla v \, \nabla u \, dx - \int_\Omega vu \, dx - \int_\Omega v \, \Delta u \, dx + \int_\Omega |\nabla u|^2 \, dx + \int_\Omega (u^2+v^2)dx=$$

$$= \int_\Omega (u^2+v^2 -uv)dx + \int_\Omega |\nabla u|^2 dx \geq 0 \ \text{(since } u^2+v^2-uv \geq \tfrac{1}{2}(u^2+v^2) \geq 0).$$

Case (3). $\Omega = R^N$. In this case the right space is $H = H^1(R^N) \times L^2(R^N)$. The operator A is defined as in (4.52). The norm of H is that given by (4.67) i.e.

$$\|U\|^2 = \int_{R^N} (|\nabla u|^2 + |u|^2 + |v|^2)dx, \text{ where } U = \binom{u}{v} \in H \qquad (4.68)$$

The operator A is defined by

$$AU = \binom{-v}{-\nabla u}, \ U = \binom{u}{v} \in D(A) = H^2(R^N) \cap H^1(R^N) \qquad (4.69)$$

By using the Fourier transform $\hat{f}$ of $f \in L^2(R^N)$, i.e.

$$(\zeta f)(\xi) \equiv \hat{f}(\xi) = (2\pi)^{-N/2} \int_{R^N} e^{-i <x,\xi>} f(x)dx \qquad (4.70)$$

one proves that

$$\|(I + A)^{-1}F\| \le (1-2|\lambda|)^{-1}\|F\|, \ F \in H, \ \lambda \in R, \ 0 < |\lambda| < 1/2 \qquad (4.71)$$

which implies (cf. Pazy [4], pp. 219-223).

Theorem 4.3. The operator A given by (4.69) is the infinitesimal generator of a C_o-group $S(t)$ on $H^1(R^N) \times L^2(R^N)$ (with $\|S(t)\| \le e^{2t}$, $t \in R$).

Example 4.8. (Schrödinger Operator). Define

$$Au = i\Delta u, \text{ for } u \in D(A) = H^2(R^N). \qquad (4.72)$$

this operator is called a Schrödinger operator, since it arises in the Schrödinger equation

$$\frac{1}{i} \frac{\partial u}{\partial t} = \Delta u - Vu \qquad (4.73)$$

where V is a potential in $L^2(R^N)$ (cf. Pazy [4] p. 224).

Corollary 4.5. The operator iA is selfadjoint in $L^2(R^N)$ (and therefore by Stone's theorem, A is the infinitesimal generator of a group of unitary operators $S(t)$, $t \in R$ in $L^2(R^N)$).

Proof. Let $u, v \in H^2(R^N)$. Then according to the definition of inner product in $L^2(R^N)$ and the integration by parts,

$$< \Delta u, v > = \int_{R^N} (\Delta u)\bar{v} \, dx = \int_{R^N} u\Delta\bar{v} \, dx = < u, \Delta v >$$

so $iA = -\Delta u$ is symmetric ($iA \subset (iA)^*$). In order to have $iA = (iA)^*$ it suffices to prove that $R(\lambda I - iA)$ is dense in $L^2(R^N)$ (for Im $\lambda \ne 0$). Take $f \in C_o(R^N)$. Then the solution u to the equation $(\lambda I - IA)u = f$ is given by $\lambda\hat{u} + |\xi|^2\hat{u} = \hat{f}(\xi)$ i.e.

$$u(x) = (2\pi)^{-N/2} \int_{R^N} \frac{f(\xi)e^{i<x,\xi>}}{\lambda + |\xi|^2} d\xi , \quad x \in R^N$$

so $R(\lambda I - iA) \supset C_o(R^N)$ (which is dense in $L^2(R^N)$). This completes the proof. Note that in the case N=2, the group S(t) of unitary operators in $L^2(R^2)$ has the following remarkable explicit form

$$S(t)u_o(x) = \frac{1}{4\pi ti} \int_{R^2} e^{i\frac{\|x-y\|^2}{4t}} u_o(y)dy, \quad u_o \in L^2(R^2), t \in R, \ x \in R^2 \tag{4.74}$$

The integral representation (4.74) can be easily found taking into account that $u(t,x) = S(t)u_o)(x)$ is the solution of the problem

$$\frac{\partial u}{\partial t} = i\Delta u, \ u(0,x) = u_o(x), \ x \in R^2, \ t \in R, \ u_o \in L^2(R^2) \tag{4.75}$$

This (classical) solution can be determined by the Fourier transform method. A direct verification of the fact that $u(t) = S(t)u_o$ (given by (4.74)) is the solution of (4.75), is also a useful exercice (which is left to the reader).

Example 4.9. Let us consider the space

$$X = \{u; \ u \text{ continuous from } [0,1] \text{ into } R, \ u(0)=u(1)\} \tag{4.76}$$

endowed with the supremum norm. Set

$$D(A) = \{u \in C^2([0,1]); \ u,u',u'' \in X\}, \ Au = u'' \text{ for } u \in D(A) \tag{4.77}$$

Theorem 4.4. The operator A defined by (4.77) is the infinitesimal generator of a compact (and analytic) semigroup on X.

The proof can be found in Pazy [4, p. 234]. A more detailed proof is given in the author's book [15, p. 220].

Example 4.10. Let $V \in L^\infty(R^N) + L^{N/2}(R^N)$, $N \geq 3$ be a real potential. Set $H = L^2(R^N, C)$ (the square integrable complex-valued functions on R^N) and

$$D(A) = \{u \in H \cap H^1_{loc}(R^N); \ u + Vu + ku \log(|u|^2) \in H\} \tag{4.78}$$

$$Au = -i(\Delta u + Vu + ku \log(|u|^2)), \ u \in D(A), \ k \in R \tag{4.79}$$

Theorem 4.5. The operator A + 2 k I is maximal monotone in H.

The proof of this surprising result is carried out in the book of Haraux [1, pp. 118-125]. Recall also a result on "convex integrands" which is useful in convex analysis. Let $g: R \to [-\infty, +\infty]$ be a lower semicontinuous convex proper (l.s.c.) function. Define

$$f(u)=\begin{cases}\int_\Omega g(u(x))dx, & \text{if } g(u)\in L^1(\Omega)\\ +\infty & \text{, otherwise}\end{cases}$$

Then we have

Proposition 4.2. Let Ω be a bounded domain in R^N. The functional $f: L^2(\Omega) \to]-\infty, +\infty]$ is l.s.c. and $F\in \partial f(u)$ iff $F(x)\in \partial g(x)$ a.e. in Ω. Moreover

$$D(f) = \{u\in L^2(\Omega);\ u(x)\in D_e(g),\ \text{a.e. in } \Omega\}$$

5. Partial differential equations of parabolic type

5.1. The heat equation

Let us recall the abstract framework. Denote by $A:D(A)\subset H\to H$ the infinitesimal generator of the C_o-semigroup $S(t)$ with $\|s(t)\| \le 1$. This means that A is maximal dissipative in the real Hilbert space H (of inner product $<\ ,\ >$ and norm $\|.\|$). It is well-known that $S(t):D(A)\to D(A)$ and that the function

$$u(t) = S(t)u_o,\ u_o\in D(A),\ t \ge 0 \tag{5.1}$$

is of class $C^1([0, +\infty[;H)$. Namely,

$$u'(t) = S(t)Au_o = AS(t)u_o,\ t \ge o,\ u_o\in D(A) \qquad (5.1)'$$

so u is the unique strong solution of the problem

$$u'(t) = Au(t),\ u(0) = u_o\in D(A),\ t > 0 \tag{5.2}$$

Inasmuch as A is closed, it follows that D(A) is a Banach space with respect to the graph norm

$$\|u\|_{D(A)} = \|u\| + \|Au\|,\ u\in D(A) \tag{5.3}$$

(See (6.31) and (6.32) in Ch. 1 for the relationship between D(A) and S(t)). Clearly, a D(A)-valued continuous function $u:[0,\infty[\to H$ is continuous in the norm of D(A) iff $t\to Au(t)$ is continuous in H. Set

$$C(R_+, D(A)) = \{u; u: R_+ \to D(A),\ u \text{ continuous in the norm } \|.\|_{D(A)}\} \qquad (5.3)'$$

$$D(A^2) = \{x \in D(A);\ Ax \in D(A)\},\ A^2x = A(Ax),\ x \in D(A^2) \qquad (5.4)$$

Recursively one defines $D(A^m)$, $m = 3,4,\ldots$ i.e.

$$D(A^m) = \{u \in D(A^{m-1}),\ A^{m-1}(u) \in D(A)\},\ A^m u = A(A^{m-1}u),\ u \in D(A^m) \qquad (5.4)'$$

We are in a position to state the following result (which holds in every real Banach space X).

Theorem 5.1. Let A be the infinitesimal generator of the C_o-contraction semigroup $S(t)$ (on the real Banach space X). Then for every $u_o \in D(A)$ the problem (5.2) admits a unique solution u (given by (5.1)) with the properties:

1°) $u(t) \in D(A)$, $\forall t \geq 0$, 2°) $u \in C^1(R_+;X) \cap C(R_+, D(A))$

3°) $\|u(t\| \leq \|u_o\|$, $t \geq 0$, 4°) $\|u'(t)\| \leq \|Au_o\|$, $\forall\ t \geq 0$

5°) If in addition $u_o \in D(A^m)$, then $u^{(k)} \in C(R_+, D(A))$, $k=1,\ldots,m-1$ and $u \in C^m(R_+;X)$ for every positive integer $m \geq 1$. Moreover $\|u^{(k)}(t)\| \leq \|A^k u_o\|$, $k = \overline{1,m}$.

Proof. 1°) We have $Au(t) = S(t)Au_o$, so $t \to Au(t)$ is also continuous. Therefore $u \in C(R_+, D(A))$. Estimates 3°) and 4°) follows trivially from (5.1) and (5.1)'. Finally, if $u_o \in D(A^m)$ then

$$u^{(k)}(t) = Au^{(k-1)}(t) = S(t)A^k u_o,\ k=1,2,\ldots,m \qquad (5.5)$$

which implies 5°), q.e.d.

Note that

Lemma 5.1. $D(A^2)$ is dense D(A) (in the graph norm), i.e. for every $x \in D(A)$ there is $x_n \in D(A^2)$ such that

$$\|x_n - x\|_{D(A)} \to 0 \text{ as } n \to \infty$$

Proof. For $x \in D(A)$ set $x_n = J_{1/n}x$. Then $Ax_n = AJ_{1/n}\,x = J_{1/n}Ax$ and $x_n \to x$, $Ax_n \to Ax$ in X (i.e. $\|x_n - x\|_{D(A)} \to 0$ as $n \to \infty$) q.e.d.

Theorem 5.2. Suppose (in addition to Hypotheses of Theorem 5.1) that $X=H$ and $A=A^*$ (i.e. A is selfadjoint). Then in addition to the properties 1°)-5°) the solution $u(t)=S(t)u_o$ of $u' = Au$ satisfies

(a) for $u_o \in H$, $u\ \ C([0,+\infty[;H) \cap C^1([0,+\infty[;H) \cap C([0,+\infty[;D(A))$

(b) $S(t): H \to D(A)$, $\forall t > 0$, $t \to S(t)$ is continuous at $t > 0$ in the uniform operator topology,

(c) $\|u'(t)\| = \|AS(t)u_o\| \le \frac{1}{t}\|u_o\|$, $\forall\, t>0$, $u_o \in H$ (5.6)

(d) $u \in C^k([0,+\infty[;H)$, $u^{(k)} \in C([0,+\infty[;D(A))$, $k=1,2,\ldots$ with $u^{(k)}(t) = Au^{(k-1)}(t)$ $(u \in C^k(]0,\infty[;D(A^k))$, $t>0$

(e) $S(t)u_o \rightharpoonup a$ (weakly) as $t \to \infty$, with $a \in A^{-1}o$ (i.e. $Aa=0$)

Proof. In this case $A = -\partial f$ with f given by (1.27) in Appendix. Therefore, on the basis of Theorem 1.7 in Ch.2, $S(t):\overline{D(A)} \to D(A)$. Or in our case here $\overline{D(A)} = H$. Moreover since $A_o=A$, the inequality (7.25) in Ch. 2 becomes

$$\|AS(t)u_o\| \le \frac{1}{t}\|u_o-y\| + \|Ay\|,\ \forall y \in D(A),\ u_o \in H \tag{5.7}$$

Thus Properties (b) and (c) are proved (see also (5.30)'"). Now let $t>0$. Let us show for example that $u''(t)$ exists. To this goal, take r such that $0 < 2r < t$. Then for $u_o \in H$,

$$u(t) = S(t)u_o = S(t-2r)S(2r)u_o,$$
$$u'(t)= AS(t)u_o = S(t-2r)S(r)AS(r)u_o = S(t-r)AS(r)u_o \in D(A)$$
$$\text{i.e. } u'(t) \in D(A^2)$$
$$u''(t)=Au'(t)=S(t-2r)AS(r)AS(r)u_o \in D(A), \text{ so } u''(t) \in D(A^3),\ t>2r$$

It is now clear that $u(t)$ and $Au'(t)$ are continuous in t so $u' \in C([0,+\infty[;D(A))$. In such a manner we conclude that (a) and (d) are also valid. Property (e) is a consequence of Bruck's theorem (Theorem 9.3 in Ch. 2).

Remark 5.1. Here by $u'(t)$ we mean the strong derivative of u evaluated at $t>0$, i.e.

$$u'(t) = \lim_{h\to 0} (u(t+h)-u(t))/h = \frac{d}{dt}S(t)u_o$$

If $t=0$, then $u'(0)$ denotes the right derivative of u at $t=0$ i.e.

$$u'(0)=\lim_{h\downarrow 0} \frac{u(h)-u(0)}{h} = \frac{d^+}{dt}S(t)u_o\Big|_{t=0}$$

(see Remark 6.8 in Ch. 2 for details). Now let us consider the heat equation

$$\frac{\partial u}{\partial t}(t,x) = \Delta u(t,x) \text{ in } Q \tag{5.8}$$

where Q (and F_ℓ) is defined by (4.16).

Other significance of (5.8) were given in § 3. We are going to recall some fundamental results on the existence, uniqueness and regularity of solution (by semigroup approach) to (5.8) under the following initial-boundary conditions

$$u(t,x) = 0 \quad , \text{ on } F_\ell \tag{5.9}$$

$$u(0,x) = u_o(x) \quad , \text{ in } \Omega \tag{5.10}$$

Instead of Dirichlet condition (5.9), we can also impose the Neumann condition

$$\frac{\partial u}{\partial n} = 0 \text{ on } F_\ell \tag{5.11}$$

or more general

$$- \frac{\partial u}{\partial n}(t,x) \in b(u(t,x)), \text{ a.e. on } F_\ell \tag{5.12}$$

(see (4.1) for notations). Actually "on F_ℓ" 'or "in Ω") means in general "a.e. on F_ℓ" (and "a.e. in Ω"). Let $S_\Delta = S(t)$ be the semigroup generated by Δ as indicated by Corollary 4.1.

As an applicaiton of Theorem 5.2 we can prove

Theorem 5.3. Let $u_o \in L^2(\Omega)$. Then there exists a unique function $u(t,x) = (S_\Delta(t)u_o)(x)$, $x \in \Omega$, $t \geq 0$ verifying (5.8), (5.9) and (5.10) and (with $\infty \equiv +\infty$)

$$u \in C([0,\infty[;\ L^2(\Omega)) \cap C(]0,\infty[;H^2(\Omega) \cap H_o^1(\Omega)) \tag{5.13}$$

$$u \in C^k(]0,\infty[;L^2(\Omega)),\ \frac{\partial^k u}{\partial_t^k} \in C(]0,\infty[;H^2(\Omega) \cap H_o^1(\Omega))\ k=1,2,\ldots \tag{5.14}$$

$$\|\frac{\partial u}{\partial t}(t)\| = \|\Delta u(t)\|_2 \leq \frac{1}{t}\|u_o\|_2, \quad t > 0 \tag{5.15}$$

$$u \in C^\infty([a,+\infty[\times\bar\Omega) \text{ for all } a>0 \text{ (if } u_o \in C_o^\infty(\Omega)\text{, then } a=0 \tag{5.16}$$

$$S(t)u_o \to 0 \text{ in } L^2(\Omega) \text{ as } t \to \infty \text{ (i.e. } \int_\Omega |u(t,x)|^2 dx \to 0 \text{ as } t \to \infty) \tag{5.17}$$

Proof. One applies Theorem 5.2 with $H=L^2(\Omega)$ and $A=\Delta$ as defined by (4.14). It is well-known that A is self-adjoint in $L^2(\Omega)$. Indeed, for all $u,v \in D(A)$ we have

$$\langle Au, v\rangle = \int_\Omega (\Delta u)v\,dx = -\int_\Omega (\nabla u)(\nabla v)dx = \int_\Omega u\,\Delta v\,dx = \langle u, Av\rangle \tag{5.17'}$$

that is A is symmetric (and maximal monotone in L^2) hence $A=A^*$ so $u(t) \in D(A^k)$, $k=0,1,\ldots$, $t>0$.

According to Theorem 5.2, the solution $u(t)=S(t)u_o$ (with $S(t) \equiv S_\Delta(t)$) has the property $u \in C^k(]0,\infty;D(A^\ell))$ for every $k, \ell = 0,1,\ldots$ with $D(A^o) = D(A)$. On the other hand $D(A^k) \subsetneq H^{2k}$ (Brézis [5, p. 206) and

$$H^{k+m}(\Omega) \subsetneq C^m(\Omega),\ k > \frac{N}{2},\ m=0,1,\ldots \tag{5.18}$$

Therefore $u \in C^k(]0,\infty[;C^k(\Omega))$, $k=0,1,\ldots$ so (5.16) holds. If $u_o \in C_o^\infty(\Omega)$, then $u_o \in D(A^\ell)$ for all positive integers , so we have

$$u'(t) = S(t)Au_o,\ u''(t) = S(t)A^2u_o = Au'(t),\ \forall\ t \geq 0,\ \text{a.s.o.}$$

It follows $u \in C^k(]0,\infty[;D(A^\ell))$ on which in turn yields $u \in C^k([0,\infty[;C^k(\Omega))$. Finally, the fact that the solution $u(t)=S(t)u_o$ tends in $L^2(\Omega)$ to an equilibrium point $y \in H^2 \cap H_o^1$ (with $\Delta y=0$ (i.e. (5.17)) follows from Corollary 9.2 in Ch. 2. Clearly $\Delta y=0$ with $y \in H^2 \cap H_o^1$ yields $y=0$ in Ω q.e.d.

<u>Theorem 5.4.</u> For $u_o \in L^2(\Omega)$, the corresponding solution $u(t) = S(t)u_o$ of the problem (5.8), (5.9), (5.10) satisfies: $u \in L^2(0,\infty;H_o^1(\Omega))$ and

$$\frac{1}{2}\|u(T)\|^2 + \int_0^T \|\nabla u(t)\|_2^2\,dt = \frac{1}{2}\|u_o\|_2^2,\ \forall\ \ T>0 \tag{5.19}$$

$$\|S(t)u_o\|_p \leq \|u_o\|_p,\ \forall u_o \in L^p(\Omega),\ 1 \leq p \leq \infty \tag{5.20}$$

$$\text{Min}\,\{\inf_\Omega u_o, 0\} \leq u(t,x) \leq \text{Max}\,\{0, \sup_\Omega u_o\},\ \forall\ (t,x) \in Q \tag{5.21}$$

$S(t): D_+ \to D_+$, $t \geq 0$ (see (4.18)), i.e. the semigroup generated by Δ (as in (4.14)) is a "positive semigroup".

For each $t>0$, $S(t)$ is compact in $L^2(\Omega)$.

<u>Proof.</u> One consideres the function $f(t)= \frac{1}{2}\|u(t)\|_2^2$ (which is of class $C^1(]0,\infty[)$) We have

$$f'(t) = \langle u(t),u'(t) \rangle = \int_\Omega u\,\Delta u\,dx = -\int_\Omega |\nabla u(t,x)^2|\,dx$$

Integrating over $[a,T]$ with $0 < a < T$ and then letting $a \downarrow 0$ one obtains (5.19). In order to get (5.20) we multiply (5.8) by $|u(t,x)|^{p-2}u(t,x)$ and take into account that

$$\nabla(|u|^{p-2}u) = (p-1)|u|^{p-2}\,\nabla u \tag{5.22}$$

Then

$$\frac{1}{p}\frac{d}{dt}\|u(t)\|_p^p = \int_\Omega (\Delta u)u|u|^{p-2}\,dx = -(p-1)\int_\Omega |\nabla u|^2\,|u(t,x)|^{p-2}dx \leq 0$$

i.e. $t \to \|S(t)u_o\|_p$ is decreasing and (5.20) follows.

Clearly, the case $p=+\infty$ in (5.20) is obtained letting $p \to +\infty$ in 5.20 (with p-finite). The maximum principle given by (5.21) is proved e.g.

in Brézis [5, p.211]. The compactness of $S(t)$ as well as its positivity have already been proved (Corollaries 4.1,4.2).

Remark 5.2. Note that in Theorem 5.1, $u_0 \in D(A)$ while in Theorem 5.2, $u_0 \in H$. It is clear that the solution $u(t)=S(t)u_0$ of the heat equation is smoother than the initial condition u_0. One says that in parabolic case, the semigroup $S(t)$ has a smoothing effect on the initial data. (This is not the case in the hyperbolic case). If $u_0 \in H_0^1(\Omega)$, then $u(t)= =S(t)u_0$ has additional regularities. (See e.g. Brézis [5,Ch. X]). In the case $\Omega = R^N$, the problem (5.8)-(5.10) becomes

$$\frac{\partial u}{\partial t} = \Delta u,\ t \geq 0,\ u(0,x) = u_0(x),\ x \in R^N \tag{5.23}$$

and admits the following explicit solution

$$u(t,x)=(S(t)u_0)(x) = (4\pi t)^{-N/2} \int_{R^N} e^{-\|x-y\|^2/4t} u_0(y)dy, \tag{5.24}$$

$t \geq 0$, $x \in R^N$, $u_0 \in L^2(R^N)$. Let us prove (5.24) in the case N=1 (for simplicity of writing). Applying the Fourier transform ζ (see (4.70)) the problem (5.23) becomes

$$\frac{\partial}{\partial t} \hat{u}(t,\xi) = -\ \xi^2\ \hat{u}(t,\xi),\ \hat{u}(0,\xi) = u_0(\xi),\ t \geq 0,\ \ \xi \in R \tag{5.25}$$

Obviously, the solution $\hat{u}$ of (5.25) is given by

$$\hat{u}(t,\xi) = e^{-t\ \xi^2}\ \hat{u}_0(\xi) \tag{5.26}$$

Consequently

$$\begin{aligned}\hat{u}(t,\xi) &= \zeta(u(t,x))(\xi) = \zeta((4\pi t)^{-\frac{1}{2}}\ e^{-x^2/4t})(\xi)(\zeta u_0)(\xi) = \\ &= \zeta((4\pi t)^{-\frac{1}{2}}\ e^{-x^2/4t} * u_o)(\xi)\end{aligned} \tag{5.27}$$

where $f*g$ denotes the convolution of the function f and g, i.e.

$$(f*g)(x) = \int_R f(x-y)g(y)dy \tag{5.28}$$

Thus, (5.27) implies (5.24) with N = 1. Recall also that the solution $u(t,x)=(S(t)u_0(x)$ of the problem

$$\frac{\partial u}{\partial t} = \frac{\partial^2 u}{\partial x^2},\ u(t,0)=u(t,1) = 0,\ u(o,x) = u_0(x),\ 0 \leq x \leq 1,\ t \geq 0 \tag{5.29}$$

is given by

$$\begin{aligned}u(t,x) &= \int_0^1 (2 \sum_{k=1}^{\infty} e^{-k^2\pi^2 t} (\sin k\pi x)\sin k\pi y)u_0(y)dy = \\ &= (S(t)u_0)(x),\ t \geq 0,\ 0 \leq x \leq 1\end{aligned} \tag{5.30}$$

Remark 5.3. Condition $S(t):H \to D(A)$ in Theorem 5.2 implies the fact that for every $u_o \in H$, the function $t \to S(t)u_o$ is of class C^∞ on $]0,+\infty[$. Since one has in addition the estimate (5.6), the function $t \to S(t)u_o$ can be extended in a sector

$$D_d = \{z \in R^2, \ |\arg z| < d, \ d \in]0, \tfrac{\pi}{2}[\ \} \tag{5.30)'$$

(of the complex plane R^2) as an anlytic function (analytic semigroup, cf. Pazy [4]. Therefore, we can say that a maximal monotone selfadjoint (linear) operator A acting in a real Hilbert space H, is the infinitesimal generator of a C_o-analytic semigroup of contractions. In particular the semigroup generated by Laplace operator Δ in $L^2(\Omega)$ (with $D(\Delta) = H_o^1(\Omega) \cap H^2(\Omega)$) is analytic (this is because Δ is selfadjoint in $L^2(\Omega)$, according to (5.17)'). Moreover, the semigroup $S_A(t)$ generated by the maximal monotone selfadjoint operator $A(A=A^*)$ is continuous (being analytic) in the uniform operator topology (at $t>0$). Indeed, for $u_o \in H$ and $d>0$, (7.38)' in Ch. 2 with $y=0$, becomes (see (5.6))

$$\|S(t)u_o - S(s)u_o\| \le |t-s| \ \|AS(d)u_o\| \le |t-s|\frac{1}{d} \ \|u_o\|,$$

$$\text{for all } t,s \ge d > 0 \tag{5.30)''$$

i.e.

$$\|S(t) - S(s)\| \le \frac{1}{d} \ |t-s|, \ \forall \ t,s \ge d \tag{5.30)'''$$

5.2. Some nonlinear parabolic equations

A first example of partial differential equations of parabolic type we have treated via semigroup approach is that given by (3.26). In what follows we will give some other examples, as applications of the results in Ch. 2 (e.g. Corollary 5.2, Theorem 7.3, Theorem 10.3) in conjunction with some results in § 4. (Recall the notation $\|u\|_2 = \{\int_\Omega |u(x)|^2 dx\}^{1/2}$).

Theorem 5.5. Let Ω be a bounded domain in R^N, with smooth boundary Γ, and $j:R \to [0,+\infty]$ l.s.c. with $j(0)=0$. Let us consider the problem

$$\frac{\partial u}{\partial t}(t,x) = \Delta u(t,x), \text{ a.e. in } Q \text{ (see (4.16))} \tag{5.31}$$

$$- \frac{\partial u}{\partial n}(t,x) \in \partial j(y(t,x)), \text{ a.e. on } F_\ell \qquad (5.32)$$

$$u(0,x) = u_o(x), \text{ a.e. in } \Omega \qquad (5.33)$$

Then for each $u_o \in L^2(\Omega)$, the problem (5.31)-(5.33) admits a unique L^2-strong solution $u = u(t,x) = u(t))(x)$ with the properties

(1) $u(t) \equiv u(t,.) \in H^2(\Omega)$, $x \to j(u(t,x))$ is in $L^1(\Gamma)$ for every $t > 0$, u_o, $u \in C(R_+, L^2(\Omega))$ and $\| \Delta u(t) \|_2 \leq \frac{1}{t} \|u_o\|_2$, $\forall\ t > 0$

(2) For every $d > 0$,

$$\|u(t) - u(s)\|_2 \leq |(t-s|\ \|\Delta u(d)\|_2 \leq \frac{|t-s|}{d} \|u_o\|_2 \qquad (5.34)$$

(where $\|u(t)\|_2^2 = \int_\Omega |u(t,x)|^2 dx$), for all $t,s \geq d > 0$.

(3) $t \to u(t)$ is a.e. differentiable on $]0,+\infty[$ (in the sense of $L^2(\Omega)$), it satisfies (5.31)-(5.33) and (with $u_t^+ = \frac{\partial u}{\partial t} = u'$).

$$\|u_t'(t)\|_2 \leq \frac{1}{d} \|u_o\|_2, \text{ for every } d > 0 \text{ and } t > d \qquad (5.35)$$

$u_t^+ = \frac{d^+u}{dt}(t)$ exists at every $t > 0$ (in L^2-norm)

(4) $\int_0^\infty f(u(t))dt \leq \frac{1}{2}\|u_o\|_2^2$, $\int_0^\infty t\|u'(t)\|_2^2\, dt \leq \|u_o\|_2^2$

(i.e. $\sqrt{t}\, u' \in L^2(R_+; L^2(\Omega))$ and $u' \in L^2([d,\infty[; L^2(\Omega))$, $d > 0$) where f is defined by (4.2) and $t \to f(u(t))$ is Lipschitz continuous convex and decreasing on $[d,\infty[$

(5) If $u_o \in H^1(\Omega)$ and $x \to j(u_o(x))$ is in $L^1(\Gamma)$, then

$$\frac{1}{2}\int_\Omega |\text{grad } u(t,x)|^2\, dx + \int_\Gamma j(u(t,x))d\sigma \leq \frac{1}{2}\int_\Omega |\text{grad } u_o(x)|^2\, dx + \int_\Gamma j(u_o(x))dx,\ \forall\ t \geq 0, \qquad (5.36)$$

$$\int_0^t \|u'(s)\|_2^2\, ds + f(u(t)) = f(u_o),\ \forall\ t \geq 0 \qquad (5.37)$$

$$\int_0^\infty \|u'(s)\|_2^2\, ds = f(u_o) \qquad (5.38)$$

(6) If $u_o \in H^2(\Omega)$ and $-\frac{\partial u_o}{\partial n}(x) \in \partial j(u_o(x))$ a.e. on Γ then

$$\|u(t) - u(s)\|_2 \leq |t-s|\ \|\Delta u_o\|_2,\ \forall\ t,s \geq 0 \qquad (5.39)$$

$$\|u'(t)\|_2 \leq \|u_o\|_2, \text{ a.e. on }]0, +\infty[\qquad (5.40)$$

$u_t^+(t)$ exists at every $t \geq 0$ and $t \to f(u(t))$ is a convex, Lipschitz continuous, decreasing function on $[0,+\infty[$

(7) For every $u_o \in L^2(\Omega)$, the corresponding solution u(t) of (5.31)-(5.33) has the property that $u(t) \to y$ in $L^2(\Omega)$ as $t \to +\infty$, with $y \in H^2(\Omega)$, $-\frac{\partial y}{\partial n} \in \partial j(y(x))$ a.e. on Γ and $\Delta y = 0$ (i.e. u(t) tends in $L^2(\Omega)$ as $t \to +\infty$, to an equilibrium point y. Precisely,

$$\int_\Omega |u(t,x) - y(x)|^2 \, dx \to 0 \text{ as } t \to +\infty \qquad (5.40)'$$

and

$$\Delta y = 0 \text{ in } L^2(\Omega), \; j(y(x)) = 0 \text{ a.e. on } \Gamma \qquad (5.41)$$

If j is strictly increasing then y = 0 a.e. in Ω (is the only equilibrium point).

<u>Proof.</u> The crucial remark is to observe that the problem (5.31)-(5.33) can be reduced to an abstract Cauchy problem in $L^2(\Omega)$, namely,

$$u'(t) = -\partial f(u(t)), \; u(0) = u_o \in L^2(\Omega), \; t \geq 0 \qquad (5.42)$$

with f defined by (4.2). Indeed, in view of Theorem 4.1 we have $\Delta = -\partial f = A$, with D(A) given by (4.1). In this case the l.s.c. function f is positively valued

$$D_e(f) = \{ v \in H^1(\Omega); \; j(u|_\Gamma) \in L^1(\Gamma) \} \qquad (5.43)$$

Let $D(\Delta) = D(\partial f) \equiv D(A)$ given by (4.1). Then, by Theorem 4.1 and (2.35) in Appendix, we have

$$D(\partial f) \subset D_e(f) \subset \overline{D_e(f)} = \overline{D(\Delta)} = L^2(\Omega) \qquad (5.44)$$

In our case $f: H \to [0,+\infty]$, $f(0)=0$, $\partial f(0)=0$

It is now easy to check that Properties (1)-(7) above of the strong solution $u(t)=S(t)u_o$ of (5.42) are direct consequences of Theorem 7.3 and Corollary 7.1 in Ch. 2. Indeed, in this case, the level sets C_r given by (4.11) are precompact in $L^2(\Omega)$ and therefore ae are in Hypotheses of Corollary 7.1 in Ch. 2. Clearly, the fact that the equilibrium point $y \in D(\Delta) \equiv D(A)$ given by (4.1) and 0 $\partial f(y)$ means

$$y \in H^2(\Omega), \; -\frac{\partial y}{\partial n}(x) \in \partial j(u(x)), \text{ a.e. on } \Gamma \;, \; \Delta y = 0 \qquad (5.45)$$

Finally, inasmuch as y is a minimum point of f given by (4.2) (whose minimum value is zero) it follows that f(y)=0, which is equivalent to (5.41). If j is striclty increasing, then $y|_\Gamma = o$. This fact in conjuction with $\nabla y = 0$ and Poincaré inequality

$$\|y\|_p \leq C\|\nabla y\|_p, \quad y\in W_0^{1,p}(\Omega), \quad 1 \leq p < \infty \tag{5.46}$$

yields y = 0 in Ω, q.e.d.

Remark 5.4. In the case

$$\partial j(s) = \begin{cases} r(s^4-v^4), & \text{for } s \geq 0 \\ -rv^4, & \text{for } s < 0 \end{cases} \tag{5.47}$$

the boundary condition (5.32) is just the Boltzman law of the black body radiation heat emission on Γ, where v is the surrounding temperature and r> 0. Clearly, (5.47) corresponds to

$$j(s) = \begin{cases} \frac{r}{5} s^5 - rsv^4, & \text{for } s \geq 0 \\ -rsv^4, & \text{for } s < 0 \end{cases} \tag{5.48}$$

In this case, for v=0 we are in Hypotheses of Theorem 5.5 and therefore the equilibrium point is y=0. This is not the case if v≠0. Let us consider the natural convection which correspond to (5.32) with

$$j(s) = \begin{cases} r\, s^{5/4}, & \text{for } s \geq 0 \\ 0, & \text{for } s < 0 \end{cases} \tag{5.49}$$

We take for example

$$j(s) = \begin{cases} \frac{4r}{9} s^{9/4}, & \text{if } s \geq 0 \\ 0, & \text{if } s < 0 \end{cases} \tag{5.50}$$

In this case, the equilibrium point y in Theorem 5.5 is again y=0 in Ω. A generalization of the problem (5.31)-(5.33) is the following one (with $p \geq 2$, $r \geq 0$)

$$\frac{\partial u}{\partial t}(t,x) = \sum_{i=1}^{N} \frac{\partial}{\partial x_i}\left(\left|\frac{\partial u}{\partial x_i}\right|^{p-2} \frac{\partial u}{\partial x_i}\right) - ru|u|^{p-2}, \quad \text{a.e. in } Q \tag{5.51}$$

$$-\sum_{i=1}^{N} \left|\frac{\partial u}{\partial x_i}\right|^{p-2} \frac{\partial u}{\partial x_i} \cos(n,e_i) \in \partial j(u(t,x)), \quad \text{a.e. on } F_\ell \tag{5.52}$$

(see (4.24) for notations) $u(0,x)=u_0(x)$, a.e. in Ω (5.53)

In this case the right hand side is the subdifferential of the function $f=f_p:L^2(\Omega) \to [0,+\infty[$ given by

$$f_p(u) = \begin{cases} \int_\Omega \frac{1}{p} \sum_{i=1}^{N} \left|\frac{\partial u}{\partial x_i}\right|^p dx + \frac{r}{p}\int_\Omega |u|^p dx + \int_\Gamma j(u(x))d\sigma, \\ \qquad \text{if } u\in W^{1,p}(\Omega) \text{ and } j(u)\in L^1(\Gamma) \\ +\infty, \qquad \text{otherwise} \end{cases} \tag{5.54}$$

(see (4.25) and (4.28)). The domain $D(\partial f_p)$ is given by (4.23) in Example 4.4 with $\Delta_p u - u|u|^{p-2} \equiv L_p u$ in place of Δ_p (see 4.27) i.e.

$$D(\partial f_p) \equiv D(L_p) = \{u \in W^{1,p}(\Omega);\ \Delta_p u - u|u|^{p-2} \in L^2(\Omega); \tag{5.55}$$

$$-\frac{\partial u}{\partial n_p}(x) \in \partial j(u(x)), \text{ on } \Gamma\} \text{ with } \frac{\partial u}{\partial n_p} \text{ given by (4.24).}$$

Theorem 5.6. Let Ω and j be as in Theorem 5.5 (and $p \geq 2$). Then the semigroup $S_p(t)$ generated by L_p (as given by (4.27)) in $L^2(\Omega)$ is compact for $t > 0$ (for every $p \geq 2$). Moreover, $L_p = \partial f_p$ is densely defined in $L^2(\Omega)$ and for every $u_o \in L^2(\Omega)$, the function $u_p(t) = S_p(t)u_o$ is the unique L^2-strong solution of the problem (5.51)-(5.53). Finally, the solution u_p has the properties (1)-(4), (5.37), (5.38), (5.40)' and (5.41) with f_p in place of f. If j is strictly increasing, then $y=0$ in Ω is the only equilibrium point.

Proof. Since $p \geq 2 > \frac{2N}{N+2}$, the imbedding $W^{1,p}(\Omega) \subset L^2(\Omega)$ is compact. Accordingly the level sets

$$C_k = \{u \in L^2(\Omega),\ \|u\|_2^2 + f_p(u) \leq k\}\ ,\ k > 0 \tag{5.56}$$

are compact in $L^2(\Omega)$ hence $S_p(t)$ is compact in $L^2(\Omega)$ for $t > 0$. Clearly, $D(\partial f_p) \supset C_o^\infty(\Omega)$ which is dense in $L^2(\Omega)$. Thus the conclusions of Theorem 5.5 hold with f_p in place of f and L_p in place of Δ (and other corresponding modifications). In (5.37), $f = f_p$ and $u_o \in W^{1,p}(\Omega)$ with $j(u) \in L^1(\Gamma)$ (i.e. $u_o \in D_e(f_p)$). Clearly, $S_p(t)u_o \to y$ as $t \to \infty$ (in $L^2(\Omega)$) and if j is strictly increasing then $y|_\Gamma = 0$. In view of Poincaré inequality (5.46), it follows $y=0$ in Ω (since we also have $y=0$ in $L^2(\Omega)$). As an application of Theorem 10.3 in Ch. 2, we can prove

Theorem 5.7. Let Ω and j be as in Theorem 5.5 and let $g \in L^2(0,T;L^2(\Omega))$ with $T > 0$. Then for every $u_o \in L^2(\Omega)$ the equation

$$\frac{\partial u}{\partial t}(t,x) = \Delta u(t,x) + g(t,x), \text{ a.e. in } Q \tag{5.57}$$

with the boundary condition (5.32) and initial condition (5.33) admits a unique $L^2(\Omega)$-strong solution u on $[0,T]$ with the additional properties

(1) $\sqrt{t}\, u' \in L^2(0,T;L^2(\Omega))$, $t \to f(u(t))$ is in $L^1(0,T;L^2(\Omega))$

(2) $t \to f(u(t))$ is absolutely continuous on every compact $[d,T]$, $u \in W^{1,2}(d,T;L^2(\Omega))$ with $0 < d < T$ and f given by (4.2) (with $u' = \frac{du}{dt}$ - the strong derivative in $H = L^2(\Omega)$).

(3) If $u_0 \in H^1(\Omega)$ and $j(u_0) \in L^1(\Gamma)$, then $u \in W^{1,2}(0,T;L^2(\Omega))$ and $t \to f(u(t))$ is absolutely continuous on $[0,T]$.

(4) If in addition $g \in W^{1,1}(0,T;L^2(\Omega))$, then $u(t) \in D(\partial f)$ for all $t \in]0,T[$, i.e. for all $t \in]0,T[$, $u(t) \in H^2(\Omega)$;

$-\frac{\partial u}{\partial n}(t,x) \in \partial j(t,x)$ a.e. on Γ;

$\frac{d^+u}{dt}(t) = Au(t)+g(t)$ in $L^2(\Omega)$, $\forall t \in]0,T[$

where $A=\Delta$ is defined by (4.1).

Proof. One applies Theorem 10.3 and 10.6 with $H=L^2(\Omega)$ and f given by (4.2) (i.e. the abstract scheme is

$$u'(t) \in -\partial f(u(t))+g(t),\ u(0)=u_0 \in L^2(\Omega),\ t \in]0,T[\tag{5.58}$$

Conditions $u_0 \in F^1(\Omega)$ and $j(u_0) \in L^1(\Gamma)$ mean that $u_0 \in D_e(f)$, with f given by (4.2).

5.3. A class of semilinear parabolic equations.

We shall be concerned with some results on the semilinear equation

$$u'(t)=Au(t)+F(u(t)),\ 0 \le t \le T,\ u(0) = u_0 \in H \tag{5.59}$$

where $A:D(A) \subset H \to H$ is the infinitesimal generator of the C_0-contraction semigroup $S(t)$ and $F:H \to H$ is continuous.

Recall that if $g:[0,T] \to H$ is Lipschitz continuous, then the "mild" solution $u:[0,T] \to X$ of

$$u'(t) = Au(t)+g(t),\ u(0) = u_0 \tag{5.60}$$

is given by (see (10.5) in Ch. 2)

$$u(t) = S(t)u_0 + \int_0^t S(t-s)g(s)ds,\ 0 \le t \le T \tag{5.61}$$

and it has the properties (in every reflexive space):

Proposition 5.1. If $u_0 \in D(A)$ and $g:[0,T] \to H$ is Lipschitz continuous on $[0,T]$, then u given by (5.61) is in $C([0,T];D(A)) \cap C^1([0,T];H)$ (and (5.60) is satisfied for all $t \in [0,T]$ (i.e. u is a classical solution of (5.60) on $[0,T]$).

Proof. We first note that Lipschitz continuity of g on $[0,T]$ (say of Lipschitz constant L) implies Lipschitz continuity of u given by (5.61). This can be proved directly, as follows:

$$u(t+h)=S(t+h)u_0 + \int_0^t S(t-s)g(h-s)ds + \int_0^h S(t+s)g(h-s)ds,\ 0 \le t \le T-h \tag{5.62}$$

$$\|S(t+h)u_0 - S(t)u_0\| \le h\|Au_0\|,\ 0 \le t;\ h > 0$$

Therefore

$$\|u(t+h)-u(t)\| \le h(\|Au_o\| + LT + \|g\|) \qquad (5.62)'$$

for all $h>0$, $t,t+h \in [0,T]$ where

$$\|g\| = \sup\{\|g(t)\|,\ t\in[0,T]\} \equiv \|g\|_T$$

Another proof of (5.62)': The mild solution u is the unique integral solution of (5.60) so we can use the estimate (10.4) (in Ch. 2 with $v(t)=u(t+h)$ and $f(t)=g(t+h)$. Therefore

$$\|u(t+h)-u(t)\| \le \|u(h)-u(0)\| + \int_o^t \|g(s+h)-g(s)\|ds \qquad (5.63)$$

On the other hand, by (5.61),

$$\|u(h)-u(0)\| \le \|S(h)u_o - u_o\| + \int_o^h \|g(s)\|ds \qquad (5.64)$$

Clearly, (5.63) and (5.64) imply again (5.62)'. It follows that u is a.e. differentiable on [0,T] and

$$\begin{aligned} u'(t) &= S(t)Au_o + S(t)g(0) + \int_o^t S(s)g'(t-s)ds = \\ &= S(t)Au_o + S(t)g(0) + \int_o^t S(T-s)g'(s)ds \end{aligned} \qquad (5.65)$$

We have

$$\frac{1}{h}(v(t+h)-v(t)) = \int_o^t S(s)\ \frac{g(t-s+h)-g(t-s)}{h}\ ds + \frac{1}{h}\int_t^{t+h} S(s)g(t+h-s)ds \qquad (5.66)$$

Let us assume that $v'(t)$ exists for $t\in]0,T[$. Then for $0\le s\le t$ ($g'(t-s)$ exists for almost all $s\in]0,t[$ and) we have

$$v'(t) = \int_o^t S(s)g'(t-s)ds + S(t)g(0) \qquad (5.67)$$

and (5.65) follows (with $g'(t-s) = \lim_{h\downarrow 0}(g(t-s+h)-g(t-s)/h$. But the right hand side of (5.65) is continuous on [0,T], so u' can be extended by continuity to a continuous function on [0,T] (since u' is continuous almost everywhere on [0,T]). It follows that u is differentiable on [0,T] (since $u(t_1)-u(t_2) = \int_{t_1}^{t_2} u'(s)ds$, $t_1,t_2\in[0,T]$). Therefore u(t) D(A), $\forall\, t\in[0,T]$ $t\to Au(t) = u'(t) - g(t)$ is continuous on [0,T] hence $u\in C([0,T];D(A))$. The proof is complete.

<u>Definition 5.1.</u> (1) A function $u:[0,T]\to X$ is said to be a classical solution of (5.59) if $u(0)=u_o$,

$$u\in C([0,T];\ D(A))\cap C^1([0,T];H) \qquad (5.68)$$

and (5.59) is satisfied. (2) The function u is a classical solution on $[0,+\infty[$ of (5.59) if (5.68) holds for every $T>0$ and (5.59) is satisfied for every $t>0$.

Recall that by a mild solution of (5.59) we mean a continuous function $u:[0,T]\to H$ satisfying

$$u(t) = S(t)u_o + \int_0^t S(t-s)F(u(s))ds, \quad 0\leq t\leq T \tag{5.69}$$

$F:H\to H$ is said to be L-Lipschitz continuous (or globally Lipschitz continuous) with $L>0$, if

$$\|Fu-Fv\| \leq L\|u-v\|, \ \forall\ u,v\in H. \tag{5.70}$$

F is said to be locally Lipschitz continuous if for every $r>0$, there is $L_r>0$ such that

$$\|Fu-Fv\|\leq L_r\ \|u-v\|,\ u,v\in H, \text{ with } \|u\|\leq r,\ \|v\|\leq r \tag{5.71}$$

<u>Theorem 5.8.</u> (1) If $F:H\to H$ is locally Lipschitz continuous then for every $u_o\in H$, there is a unique mild solution $u:[0,T_{max}[\to H$ of (5.59) with either $T_{max}=+\infty$, or $T_{max}<+\infty$ and $\lim\limits_{t\uparrow T_{max}}\|u(t)\| = +\infty$. If $u_o\in D(A)$, the mild solution u is classical. (2) If F is Lipschitz continuous, then $T_{max}=+\infty$.

<u>Proof.</u> (1) First it follows that F maps bounded sets into bounded sets, i.e.

$$\|F(v)\| \leq M, \text{ for all } v \text{ with } \|v\|\leq r+\|u_o\| \tag{5.72}$$

with $M=M(r,u_o)>0$. Set

$$B = \{u\in C([0,T];H),\ \|u(s)-u_o\|\leq r,\ 0\leq s\leq T\} \tag{5.73}$$

with $T>0$ such that

$$\|S(t)u_o-u_o\|+TM\leq r,\ 0\leq t\leq T,\ TL_{r+\|u_o\|} < 1 \tag{5.74}$$

Define $G:C([0,T];H)\to C([0,T];H)$ by

$$(Gu)(t) = S(t)u_o + \int_0^t S(t-s)F(u(s))ds, \quad 0\leq t\leq T \tag{5.75}$$

Then $G:B\to B$ is a b-contraction with $0<b=TL_{r+\|u_o\|}<1$ so it has a unique fixed point u B. The fact that $T_{max}<+\infty$ implies $\lim\limits_{t\uparrow T_{max}}\|u(t)\| = +\infty$ is routine (see e.g. Proposition 4.3, p. 209 in the author's book [15] or even Theorem 1.3 in this chapter).

(2) If F is (even globally) Lipschitz, then it can be easily proved that u is bounded on $[0,T_{max}[$, and therefore, $T_{max} = +\infty$. In this case we can prove directly (without Part (1)) that there exists a unique solution u of (5.69) on $R_+ = [0,+\infty[$. Namely, one considers the Banach space

$$X = \{u \in C(R_+;H), \sup_{t \geq 0} \|u(t)\| e^{-rt} < \infty\} \tag{5.76}$$

with a Bielecki norm

$$\|u\| = \sup_{t \geq 0} \|u(t)\| e^{-rt}, \; r > 0 \tag{5.77}$$

It is routine to check that $G:X \to X$ is $\frac{L}{r}$- Lipschitz continuous. Accordingly, with $r=2L$, G has a unique fixed point $u \in X$.

The case $u_0 \in D(A)$. In this case the mild solution u (in both cases 51) and (2)) is classical. Indeed, set $g(t) = F(u(t))$. According to (5.63) and (5.64)

$$\|u(t+h)-u(t)\| \leq h(\|Au_0\| + \|g\|) + L\int_0^t \|u(s-h)-u(s)\|ds \tag{5.78}$$

on every compact $[0,T]$ (i.e. $0 \leq t < t+h \leq T$) which yields

$$\|u(t+h)-u(t)\| \leq h(\|Au_0\| \|g\|)e^{LT} \tag{5.79}$$

In other words, $t \to u(t)$ is Lipschitz continuous on $[0,T]$ and therefore so is $t \to F(u(t)) = g(t)$.

On the basis of Proposition 5.1, u is a classical solution of (5.59).

Remark 5.5. Theorem 5.8 is a particular case of some very general results (cf. Pavel [15, Ch. 5]. However, it is convenient to be used in D(A) endowed with the graph norm (5.3). Clearly, in Theorem 5.8, H=X can be replaced with any (real) reflexive space. Actually, for the existence of mild solution, X need not be reflexive.

Theorem 5.9. Let D(A) be the Banach space endowed with the graph norm (5.3). (1) If $F:D(A) \to D(A)$ is locally Lipschitz continuous, then for every $u_0 \in D(A)$, the problem (5.59) has a unique classical solution u, i.e.

$$u \in C([0,T_{max}[;D(A)) \cap C^1([0,T_{max}[;H) \tag{5.80}$$

with (either)

$$T_{max} = +\infty \text{ or } T_{max} < +\infty \text{ and } \lim_{t \uparrow T_{max}} (\|u(t)\| + \|Au(t)\|) = +\infty \tag{5.81}$$

(2) If $F:D(A)$ is globally Lipschitz, then $T_{max} = +\infty$

Proof. (1) Set $A_1 u = Au$, for $u \in D(A^2) \equiv D(A_1)$ (see (5.4)). Then $A_1 : D(A_1) \to D(A)$, and A_1 is the infinitesimal generator of the C_0-semigroup $S_1(t) : D(A) \to D(A)$ with $S_1(t)u = S(t)u$ for $u \in D(A)$. In words : the restriction A_1 of A on $D(A^2)$ is the infinitesimal generator of the restriction of S(t) on D(A). This was the key of the proof. Indeed, by Theorem 5.8 with D(A) in place of H, the problem (5.59) has a unique mild solution u with

$$u(t) \in D(A), \ \forall\, t \in [0, T_{max}[\tag{5.82}$$

Set

$$v(t) = \int_0^t S(t-s)F(u(s))ds, \ 0 \leq t < T_{max} \tag{5.83}$$

so $u(t) = S(t)u_0 + v(t)$. Since $S(t)u_0 \in D(A)$, (5.82) is equivalent to $v(t) \in D(A)$, for all $t \in [0, T_{max}]$. But

$$h^{-1}(S(h)v(t)-v(t)) = h^{-1}(v(t+h)-v(t)) - h^{-1}\int_t^{t+h} S(t+h-s)F(u(s))ds \tag{5.84}$$

which shows that

$$Av(t) = v'(t) - F(u)), \ \forall\, t \in [0, T_{max}[\tag{5.85}$$

i.e.

$$u'(t) = AS(t)u_0 + v'(t), \ \forall t \in [0, T_{max}[\tag{5.85'}$$

which concludes the proof.

Remark 5.6. In Thoerem 5.9, we need not the reflexivity of D(A). This is because $u(t) \in D(A)$ is equivalent to the differentiability of the mild solution u at t (in the case $u(0) = u_0 \in D(A)$). This equivalence is due to (5.84). We can say that the lack of relfexivity of $D(A) \subset X$ is compensated by the additional assumption $F : D(A) \to D(A)$.

Theorem 5.9 can be applied to concrete partial differential equations, e.g.

$$\frac{\partial u}{\partial t} = \Delta u + f(u), \text{ in }]0,T[\times \Omega \tag{5.86}$$

$$u(t,x) = 0, \quad \text{on } [0,T] \times \Gamma \tag{5.87}$$

$$u(0,x) = u_0(x), \text{ in } \Omega \tag{5.88}$$

where $f : R \to R$.

A simple (and restrictive) result is

Theorem 5.10. Assume that $f\in C^1(R)$ with f' bounded on R. Then for every $u_o\in L^2(\Omega)$, there exists (in $L^2(\Omega)\equiv L^2$), a unique L^2-classical solution u on $]0,+\infty[$ of (5.86)-(5.88), namely

$$u\in C([0,+\infty[;L^2)\cap C(]0,+\infty[;H^2(\Omega)\cap H_o^1(\Omega))\cap C^1(]0,+\infty[;L^2) \quad (5.89)$$

If $u_o\in H_o^1(\Omega)\cap H^2(\Omega)$, then

$$u\in C([0,+\infty[;H^2(\Omega)\cap H_o^1(\Omega))\cap C^1([0,+\infty[;L^2) \quad (5.90)$$

Theorem 5.10 is a direct consequence of the following version of Theorem 5.8.

Theorem 5.11. (1) Suppose in addition to Hypotheses of Theorem 5.8, that $A=A^*$. Then for every $u_o\in H$(a real Hilbert space) the mild solution u of (5.59) satisfies

$$u\in C^1(]0,T_{max}[;H)\cap C(]0,T_{max}[;D(A))\cap C([0,T_{max}[;H) \quad (5.91)$$

If $u_o\in D(A)$, then $u\in C^1([0,T_{max}[;H)\cap C([0,T_{max}[;D(A))$ (5.92)

(2) If $F:H\to H$ is globally Lipschitz, then $T_{max}=+\infty$.

Proof. In this case we have the important property $S(t):H\to D(A)$ (see the proof of Theorem 5.2). COnsequently for each $d>0$, and $u_o\in H$,

$$\|S(t+h)u_o - S(t)u_o\| \le h\|AS(d)u_o\|,\ d<t<t+h\le T<T_{max} \quad (5.93)$$

and it is readily see that (5.79) becomes

$$\|u(t+h)-u(t)\| \le h(\|AS(d)u_o\| + \|g\|)e^{L(T-d)},\ d\le t\le t+h\le T \quad (5.94)$$

with $\|g\| = \sup_{0\le s\le T}\|F(u(s))\|$ (and $S(t)u_o$ differentiable at $T>0$). Therefore, the conclusion of Theorem 5.8 holds on every compact $[d,T]\subset]0,T_{max}[$ i.e. (5.91) and (5.92) are fulfilled (one takes into account that $\frac{d}{dt}(S(t)u_o)=AS(t)u_o=S(t-d)AS(d)u_o$, for $0<d\le t$). The proof is complete.

Proof of Theorem 5.10. Condition $f\in C^1(R)$ with f' bounded on R implies that f is Lipschitz continuous on R and it satisfies a growth condition

$$|f(r)| \le L|r| + |f(0)|,\ \forall r\in R \quad (5.95)$$

with $L=\sup\{|f'(r)|,\ r\in R\}$. Define the realization of f in $L^2(\Omega)$ (i.e. the Nemytski operator F), $F:L^2\to L^2$, by

$$(Fu)(x) = f(u(x)),\ x\in\Omega \quad (5.96)$$

It follows that (5.86)-(5.88) can be reduced to an abstract problem of type (5.59) in $L^2(\Omega)$, with $A=\Delta$, $D(A) = H^2(\Omega)\cap H_0^1(\Omega)$ (see also (5.17)'). Therefore the conclusion of the theorem is a consequence of Theorem 5.11, (2) q.e.d.

Theorem 5.12. Suppose that $f\in C^1(R)$ and $f(0) = 0$ (e.g. $f(r)=r|r|^{\gamma-1}$, $\gamma \geq 1$, $r\in R$). Then for every $u_0\in L^\infty(\Omega)$, (5.86)-(5.88) has a unique classical solution in $L^2(\Omega)$ (i.e. (5.91) holds with $H=L^2(\Omega)$) and $u\in L^\infty([0,T]; L^\infty(\Omega))$, for all $T\in]0,T_{max}[$. Moreover, either $T_{max} = +\infty$, or $T_{max} < +\infty$ and $\lim_{t\uparrow T_{max}} \|u(t)\|_\infty = +\infty$.

Proof. In this case f is merely locally Lipschitz on R so F given by (5.95) is not defined on the whole $L^2(\Omega)$. Then, by a standard device, one considers the trunchiate $\tilde{f}$ and f

$$\tilde{f}(r) = f(r), \text{ for } |r|\leq 1+M,\ M\ \ \|u_0\|$$

$$\tilde{f}(r) = F(M+1), \text{ for } r\geq M+1,\ f(r) = f(-M-1) \text{ for } r<-M-1 \tag{5.96)'}$$

Since f is globally L-Lipschitz on R, the problem (5.86)-(5.88) with $\tilde{f}$ in place of f has a unique solution u satisfying (5.89). We now prove that the restriction of u on [0,T] with T sufficiently small satisfies

$$\|u(t)\|_\infty \leq \|u_0\|_\infty e^{LT} \leq 1+M,\ 0 \leq t \leq T \tag{5.97}$$

Indeed, let F be the realization of f in $L^2(\Omega)$ (as defined by (5.96)). Then (in $L^2(\Omega)$) we have

$$u(t)=S(t)u_0 + \int_0^t S(t-s)F(u(s))ds,\ t\geq 0, \tag{5.98}$$

Since $\|S(t)u_0\|_\infty \leq \|u_0\|_\infty$ (see (5.20)), we derive

$$\|u(t)\|_\infty \leq \|u_0\|_\infty + L\int_0^t \|u(s)\|_\infty\, ds,\ 0\leq t$$

which implies (5.97). Finally, (5.97) shows that $f(u(t,x))=f(u(t,x))$a.e. in Ω (and $F(u(t))=F(u(t))$) $0\leq t\leq T$ hence u is a classical solution of (5.86) on [0,T]. The proof of the last part of the Theorem is routine so we omit it. As we have already remarked, the equation

$$\frac{\partial u}{\partial t} = \Delta u + u|u|^{\gamma-1},\ \gamma \geq 1,\ t>0,\ x\in\Omega \tag{5.99}$$

with the conditions (5.87) and (5.88), falls into Theorem 5.12. This is because it corresponds to $f(r)=r|r|^{\gamma-1}$ (with $f'(r)=\gamma|r|^{\gamma-1}$), which is in $C^1(R)$.

Theorem 5.13. Suppose that $f \in C^3(R)$ with $f(0)=0$. Then for every $u_0 \in H^2(\Omega) \cap H_0^1(\Omega)$, there exists a unique classical solution u of (5.86)-(5.88), on $[0,T_{max}[$ i.e.

$$u \in C([0,T_{max}[;H^2(\Omega)) \cap C^1([0,T_{max}[;L^2(\Omega)) \tag{5.100}$$

Moreover,

Either $T_{max} = +\infty$, or the solution blows up in finite time, i.e.

$$T_{max} < +\infty \text{ and } \lim_{t \uparrow T_{max}} \|u(t)\|_{H^2(\Omega)} = +\infty \tag{5.101}$$

Proof. Let F be defined as indicated by (5.95), and let $A=\Delta$ with $D(A)=H_0^1(\Omega) \cap H^2(\Omega)$. We will prove that $F:D(A) \to D(A)$ and is locally Lipschitz in the norm $\|.\|_{D(A)}$ (which is equivalent to $\|.\|_{H^2}$ with $H^2 \equiv H^2(\Omega)$). Then (5.100) and (5.101) are consequences of Theorem 5.9.

Denote by D a derivative of $u \in H_0^1 \cap H^2$. Therefore $DF(u)=f'(u)\,Du$ with $f'(r)=\frac{df}{dr}(r)$. If $N \leq 3$, $H^2(\Omega) \hookrightarrow L^\infty(\Omega)$ so $x \to f'(u(x))$ is in $L^\infty(\Omega)$ while, by hypothesis, $Du \in L^2(\Omega) \equiv L^2$. Thus $DF(u) \in L^2$. Moreover $D^2F(u) = f''(u)(Du)^2+f'(u)D^2u$.

Similarly, $f''(u) \in L^\infty(\Omega)$, $D^2u \in L^2$. But $Du \in H^1(\Omega) \hookrightarrow L^4(\Omega)$(i.e. $(Du)^2 \in L^2(\Omega)$) therefore $D^2F(u) \in L^2$(Note that we actually have ($H^2(\Omega) \subset C(\bar{\Omega})$ for $N \leq 3$) and $H^1 \equiv W^{1,2}(\Omega) \hookrightarrow L^{2^*}(\Omega)=L^6(\Omega)$($\frac{1}{2^*}=\frac{1}{2}-\frac{1}{N}$ with N=3, so $2^* = 6$), $H^1 \hookrightarrow L^q$, $\forall\, q \geq 2$ if N=2 and $H^1(\Omega) \hookrightarrow L^\infty(\Omega)$. If N=1, $\Omega = I$ - a bounded interval, then $H^1(I) \hookrightarrow C(I)$ with compact injection). Inasmuch as $f(0) = 0$ we conclude that $F:D(A) \to D(A)$ (for this fact we have merely used that $f \in C^2(R)$).

We now prove that for every $M > 0$, there is $L_M > 0$ such that

$$\|F(u)-F(v)\|_{H^2} \leq L_M \|u-v\|_{H^2} \tag{5.102}$$

for all u,v with

$$\|u\|_{H^2} \leq M, \ \|v\|_{H^2} \leq M \tag{5.103}$$

Indeed, (5.103) implies $\|u\|_\infty \leq k$, $\|v\|_\infty \leq k$(k - a positive constant depending on M), so (5.103) gives

$$|f(u(x)) - f(v(x))| \leq |u(x) - v(x)| \sup_{|r| \leq 3k} |f'(r)|, \text{ a.e. in } \Omega$$

This yields

$\|F(u)-F(v)\|_2 \leq k_1 \|u-v\|_2$, for u,v satisfying (5.103). Furthermore

$$D(F(u) - F(v)) = (f'(u)-f'(v))Du + (D(u-v))f'(v) \tag{5.104}$$

Since

$$|f'(u(x)) - f'(v(x))| \leq |u(x)-v(x)| \sup_{|r|\leq 3k} |f''(r)|, \quad \text{a.e. in } \Omega$$

and $\|u-v\|_2 \leq \|u-v\|_{H^2}$, $\|D(u-v)\|_2 \leq \|u-v\|_{H^2}$
by (5.104) we derive the existence of a constant $k_2 = k_2(M)$ such that

$$\|D(F(u)-F(v))\|_2 \leq k_2\|u-v\|_{H^2} \tag{5.105}$$

for all u and v satisfying (5.103).

Similarly one proves an inequality of type (5.105) with D^2 In place of D and k_3 in place of k_2. In such a manner it folllows that (5.102) holds, which concludes the proof.

As a simple application of Theorem 5.12 we give

Coroolary 5.1. For every $u_0 \in H_0^1(\Omega) \cap H^2(\Omega)$, the problem

$$\frac{\partial u}{\partial t} = \Delta u - u^3, \text{ in }]0,T[\times \Omega \tag{5.106}$$

$$u(t,x) = 0, \text{ on }]0,T[\times \Gamma \ ; \ u(0,x)=u_0(x), \text{ in } \Omega \tag{5.107}$$

has a unique solution $u \in C([0,+\infty[;H^2(\Omega)) \cap C^1([0,+\infty[;L^2(\Omega))$.

Proof. We will prove that in this case $T_{max} = +\infty$ (with T_{max} given by Theorem 5.13). Indeed, multiplying (5.106) by $u|u|^{p-2}$ and integrating over Ω (see (5.22)) se get

$$\frac{1}{p}\frac{d}{dt}\|u(t)\|_p^p = -(p-1)\int_\Omega |\nabla u|^2 |u(t,x)|^{p-2}\,dx - \int_\Omega u^4|u|^{p-2}dx \leq 0, \ p \geq 2 \tag{5.108}$$

hence

$$\|u(t)\|_p \leq \|u_0\|_p, \ \forall\, t \in [0,T_{max}[, \quad p \geq 2 \tag{5.109}$$

Letting $p \to \infty$, one obtains $\|u(t)\|_\infty \leq \|u_0\|_\infty$ for all t $[0,T_{max}[$ so according to Theorem 5.12, $T_{max} = +\infty$ (this is because $\|u(t)\|_\infty$ is bounded on $[0,T_{max}[$, so it cannot blow up in finite time). The formula (5.108) is correct since $u(t,.) \in H_0^1 \cap H^2$, so $\nabla u(t,.) \in L^2(\Omega)$ and $u(t,.) \in L^\infty(\Omega)$ for all $t \in [0,T_{max}[$ (see the proof of Theorem 5.13). The proof is complete.

Remark 5.7. Equation (5.106) corresponds to (5.86) with (I) : $f(u)=-u^3$ In this case the property of f which guaranteed $T_{max} = +\infty$ is the dissipativity property

$$(f(r_1)-f(r_2))\ (r_1-r_2) \leq 0,\ \forall\ r_1,r_2 \in R \tag{5.110}$$

or $f(r)\ r \leq 0$, $\forall r \in R$ (since $f(0)=0$).

One can also use the fact that the realization F of f in $L^{4/3}(\Omega)$, i.e. $F(u) = -u^3$, $u \in L^4(\Omega)$ is just $-\partial g_4(u)$ where $g_4(u) = \frac{1}{4}\|u\|_4^4$, $u \in L^4(\Omega)$ (see 4.29).

The case (II) :$f(u) = u^3$ is essentially different from the case (I) above, in the sense that in the case (II) the blow-up phenonomenon can occur. Precisely, we have

<u>Corollary 5.2.</u> Let $u_o \in H_o^1 \cap H^2$ be such that

$$E(u_o) \equiv \frac{1}{2}\int_\Omega |\nabla u_o(x)|^2 dx - \frac{1}{4}\int_\Omega (u_o(x))^4 dx \leq 0,\ u_o \neq 0 \tag{5.111}$$

Then the problem

$$\frac{\partial u}{\partial t} = \Delta u + u^3,\ \text{in}\ \ ]0,T[\times\Omega \tag{5.112}$$

$$u(t,x) = 0\ \text{on}\ ]0,T[\times\Gamma,\ u(0,x) = u_o(x)\ \text{in}\ \Omega \tag{5.113}$$

has a unique solution

$$u \in C([0,T_{max}[;H^2(\Omega)) \cap C^1(]0,T_{max};\ L^2(\Omega)) \tag{5.114}$$

with $T_{max} < +\infty$ (so $\lim_{t\uparrow T_{max}} \|u(t)\|_{H^2} = -\infty$).

<u>Proof</u>. According to Theorem 5.13 (with $f(u) = u^3$) there is a unique solution u of the problem (5.12)+(5.113) with the properties (5.114). Since $u(t) \in H^2(\Omega) \subset L^\infty(\Omega) \subset L^2(\Omega)$ we have also $u^3(t,.) \in L^2(\Omega)$. Moreover, $\Delta u(t) \in L^2(\Omega)$. It follows that the problem (5.12) + (5.13) can be written in $L^2(\Omega) \equiv H$ as

$$u'(t) + \partial g u(t) = Fu(t),\ u(0)=u_o \in H_o^1 \cap H^2,\ o \leq t < T_{max} \tag{5.115}$$

with $(Fu(t))(x) = u^3(t,x)$, $x \in \Omega$ and

$$g(v) = \frac{1}{2}\int_\Omega |\nabla v(x)|^2\ dx\ \text{for}\ v \in H_o^1(\Omega),\ g(v)=+\infty\ \text{,otherwise} \tag{5.116}$$

(see (4.1115)). On the basis of the above remarks, Hypotheses of Lemma 2.2. in Appendix are fulfilled. Thus, multiplying (in $L^2(\Omega)$) with $u'(t)$, and taking into account (2.59) in Appendix, we obtain

$$\|u'(t)\|_2^2 + \frac{d}{dt}\ g(u(t)) = \int_\Omega u'(t,x)u^3(t,x)dx,\ \text{on}\ [0,T_{max}[\tag{5.117}$$

Integrating over [0,t] one derives

$$E(u(t)) \leq E(u(0)) \equiv E(u_o), \text{ for all } t \in [0,T_{max}[\tag{5.118}$$

which yields (if $E(u_o) \leq 0$)

$$-\int_\Omega |\nabla u(t,x)|^2 \, dx \geq -\frac{1}{2}\int_\Omega u^4(t,x)dx, \quad t \in [0,T_{max}[\tag{5.119}$$

Finally, set $w(t) = \|u(t)\|_2^2 = \int_\Omega u^2(t,x)dx$. Then

$$w'(t) = 2\int_\Omega u(t,x)u_t'(t,x)dx = 2\int_\Omega u \Delta u \, dx + 2\int_\Omega u^4 dx =$$
$$= -2\int_\Omega |\Delta u|^2 dx + 2\int_\Omega u^4 dx \geq \int_\Omega u^4 dx \tag{5.120}$$

Inasmuch as

$$\int_\Omega u^2 dx \leq (\text{mes}\,\Omega)^{1/2}\{\int_\Omega u^4 dx\}^{1/2}$$

we have

$$w'(t) \geq (\text{mes}\,\Omega)^{-1} w^2(t), \quad t \in [0,T_{max}[\tag{5.121}$$

Integrating (5.121), it follows

$$t \leq (\text{mes}\,\Omega)/w(0) = (\text{mes}\,\Omega)\|u_o\|_2^{-2}, \ \forall\, t \in [0,T_{max}[\tag{5.122}$$

and hence $T_{max} < +\infty$, q.e.d.

Remark 5.8. Take $u_o \in H_o^1 \cap H^2$, with $u_o \neq 0$. If $E(u_o) > 0$ then we can find $d > 0$ such that $E(du_o) \leq 0$ (e.g. $d \geq \sqrt{2}\ \|\nabla u_o\|_2 \|u_o\|_4^{-2}$). Another important aspect is that there exists $a > 0$, such that $\|u_o\|_{H_o^1} < a$ implies $T_{max} = \infty$ (see e.g. Brézis [6], p. 26). Corollary 5.2; is a "limit" case (i.e. it corresponds formally to r=3) of the following result

Theorem 5.14. Let $\Omega \subset R^N$ and j be as in Theorem 5.5 If $u_o \in H^1(\Omega)$ and $j(u) \in L^1(\Gamma)$, then there exists at least one strong solution $u: [0,T_{max}[\to L^2(\Omega)$ of the problem

$$\frac{\partial u}{\partial t}(t,x) = \Delta u(t,x) + u(t,x)|u(t,x)|^{r-1},$$
$$\text{in }]0,T_m[\times \Omega \text{ with } 2<N,\ 1 \leq r < \frac{N}{N-2},\ T_m \equiv T_{max} \tag{5.123}$$

$$-\frac{\partial u}{\partial n} \in \partial j(u(t,x)); \text{ on }]0,T_m[\times \Gamma; u(0,x) = u_o(x) \text{ in } \Omega \tag{5.124}$$

satisfying

$$u(t) \in D(\Delta) \text{ given by (4.1), a.e. on }]0,T_m[\tag{5.125}$$

$t \to u'(t)$ and $t \to u(t,.)|u(t,.)|^{r-1}$ belong to

$$L^2(0,T;L^2(\Omega)) \text{ for every } 0 < T < T_m \tag{5.126}$$

$t \to f(u(t))$ is absolutely continuous on $[0,T]$ (with f given by (4.2)). If in addition $\tilde{E}(u_o) \leq 0$, with

$$\tilde{E}(u_o) = \frac{1}{2}\int_\Omega |\nabla u_o|^2 dx + \int_\Gamma j(u_o)d\sigma - \frac{1}{r+1}\int_\Omega |u_o|^{r+1}\, dx \tag{5.127}$$

and

$$2j(s) \geq s\ \ \partial j(s),\ \forall s \in D(\partial j) \tag{5.127'}$$

then $T_{max} < +\infty$.

<u>Proof.</u> The existence of u with the properties above follows from a general theory (see Ôtani [1] and e.g. Theorem 4.2.1 in the forthcoming book of Vrabie [3, Ch. 4]) as follows: set

$$F(v)(x) = v(x)|v(x)|^{r-1},\ v \in L^{2r}(\Omega) \tag{5.128}$$

Since $1 \leq r < N/N-2$ (with $2 < N$) we have $H^1(\Omega) \hookrightarrow L^{2r}(\Omega)$ (with compact embedding) so $F: D(\partial f) \to L^2(\Omega)$ where f is given by (4.2) and $D(\partial f) = D(\Delta) \subset H^2(\Omega)$ (see (4.1)). It is not difficult to check that F is f-demicontinuous (in the sense of Definition 4.2.1 in the above book of Vrabie [3, Ch. 4].

Since $H^1 \hookrightarrow L^{2r}$, there is $c > 0$ such that

$$\|v\|_{2r} \leq c\left(\int_\Omega |v|^2 dx + \frac{1}{2}\int_\Omega |\nabla v|^2 dx\right)^{1/2},\ \forall\ v \in H^1(\Omega) \tag{5.129}$$

Now let M be a bounded subset of $L^2(\Omega)$ (say $\|v\|_2 \leq k$, $\forall v \in M$). Since $j(s) \geq 0$, $\forall s \in R$, by (5.129) we derive

$$\|F(v)\|_2 \leq \left\{\int_\Omega |(Fv)(x)|^2 dx\right\}^{1/2} = \left\{\int_\Omega |v(x)|^{2r}\right\}^{1/2} \leq$$

$$C\left(k^2 + \frac{1}{2}\int_\Omega |\nabla v|^2 dx + \int_\Gamma j(v(x))d\sigma\right)^{1/2} = c\ \hat{g}(f(v)), \forall v \in M \cap D(\partial f) \tag{5.130}$$

where $\hat{g}(r) = (k^2 + r)^{1/2}$ is a nondecreasing function from R_+ into R_+. Therefore, Hypotheses of Theorem 4.2.1 mentioned above are fulfilled. It remains to prove that $E(u_o) \leq 0$ implies $T_{max} < +\infty$. Following the proof of Corollary 5.2 with f given by (4.2) in place of g, $u|u|^{r-1}$ in place of u^3 and E given by (5.127) in place of E given by (5.111), we get

$$E(u(t)) \leq E(u_o),\ \forall\ t \in [0, T_m[\tag{5.131}$$

Set also $w(t) = \int_\Omega u^2(t,x)dx$. Then

$$w'(t) = 2\int_\Omega u\, u' dx = 2\int_\Omega (u\Delta u + |u|^{r+1})dx = \tag{5.132}$$

$$= -2\int_\Omega |\nabla u|^2 dx + 2\int_\Gamma u\,\frac{\partial u}{\partial n}\, d\sigma + 2\int_\Omega |u(t,x)|^{r+1}\, dx$$

A simple combination of (5.131) and (5.132) (eliminating $-2\int_\Omega |\nabla u|^2 dx$) yields

$$w'(t) \geq 2\,\frac{r-1}{r+1}\int_\Omega |u(t,x)|^{r+1}dx + 2\int_\Gamma (2j(u)-u\partial j(u))d\sigma - 4E(u_o) \qquad (5.133)$$

In view of (5.127)' and if $E(u_o) \leq 0$, (5.133) yields

$$w'(t) \geq 2\,\frac{r-1}{r+1}\int_\Omega |u|^{r+1}dx,\ 0 < t < T_{max}\,. \qquad (5.134)$$

with $p = (r+1)/2$ and $q = \frac{r+1}{r-1}$, by Hölder inequality we have

$$\|u\|_2^2 = \int_\Omega |u|^2 dx \leq k_1\ \|u\|_{r+1}^2,\ k_1 = (\text{mes}\,\Omega)^{(r-1)/r+1}$$

i.e. $\int_\Omega |u|^{r+1}dx \geq k_2(w(t))^{(r+1)/2},\ k_2 = (\text{mes}\,\Omega)^{2/(r-1)}$

Therefore

$$w'(t) \geq k_3(w(t))^{(r+1)/2},\ 0 \leq t < T_m,\ k_3 = \frac{2(r-1)}{r+1}\,k_2 \qquad (5.135)$$

Integrating (5.135) one obtains

$$\frac{2}{r-1}\,(w(0))^{(1-r)/2} \geq tk_3,\ \forall t \in [0,T_m[\quad (w(0)=\|u_o\|_2^2$$

so $T_{max} < +\infty$, q.e.d.

<u>Remark 5.9.</u> Let us comment the condition $E(u_o) \leq 0$. If e.g. $j(u_o(x))=0$ on Γ or if $j(s)=ks^2$ with $k>0$, $s \in R$ then we can choose $d>0$ such that $E(du_o) \geq 0$ (this is because $r>1$. See Remark 5.8). Clearly, Inequality (5.127)' means that for every $y \in \partial j(s)$ we have $2j(s) \geq s\,y$. In the case $j(s) = ks^2$ we have $\partial j(s)=2ks$ so (5.127)' holds. We give two examples of j satisfying Hypotheses of Theorem 5.14 as follows

$$j(s)= \begin{cases} 0 & ,\text{ if } s \geq 0 \\ +\infty, & \text{if } s < 0 \end{cases} \qquad (5.136)$$

In this situation

$$\partial j(s)= \begin{cases}]-\infty,0] & ,\text{ if } s = 0 \\ 0 & ,\text{ if } s > 0 \\ \emptyset & ,\text{ if } s < 0 \end{cases} \qquad (5.136)'$$

where $\emptyset$ is the empty set (i.e. $D(\partial j) = [0,+\infty[$).

Of course, (5.127)' is trivially satisfied. Moreover, in this case, the existence of a solution u to (5.123)-(5.125) implies $u(t,x)\ D(\partial j)$, on $]0,T_m[\times\Gamma$, i.e. $u(t,x) \geq 0$ so $j(u(t,x) = 0$ on $]0,T_m[\times\Gamma$. We can also choose u_o such that $u_o(x) \geq 0$ on (so $j(u_o) = 0$). Therefore in these conditions, $T_{max} < +\infty$. Another example is the following one

$$j(s)=\begin{cases} 0 & ,\text{if } s = 0 \\ +\infty & ,\text{if } s \neq 0 \end{cases} \tag{5.137}$$

Then

$$\partial j(s)=\begin{cases}]-\infty,+\infty[, & \text{if } s = 0 \\ \emptyset & , \text{is } s \neq 0 \end{cases} \tag{5.137'}$$

i.e. $D(\partial j) = \{0\}$ and $\partial j(0)= R$. In this case $D(\Delta) = H_o^1 \cap H^2$ (see (4.14)) and f is given by (4.15). Thus, the boundary condition (5.124) becomes just (5.87) and $\tilde{E}$ has to be E_r below

$$E_r(u_o) = \frac{1}{2}\int_\Omega |\nabla u_o|^2 dx - \frac{1}{r+1}\int_\Omega |u_o|^{r+1} dx, \quad u_o \in H_o^1(\Omega) \tag{5.138}$$

Therefore, the fact that the problem (5.123)-(5.124) with $E_r(u_o) \leq 0$ (and j given by (5.137) has a solution u with the properties (5.125)-5.126) (with $T_{max} < +\infty$) follows from Theorem 5.14. However, in this (semilinear) case a more precise result (due to Ball [2]) is known, as follows

<u>Theorem 5.15.</u> Let $1 \leq r < N/(N-2)$ for $N \geq 3$ and $r \geq 1$ if N=1, N=2. If $u_o \in H_o^1(\Omega)$, there is a unique solution u to the problem (5.123) with $u(t,x)=0$ on $[0,T_m[\times \Gamma$ and $u(0,x) = u_o(x)$ in Ω. Solution u has the regularity

$$\begin{aligned} &u \in C([0,T_{max}[;H_o^1(\Omega)) \cap C^1(]0,T_{max}[;L^2(\Omega)), \\ &\qquad u(t) \in H_o^1(\Omega) \cap H^2(\Omega) \quad \text{for } t \in [0,T_{max}[\end{aligned} \tag{5.139}$$

If, in addition $E_r(u_o) \leq 0$, then $T_{max} < +\infty$ and

$$\lim_{t \uparrow T_{max}} \int_\Omega |u(t,x)|^{r+1} dx = +\infty .$$

Note that in the case N=1 or N=2 the only restriction on r is $r \geq 1$ (this is because in these cases we have $H^1(\Omega) \hookrightarrow L^p(\Omega)$, $\forall p \geq 1$, with compact injection.

§ 6. Wave Equation. Schrödinger equation.

6.1. Wave equation

The wave equation below is an important example of hyperbolic equation. Precisely , we are concerned with the existence and uniqueness of a function $u:Q \to R$ such that

$$\frac{\partial^2 u}{\partial t^2} - \Delta u = 0, \quad \text{in } Q \tag{6.1}$$

$$u(t,x) = 0, \quad \text{on } F \tag{6.2}$$

$$u(0,x) = u_o(x), \ \frac{\partial u}{\partial t}(0,x) = v_o(x) \text{ in } \Omega \tag{6.3}$$

where Q and F are defined by (4.16). For simplicity of writing set,

$$\frac{\partial u}{\partial t} = u_t$$

Theorem 6.1. Let Ω be a bounded domain of R^N with smooth boundary Γ. Then for every $u_o \in H^2(\Omega) \cap H^1_o(\Omega)$ and $v_o \in H^1_o(\Omega)$, the problem (6.1)-(6.3) has a unique (classical) solution u (see Definition 5.1) on $[0,+\infty[$, with

$$u \in C([0,+\infty[;H^2 \cap H^1_o) \cap C^1([0,+\infty[;H^1_o) \cap C^2([0,+\infty[;L^2(\Omega)) \tag{6.4}$$

In addition, u satisfies the energy equation

$$E(u(t,.), u_t(t,.)) = E(u_o,v_o), \ \forall\, t \geq 0 \tag{6.5}$$

where

$$E(u_o,v_o) = \int_\Omega |\nabla u_o|^2 dx + \int_\Omega (v_o(x))^2 dx \tag{6.6}$$

(II) If the initial conditions u_o and v_o satisfy $u_o, v_o \in H^k(\Omega)$, k=1,2,... with the compatibility conditions $y = \Delta y = \dots = \Delta^j y = \dots = 0, y=u_o,v_o$ (j=1,2,...) on Γ, then the corresponding solution $u \in C^\infty([0,+\infty[\times \Omega)$.

Proof. Equation (6.1) can be written in the form

$$u_t = v, \ v_t = \Delta u \text{ in } Q \tag{6.7}$$

or

$$U' = AU, \ U(o) = \begin{pmatrix} u_o \\ v_o \end{pmatrix} = U_o; \ U = \begin{pmatrix} u \\ v \end{pmatrix}, \ t \geq 0 \tag{6.8}$$

in $H = H^1_o(\Omega) \times L^2(\Omega)$

with

$$AU = \begin{pmatrix} 0 & I \\ \Delta & 0 \end{pmatrix} \begin{pmatrix} u \\ v \end{pmatrix}, \ D(A) = (H^2 \cap H^1_o) \times H^1_o \tag{6.9}$$

According to Corollary 4.3, $-A$ is a maximal-monotone skew-adjoint operator in the Hilbert space H with inner producet (4.55). Let $S(t)$ be the semigroup generated by A. We have laready proved that $S(t)$ is unitary (i.e. $S^{-1}(t) = S(t)$) and that $S(t)$ is an isometry i.e. $\|S(t)U_o\| = \|U_o\|$, $\forall\, t \geq 0$. Since $u(t) = S(t)u_o = U(t) = \binom{u}{u_t}$ is the strong solution to (6.8) with $U_o \in D(A)$, (6.4) and (6.5) follow.

The fact that $S(t)$ is an isometry means just (6.5) (this is because the inner product on H is given by (4.55), i.e. $\|U_o\|^2 = \|u_o\|_2^2 + \|v_o\|_2^2$) Actually $S(t)$ can be extended to a group as indicated by (4.60).

(II) The regularity of u (i.e. the property $u \in C^\infty([0,+\infty[\times\Omega))$ can be proved as in the case of Theorem 5.3.

<u>Remark 6.1.</u> In the case of Hyperbolic equation (6.1) the corresponding semigroup $S(t)$ (generated by A defined as in (6.9)) is not compact. This is because $S(t)$ is a group (see Remark 6.1 in Ch. 2). Moreover, $S(t)$ has not a regularizing (smoothing) effect on the initial data as in the case of the heat equation, i.e. the solution is not smoother than the initial conditions u_o and v_o. Indeed take $\Omega = R$. Then the corresponding solution of (6.1) (6.3) has the following explicit form

$$u(t,x) = \frac{1}{2}\,[u_o(x+t)+u_o(x-t)] + \frac{1}{2}\int_{x-t}^{x+t} v_o(s)ds = S(t)\begin{pmatrix}u_o\\ v_o\end{pmatrix} \tag{6.9'}$$

For $v_o = 0$ it is clear that $u(t,x)$ cannot be more regular than u_o. If e.g. u_o is not differentiable at $x_o \in R$, then $u(t,x)$ is not differentiable along the characteristics

$$x + t = x_o,\quad x - t = x_o \tag{6.10}$$

In other words there is a propagation of singularities along the characteristics. The equation of Klein-Gordon

$$u_{tt} = \Delta u - a^2 u \text{ in } Q \tag{6.11}$$

with $a \neq 0$ can be reduced to (6.1) by a change of variable $v(t,x) = e^{rt}u(t,x)$, $r \in R$. Actually, the operator $Bu = \Delta u - a^2 u$, $u \in H^2 \cap H_o^1 = D(B)$ is strongly dissipative in $L^2(\Omega)$, i.e.

$$\langle Bu,u\rangle \leq -\,a^2\,\|u\|_2^2,\quad u \in D(B) \tag{6.12}$$

so the conclusion (6.4) of Theorem 6.1 remains also valid in the case of (6.11). In the case $\Omega = R^N$, taking into account Theorem 4.3 we have (see also Pazy [4], p. 222).

Theorem 6.2. If $u_o \in H^2(R^n)$ and $v_o \in H^1(r^N)$, then the problem (6.1)+(6.3) with $\Omega = R^N$ has a unique solution $u \in C^1([0,\infty[;\ H^2(R^N))$.

6.2. The semilinear wave equation

Similar to (5.86) we now consider the problem

$$u_{tt} = \Delta u + f(u), \text{ in }]0,T[\times\Omega \tag{6.13}$$

$$u(t,x) = 0 \quad , \text{ on }]0,T[\times\Gamma \tag{6.14}$$

$$u(0,x) = u_o(x),\ u_t(0,x) = v_o(x), \text{ in } \Omega \tag{6.15}$$

Set $U_o = \begin{pmatrix} u_o \\ v_o \end{pmatrix}$. The analogous of Theorem 5.10 and 5.13 are respectively

Theorem 6.3. Let $f: R \to R$ be globally Lipshitz (e.g. $f \in C^1(R)$ with f' bounded on R). Then for every $U_o \in H_o^1(\Omega) \cap L^2(\Omega)$, the problem (6.13)-(6.15) has a unique mild solution u in $H = H_o^1(\Omega) \cap L^2(\Omega)$ on $[0,\infty[$. If in addition $U_o \in (H^2 \cap H_o^1) \times H_o^1$, then u is a classical solution $[0,+\infty[$, i.e.

$$u \in C([0,\infty[;H^2 \cap H_o^1) \cap C^1([0,\infty[;H_o^1) \cap C^2([0,\infty[;L^2(\Omega)) \tag{6.16}$$

Theorem 6.4. If $f \in C^3(R)$ and $f(0) = 0$, then for every $u \in H^2 \cap H_o^1$ and $v_o \in H_o^1$, the problem (6.13)-(6.15) admits a unique classical u on $[0,T_{max}[$ with either $T_{max} = +\infty$ or

$$T_{max} < +\infty \text{ and } \lim_{t \uparrow T_{max}} \|u(t)\|_{H^2} + \|u_t(t)\|_{H_o^1} = \infty \tag{6.17}$$

As for the heat equation (Corollary 5.1), in the case $f(u) = -u^3$ (so called "the good case" - Brézis [6]) one has $T_{max} = +\infty$, i.e.

Corllary 6.1. For every $u_o \in H^2 \cap H_o^1$ and $v_o \in H_o^1$, the equation

$$u_{tt} = \Delta u - u^3, \text{ in } [0,\infty[\times\Omega \tag{6.18}$$

$$u(t,x) = 0 \quad , \text{ on } [0,\infty[\times\Gamma \tag{6.19}$$

with the initial conditions (6.15), has a unique classical solution u.

Proof. In view of Theorem 6.4 we have to prove that $T_{max} = +\infty$. Multiplying (6.18) in $L^2(\Omega)$ by u_t, using Green's formula and then integrating over $[0,t]$ one obtains the energy first integral

$$E(u(t),u_t(t)) = \frac{1}{2}\int_\Omega (|\nabla u(t,x)|^2 + (u_t(t,x)^2)\,dx + \tag{6.20}$$
$$+\frac{1}{4}\int_\Omega (u(t,x))^4\,dx = E(u_o,v_o) = \text{const.},\ \forall\, t\in [0,T_{max}[$$

which shows that the blow-up phenomenon cannot occur (hence $T_{max} = +\infty$). In the "hard case" $f(u)=u^3$, the energy first integral is clearly the following one

$$\tilde{E}(u(t),\ u_t(t)) = \frac{1}{2}\|\nabla u(t)\|_2^2 + \frac{1}{2}\|u_t\|_2^2 - \frac{1}{4}\|u(t)\|_4^4 = \tilde{E}(u_o,v_o)=\text{const.} \tag{6.21}$$

for $t\in [0,T_{max}[$, and we have

Corllary 6.2. Let $u_o\in H^2\cap H_o^1$ and $v_o\in H_o^1$. If $\tilde{E}(u_o,v_o)\leq 0$ then the classical solution u of the problem

$$u_{tt} = \Delta u+u^3,\quad \text{in } [0,T_{max}[\ \times\Omega \tag{6.22}$$

$$u(t,x) = 0\ ,\quad \text{on } [0,T_{max}[\ \times\Gamma \tag{6.23}$$

with the initial conditions (6.15), blows-up in finite time ($T_{max} < +\infty$ and (6.17) holds).

Proof. As in the proof of Theorem 5.14 set $w(t) = \int_\Omega |u(t,x)|^2 dx$ (where $u(t)$ means the function $x \to u(t,x)$). We have (according to (6.21) and (6.22))

$$w''(t) = 2\int_\Omega |u_t\ (t,x)|^2\,dx + 2\int_\Omega |u(t,x)|^4 dx - 2\int_\Omega |\nabla u(t,x)|^2 dx =$$
$$= \int_\Omega |u(t,x)|^4\,dx + 4\ \|u_t\|_2^2 - 4.\tilde{E}(u_o,v_o)$$

Therefore (see (5.135))

$$w''(t)\geq\int_\Omega |u(t,x)|^4\,dx \geqslant (\text{mes}\,\Omega)(w(t))^2\ ,\ t\in[0,T_{max}[\tag{6.24}$$

which leads (by integration) to $T_{max} < +\infty$

Now let us recall another semilinear case

$$u_{tt} = \Delta u+u|u|^{r-1},\ \text{in } [0,T_m[\times\Omega \tag{6.25}$$

with the boundary-initial conditions as in (6.14),(6.15) and $T_m=T_{max}$. In this case the energy conservation law (6.21) is given by

$$E_r(u(t),\ u_t(t)) = \frac{1}{2}\|\nabla u(t)\|_2^2 + \frac{1}{2}\|u_t\|_2^2 - \frac{1}{r+1}\|u(t)\|_{r+1}^{r+1} = \tag{6.26}$$
$$= E_r(u_o,v_o) = \text{const.},\ 0\leq t < T_m$$

The analogous of Theorem 5.15 is given by

Theorem 6.5. Let r be as in Theorem 5.15 . If $u_o, v_o \in H_o^1 \cap H^2$, then there exists a unique solution u to the problem (6.25)+(6.2)+(6.3) with $u, u_t \in C([0,T_m; L^2(\Omega))$. If in addition $E_r(u_o,v_o) \leq 0$, then $T_m < \infty$ and

$$\lim_{t\uparrow T_m} \|u\|_{r+1} = \lim_{t\uparrow T_m} (\|\nabla u(v)\|_2^2 + \|u_t\|_2^2) = +\infty$$

For the proof see Ball [2]. Note that in this case (6.24) becomes

$$\begin{aligned} w''(t) &= \int_\Omega (2|u_t|^2 - 2|\nabla u|^2 + 2|u|^{r+1})\, dx \geq \\ &\frac{2(r-1)}{r+1}\int_\Omega |u|^{r+1} dx - 4E(u_o,v_o) \geq k(w(t))^{(r+1)/2} \end{aligned} \qquad (6.28)$$

with a positive constant k. This implies $T_m < +\infty$.

6.3. Semilinear Schrödinger Equation

We have already seen that the unique (classical) solution $u:[0,+\infty[\times R^N \to H^2(R^N)$ of the problem

$$\frac{\partial u}{\partial t} = i\,\Delta u, \text{ on } [0,+\infty[\times R^N \qquad (6.29)$$

$$u(0,x) = u_o(x) \text{ on } R^N \qquad (6.29)'$$

with N = 2 is given by the convolution (see (4.74)-(4.75))

$$u(t,x) \equiv (S(t)u_o)(x) = (g_t * u_o)x \equiv u(t)(x) \qquad (6.30)$$

where

$$g_t(x) = \frac{1}{4\pi t i} e^{i\|x\|^2/4t} \qquad (6.30)'$$

Moreover, S(t) is a group of unitary operators so

$$\|S(t)u_o\|_{L^2(R^2)} = \|u_o\|_{L^2(R^2)}, \ \forall t \geq 0,\ u_o \in L^2(R^2) \qquad (6.31)$$

In the general case of R^N, (6.30) is valid with $(4\pi t i)^{N/2}$ in place of $4\pi t i$. Now let us recall briefly how one proves via semigroups approach the existence and uniqueness of solution to (6.29) with Ω in place of R^N. Therefore, let Ω be a bounded domain of R^N with smooth boundary Γ and let us consider the problem

$$\frac{\partial u}{\partial t} = i\,\Delta u, \text{ in } [0,\infty[\times\Omega \qquad (6.32)$$

$$u(t,x)=0 \text{ , on } [0,\infty[\times\Gamma \qquad (6.33)$$

$$u(0,x) = u_o(x), \text{ in } \Omega \qquad (6.34)$$

with $u: [0,\infty[\times \bar{\Omega} \to C$ (the complex field). If we set $u=u_1+iu_2$ with $u_1 = \mathrm{Re}(u)$, $u_2=\mathrm{Im}(u)$ then (6.32)-(6.34) become

$$\frac{\partial u_1}{\partial t} = -\Delta u_2, \quad \frac{\partial u_2}{\partial t} = \Delta u_1 \quad \text{in } [0,\infty[\times \Omega \tag{6.35}$$

$$u_1 = u_2 = 0 \qquad \text{on } [0,\infty[\times \Gamma \tag{6.36}$$

$$u_1(0,x) = u_1^o(x), \; u_2(0,x) = x_2^o(x), \text{ in } \Omega \tag{6.37}$$

where $u_j: [0,\infty[\times \Omega \to R$ ($j=1,2$) are real valued).

Consequently the problem (6.32)-(6.34), can be reduced (via (6.35)-(6.37)) to the abstract evolution equation

$$\frac{dU}{dt} = AU, \; U(o) = U_o = \begin{pmatrix} u_1^o \\ u_2^o \end{pmatrix} \tag{6.38}$$

in the real Hilbert space $H = L^2(\Omega)\times L^2(\Omega)$ where

$$AU = \begin{pmatrix} -\Delta u_2 \\ \Delta u_1 \end{pmatrix}, \text{ for } U = \begin{pmatrix} u_1 \\ u_2 \end{pmatrix} \in D(A) = (H^2 \cap H_o^1) \times (H^2 \cap H_o^1) \tag{6.39}$$

As usual, the inner product of H is defined by $\langle U,V \rangle_H = \langle u_1,v_1 \rangle_{L^2} + \langle u_2, v_2 \rangle_{L^2}$ for $V = \begin{pmatrix} v_1 \\ v_2 \end{pmatrix}$ and U as above.

In view of Green formula it follows at once $\langle AU,U \rangle_H = 0$, $\forall U \in D(A)$, hence A and -A are dissipative in H. It can be proved that A and -A are maximal dissipative in H and therefore A generates a group S(t) of unitary operators in H. Thus <u>for every</u> $U_o \in D(A)$, (6.38) <u>has a unique classical solution</u> $U(t) = S(t)U_o$, $t \geq 0$ in H, with $\|U(t)\|_{L^2} = \|U_o\|_2$, $t \geq 0$.

Let us proceed to study (6.32) with the nonlinear perturbation of the form $u\tilde{f}(|u|^2)$. In other words we are now concerned with the semilinear problem

$$\frac{1}{i}\frac{\partial u}{\partial t} = u + u\tilde{f}(|u|^2), \text{ in } [0,T[\times \Omega \tag{6.40}$$

$$u(t,x) = 0 \qquad , \text{ on } [0,T[\times \Gamma \tag{6.41}$$

$$u(0,x) = u_o(x) \qquad , \text{ in } \Omega \tag{6.42}$$

where $\tilde{f}: [0,+\infty[\to R$ is a real-valued function.

Set

$$F(U) = \begin{pmatrix} -u_2\tilde{f}(u_1^2+u_2^2) \\ \\ u_1\tilde{f}(u_1^2+u_2^2) \end{pmatrix}, \text{ for } U = \begin{pmatrix} u_1 \\ u_2 \end{pmatrix} \in H \qquad (6.43)$$

where $u_1^2+u_2^2 = ||U||^2 = |u|^2$, $u = u_1 + iu_2$.

It is now clear that the problem (6.40)-(6.42) reduces to (5.59) with A and F given by (6.39) and (6.43) and U_o as in (6.38). On the basis of Theorem 5.8 (part (2) one easily derives

Theorem 6.6. Suppose that: $\tilde{f}\in C^1(R_+)$, $|r\tilde{f}'(r)| \le M_1$ and $|\tilde{f}(r) \le M_2$ for every $r \ge 0$ (where M_1, M_2 are positive constants). Then for every $u_o = u_o^1 + iu_o^2$ with u_o^1, $u_o^2 \in H^2(\Omega) \cap H_o^1(\Omega)$, the problem (6.40)-(6.41) admits a unique classical solution $u=u_1+iu_2$, $u\in C([0,\infty[;H^2 \cap H_o^1) \cap L^1([0,\infty[;L^2(\Omega))$.

Theorem 6.6 is similar to Theorem 5.10. The result corresponding to Theorem 5.13 is the following one

Theorem 6.7. Suppose that $\tilde{f}\in C^3(R_+)$. Then for every u_o as in Theorem 6.6 there exists a unique (local) classical solution u on $[0,T_{max}[$ to the problem (6.40)-(6.42) with, either $T_{max} = +\infty$ or

$$T_{max} < +\infty \text{ and } \lim_{t\uparrow T_{max}} ||u(t)||_{H^2} = \infty \quad (\text{which implies}) \; ||u(t)||_\infty \underset{t\uparrow T_{max}}{\to} \infty)$$

Proof. One applies Theorem 5.9 and the technique in the proof of Theorem 5.12. If $\limsup_{t\uparrow T_{max}} ||u(t||_\infty < \infty$, it follows that $||u(t)||_\infty$ is bounded on $[0,T_{max}[$ so we cannot have a blow-up in $H^2(\Omega)$.

Remark 6.2. 1) In the case $\tilde{f}(r) = -r$ (i.e.

$$\frac{\partial u}{\partial t} = i\,\Delta u - iu|u|^2 \qquad (6.44)$$

and N = 1,2,3, then $T_{max} = +\infty$. 2) In the "hard" case $\tilde{f}(r) = r$, the conclusions are similar to those in Corollary 5.2 and remark 5.8.

For the proof we take into account that we have the following two first integrals

$$E(u(t)) = \int_\Omega |\nabla u|^2 - \frac{1}{2}\int_\Omega |u|^4\, dx = E(u_o) = \text{const.}, \; \forall\, t \ge 0$$

Appendix

In what follows we will recall some of the basic results (most of nonlinear anlysis) which have been used (more or less) in this work.

§ 1. Duality mapping. Subdifferentials.

Let X be a Banach space of norm $\|.\|$ and let X* be its dual (of norm $\|.\|_*$). The mapping $F : X \to 2^{X^*}$ defined by

$$F(x) = \{x^* \in X^*;\ x^*(x) = \|x\|^2 = \|x^*\|_*^2\} \tag{1.1}$$

is called the duality mapping of X.

Several properties of F can be found in author's books [12],[15]. See also Remark 1.6.

The function $f:X \to]-\infty, +\infty]$ is said to be proper if it is nonidentically $+\infty$. Denote by

$$D_e(f) = \{x \in X,\ f(x) < +\infty\} \equiv D(f) \tag{1.2}$$

the effective domain of f.

The reason for which we are not interested here in the case $f(x)=-\infty$ for some $x \in X$ is pointed out by

Proposition 1.1. If $f:X \to [- \ ,+\infty]$ is a lower semicontinuous convex (l.s.c.) function such that $f(x_o)=-\infty$ form some $x_o \in X$, then f is nowhere finite.

Proof. Assume by contradiction that there is $x_1 \in X$, such that $f(x_1)$ is a finite number. The convexity of f implies obviously

$$f(\lambda x_o+(1-\lambda)x_1)=-\infty \quad \text{for every } \lambda \in]o,1[$$

Letting $\lambda \downarrow o$ and taking into account the lower semicontinuity of f at x_1 one obtains $f(x_1)=-\infty$ which concludes the proof. Recall that a proper (l.s.c.) $f:X \to]- \ ,+\infty]$ is bounded from below by an affine function, i.e. there exists $x_o^1 \in X^*$ and $b \in R$ such that $f(x) \geq x_o^*(x)+b$ for all $x \in D_e(f)$ (see e.g. Barbu-Precupanu [1] or Pavel [12]).

Definition 1.1. Let $f:X \to]-\infty,+\infty]$ be a proper function. 1) The subdifferentials of f at $x \in X$ (denoted by $\partial f(x)$) are defined as below

$$\partial f(x) = \{x^* \in X\ ,\ f(y)-f(x) \geq \langle x^*,y-x\rangle\ ,\forall y \in X\} \tag{1.3}$$

where $\langle x^*,y\rangle$ denotes the value of x* at y (i.e. x*(y)).

2) $\partial f: D(\partial f) \subset X \to 2^{X^*}$, is called the subdifferential of f and the elements $x^* \in \partial f(x)$ are also called subgradients of f at x.

It is easy to see that $\partial f(x)$ may be empty or it may contain more than one element. Indeed, let K be a nonempty convex subset of X. The indicator function I_K of K i.e.

$$I_K(x)=0 \text{ if } x \in K \text{ and } I_K(x)= +\infty \text{ if } x \in X \setminus K \tag{1.4}$$

is a proper l.s.c. function. Clearly,

$$D(\partial I_K) = D_e(I_K)=K, \ \partial I_K(x)=\{ x^* \in X^*, \ \langle x^*, x-y \rangle \geq 0, \forall y \in K\} \tag{1.5}$$

for all $x \in K$.

Moreover, if $x \in X \setminus K = \{ z \in X, \ z \bar{\in} K\}$, $\partial I_K(x) = \emptyset$ and if $x_o \in \text{Int}K$ (the interior of K), then $\partial I_K(x_o)=\{o\}$. Finally, if y is tangent to K at $x \in K$ (in the sense of (7.15) in Cap. 1) i.e.

$$x+hy+hr(h) \in K \quad \text{for all } h > o \tag{1.6}$$

with some $r(h) \to o$ as $h \downarrow o$, then for each $x^* \in \partial I_K(x)$, it follows $\langle x^*, y \rangle \leq o$. If (1.6) holds for all sufficiently small $h \in R$, then $\langle x^*, y \rangle = o$. Therefore, it is natural to say that $\partial I_K(x)$ is the cone of "normals" to K at x, $T_K(x)$- the tangent cone to K at x, where

$$T_K(x) = \{ y \in X, \lim_{h \downarrow o} d[x+hy;K]/h = o\} = \{ y \in X, \ y \text{ satisfies } (1.6)\} \tag{1.7}$$

For other aspects in this direction we refer to Clarke [1], Pavel [15] and Ursescu [1]).

Given the convex (proper) functional $f: X \to R$ set

$$f'_+(x,y) = \lim_{t \downarrow o} \frac{f(x+ty)-f(x)}{t}$$
$$f'_-(x,y) = \lim_{t \uparrow o} \frac{f(x+ty)-f(x)}{t}, \ x, y \in X \tag{1.8}$$

The existence of the directional derivatives $f'_+(x,y)$ and $f'_-(x,y)$ is a consequence of convexity of f. In the case $f'_+(x,y)=f'_-(x,y) \equiv f'(x,y)$ $\langle \text{grad } f(x), y \rangle$ we say that f is Gâteaux differentiable at x in the direction of y (in short, f is G-differentiable at x). Clearly $f'_-(x,y) \leq f'_+(x,y)$ and

$$f'_+(x,ty) = tf'_+(x,y),\ \forall t \geq 0,\ f'_+(x,-y) = -f'_-(x,y) \tag{1.9}$$

$$f'_+(x,y_1-y_2) \leq f'_-(x,y_1)+f'_+(x,y_2),\ y_1,y_2,\ x \in X \tag{1.10}$$

Note that for proving (1.10) one observes that the convexity of f implies

$$t^{-1}(f(x+t(y_1+y_2))-f(x) \leq (2t)^{-1}\ (f(x+2ty_1)-f(x)) + \\ + (2t)^{-1}(f(x+2ty_2) - f(x))$$

An important result is

Theorem 1.1. Let $f: X \to R$ be a convex functional which is continuous at $x \in X$. Then for each $y_o \in X$, there exists $x_o^* \in X^*$(depending on x and y_o) with the properties

(1) $x_o^* \in \partial f(x), \langle x_o^*, y_o\rangle = f'_+(x,y_o)$ (in other words, f is subdifferentiable at x).

(2) $f(x+y)-f(x) \geq f'_+(x,y) \geq \langle x_o^*,y\rangle \geq f'_-(x-y) \geq -(f(x-y)-f(x))$ for all $y \in X$.

Note that $f'_+(x,y) \geq \langle x_o^*,y\rangle$ implies all of the inequalities in (2) (but it is useful to have (2) in this form).

Proof of Theorem 1.1. Set $X_o = \{ay_o,\ a \in R\}$ and define $g:X_o \to R$ by $g(ay_o) = af'_+(x,y_o)$. Therefore $g(y_o)=f'_+(x,y_o)$ and

$$g(z) \leq f'_+(x,z),\ \forall\ z \in X_o \tag{1.10)'$$

Indeed, if $z=ay_o$ with $a \geq 0$, then (1.10) holds (with $g(z)=f'_+x,z)$). If $a < 0$, then $g(z)=g(ay) \equiv af'_+(x,y_o)=-f'_+(x,-ay_o)=f'_-(x,ay_o) \leq f'_+(x,ay_o)= =f'_+(x,z)$ which proves (1.10). In view of Hahn-Banach theorem, the linear functional g can be extended from X_o to X. Denote by x_o^* such an extension of g. Clearly x_o^* satisfies both (1) and (2). The continuity of f at x implies (in view of (2)) the continuity of x_o^* as well as $x_o^* \in \partial f(x)$. This completes the proof.

Corollary 1.1. Let $f:X \to R$ be a convex functional which is continuous at x. Then f is differentiable at x and for each $y \in X$ we have

$$f'_+(x,y) = \sup\{x^*(y),\ x^* \in \partial f(x)\} = x_1^*(y) \equiv \langle x_1^*,y\rangle \tag{1.11}$$

$$f'_-(x,y) = \inf\{x^*(y),\ x^* \in \partial f(x)\} = x_2^*(y) \equiv \langle x_2^*,y\rangle \tag{1.11)'$$

for some x_1^* and x_2^* in $\partial f(x)$.

Proof. Fix $y \in X$ and $x^* \in \partial f(x)$. Then $f(x+ty)-f(x) \geq x^*(ty)$, for all $t > 0$. This yields $f'_+(x,y) \geq x^*(y)$. On the other hand, by Theorem 1.1, there exists $x_1^* \in \partial f(x)$ such that $x_1^*(y) = f'_+(x,y)$. Consequently (1.11) holds. Clearly (1.11) and $f'_-(x,y) = -f'_+(x,-y)$ imply (1.11)'.

Proposition 1.2. The duality mapping F is the subdifferential of the function f given by $f(x) = \frac{1}{2}\|x\|^2$, $x \in X$

Proof. Let $x \in X$ and $x^* \in F(x)$. then $\frac{1}{2}\|x+y\|^2 - \frac{1}{2}\|x\|^2 \geq \|x+y\| \ \|x\| - \langle x^*,x\rangle \geq \langle x^*,y\rangle$, $\forall y \in X$. Hence $F(x) \subset \partial f(x)$. Conversely, take $x^* \in \partial f(x)$ i.e.

$$\tfrac{1}{2}\|x+y\|^2 - \tfrac{1}{2}\|x\|^2 \geq \langle x^*,y\rangle, \forall\, y \in X. \tag{1.12}$$

Replacing y by ty with $t > 0$, it follows

$$\langle x^*,y\rangle \leq \lim_{t\downarrow 0} \frac{\|x+ty\|^2 - \|x\|^2}{2t} = \|x\| \lim_{t\downarrow 0} \frac{\|x+ty\| - \|x\|}{t} =$$

$$= \|x\|\langle y,x\rangle_+ \leq \|x\| \ \|y\|, \forall y \in X \tag{1.13}$$

Therefore $\|x^*\|_* \leq \|x\|$. Finally, for $y=tx$, with $t < 0$, (1.12) yields $\langle x^*,x\rangle \geq \|x\|^2$, hence $\|x^*\|_* = \|x\|$ and $x^*(x) = \|x\|^2$ (i.e. $x^* \in F(x)$), q.e.d.

Remark 1.1. If $f(x) = \frac{1}{2}\|x\|^2$, then $f'_+(x,y) \leq \langle y,x\rangle_s$ (see (2.1) in Ch. 1) and Corollary 1.1 becomes Proposition 2.1 in Ch. 1. Another useful result in nonlinear analysis is given by

Proposition 1.3 Assume that X is reflexive and $f:X \to]-\infty,+\infty]$ is a proper l.s.c. function with $f(x) \to +\infty$ as $\|x\| \to +\infty$. Then there exists $x_0 \in X$ such that

$$\inf_{x \in X} f(x) = f(x_0) \tag{1.14}$$

Proof. There exist a sequence $\{x_n\} \subset X$ with the property that $f(x_n) \to \inf_{x\in X} f(x)$, as $n \to \infty$. Because f is proper, it is easily seen that x_n is bounded and therefore we may assume that x_n is weakly convergent to an element $x_0 \in X$. Since f is weakly lower semicontinuous, we have $\lim_{n\to\infty} f(x_n) \geq f(x_0)$ which yields (1.14). q.e.d.

Clearly, if in addition f is strictly convex, then x_0 in (1.14) is unique.

Definition 1.2 The (possible multivalued) mapping $A:D(A) \subset X \to 2^{X^*}$ is said to be monotone if $\langle y_1-y_2, x_1-x_2\rangle \geq 0$, $\forall$ $x_i \in D(A)$ and $y_i \in Ax_i$ $i=1,2$.

A is said to be maximal monotone if it is not properly contained in any other monotone set of $X \times X^*$.

A direct consequence of Definition 1.1 is the fact that ∂f is monotone. Moreover,

Theorem 1.2 Let X be a real Banach space and let $f: X \to]-\infty,+\infty]$ be a proper, l.s.c. function. Then ∂f is maximal monotone.

For the proof see e.g. Rockafellar[1].

In this work we have used Theorem 1.2 in the case in which X is a real Hilbert space H. In this case, Theorem 1.2 can be restated as follows (with H^* identified with H).

Theorem 1.3. If $f:H \to]-\infty,+\infty]$ is a proper l.s.c. function, then f is subdifferentiable at least in one point $x_o \in D_e(f)$ (i.e. $D(\partial f) \neq \emptyset$) and ∂f is maximal monotone (which is equivalent to $R(I+\partial f)=H$).

Proof. Let $y \in H$. Define $g:H \;]-\infty,+\infty]$ by

$$g(x) = f(x) + \tfrac{1}{2} \|x-y\|^2, \quad x \in H \tag{1.14)'$$

Clearly, g is proper, l.s.c. and $g(x) \to +\infty$ as $\|x\| \to +\infty$ (the latter property is due to the fact that f is bounded from below by an affine function). On the basis of Proposition 1.3, there exists $x_o \in H$, such that $g(x) \geq g(x_o)$, $\forall x \in H$, i.e.

$$f(x) - f(x_o) \geq \tfrac{1}{2}\|x_o-y\|^2 - \tfrac{1}{2}\|x-y\|^2 \tag{1.15}$$

Replace $x= x_o+t(z-x_o)$ in (1.15) (with $o \leq t \leq 1$ and $z \in H$).

Convexity of f and an elementary computation lead us to

$$f(z) - f(x_o) \geq \langle y-x_o, z-x_o \rangle, \quad \forall z \in H$$

i.e. $x_o \in D(\partial f)$, and $y-x_o \in \partial f(x_o)$, q.e.d.

Remark 1.2. Obviously, (1.14) holds iff $o \in \partial f(x_o)$. In Theorem 8.10, Ch. 2 one points out that if $o \in \mathrm{Int}(D_e(f)-D_e(g))$, then $\partial f+\partial g=\partial(f+g)$. Following the proof of Theorem 1.3 one conlcudes that we have

Proposition 1.4. Let $f:H \to]-\infty, +\infty]$ be a convex proper function, $a \geq o$ and

$$g(x) = f(x)+ \frac{a}{2}\|x-y\|^2, \quad x, y \in H \tag{1.16}$$

Then

$$\min_{x \in H} g(x) = g(x_o) \quad \text{iff } a(y-x_o) \in \partial f(x_o) \tag{1.17}$$

(i.e. iff $x_0 = (I + \frac{1}{a}\partial f)^{-1} y \equiv J_{\frac{1}{a}} y$).

Another important consequence of Corollary 1.1 is the result below.

Proposition 1.5. Let $f: X \to R$ be a convex functional which is continuous at $x \in X$. If f is Gâteaux differentiable at x (i.e. $f'_+(x,y) = f'_-(x,y) \equiv f'(x,y)$ for all $y \in X$) then $\partial f(x)$ consists of a single element, namely $\partial f(x) = f'_+(x,.) \in X^*$. Conversely, if $\partial f(x)$ consists of a single element then f is G-differentiable at x, and $f'(x,.) = \partial f(x)$.

Obviously the conclusion of the proposition is a direct consequence of (1.11) and (1.11)'.

In particular, Proposition 1.5 yields

Proposition 1.6. If $f: X \to R$ is convex and Frechet differentiable at $x \in X$, then f is subdifferentiable at x, and $\partial f(x) = \dot{f}(x)$ (where $\dot{f}(x)$ denotes the Frechet derivative of f at x).

Proposition 1.5 can be restated as

Proposition 1.5.' Let $f: X \to]-\infty, +\infty]$ be convex, and let $x \in X$ be such that f is (finite and) continuous at x. Then $x \in D(\partial f)$. Moreover, $\partial f(x)$ consists of a single element if and only if f is G-differentiable at x and $\partial f(x) = f'(x,.) \equiv \text{grad } f(x)$.

Following the proof of Proposition 1.3 and taking into account that the weak closure of a convex set equals strong closure one can prove

Proposition 1.7. Let X be reflexive, $f: X \to]-\infty, +\infty]$ l.s.c. proper and M a bounded closed and convex subset of X. Then there exists $x_0 \in M$, such that

$$\inf_{x \in M} f(x) = f(x_0) \tag{1.18}$$

Actually, Proposition 1.7 is a consequence of the following general results

(I) a proper, lower semicontinuous functional $f: X \to]-\infty, +\infty]$ with X a topological separated space, attains its minimum on every sequentially compact set $M \subset X$.

(II) A convex lower semicontinuous function from the Banach space X into R is weakly lower semicontinuous.

(III) A bounded, convex and closed subset M of a reflexive space X is weakly compact.

We now give some otehr results on monotone operators (sets). To this goal , some notions are needed first.

Definition 1.3. (1) The (single valued) operator $A:X\to X^*$ is said to be hemicontinuous at $x\in D(A)$ if for all $y\in X$ with $x+ty\in D(A)$ (for sufficiently small $t\in R$) we have

$$A(t+ty) \rightharpoonup Ax \quad \text{as } t\to o \tag{1.19}$$

(2) A is said to be demicontinuous at $x\in D(A)$ if for every $x_n\in D(A)$ with $x_n\to x$, $w^* - \lim Ax_n = Ax$ as $n\to\infty$.

(3) The set $B\subset X\times X^*$ (see § 2) is said to be coercive, if

$$\lim \langle x_n, x_n^*\rangle / \|x_n\| = \infty$$

for every $[x_n, x^*_n]\in B$ with $\|x_n\|\to +\infty$ as $n\to\infty$.

Clearly, A is said to be hemicontinuous (demicontinuous) on D(A) if it is hemicontinuous (demicontinuous) at every $x\in D(A)$. Obviously, every demicontinuous operator is hemicontinuous, but the converse assertion fails. For example a linear unbounded operator is hemicontinuous but not necessarily demicontinuous.

The duality mapping F is coercive and if X^* is strictly convex, F is also demicontinuous (the proof of the latter property can be found e.g. in author's books [12,14].

Proposition 1.8. Every monotone and hemicontinuous operator $B:X\to X^*$ with $D(B)=X$ is maximal monotone.

Proof. Let $x_o\in X$ and $x_o^*\in X^*$ be such that

$$\langle Bx-x_o^*, x-x_o\rangle \geq 0, \ \forall\, x\in X \tag{1.20}$$

The conclusion of the proposition means to prove that $x_o^*=Bx_o$. Indeed, for each $u\in X$, set $x_t=x_o+t(u-x_o)$ with $t>0$. Replacing in (1.20), $x=x_t$ we have

$$\langle u-x_o, Bx_t - x_o^*\rangle \geq o, \ \forall\, t>o \tag{1.21}$$

Letting $t\downarrow o$, (1.21) and hemicontinuity of B give

$$\langle u-x_o, Bx_o-x_o^*\rangle \geq o, \ \forall u\in X \tag{1.22}$$

which implies $Bx_o = x_o^*$, q.e.d.

We have used the notation $\langle x^*,x\rangle = \langle x,x^*\rangle = x^*(x)$.

Proposition 1.9. If X is a finite dimensional Banach space, then every monotone and hemicontinuous operator $B:X\to X^*$ (with $D(B) = X$) is continuous.

Proof. First one proves that B is bounded on bounded sets. Indeed, if this were not the case then would exist $x_n \to x_o$ such that $\|Bx_n\| \to +\infty$. Let $\frac{Bx_n}{\|Bx_n\|} \to y$ as $n\to\infty$ (or a subsequence). Then

$$\langle x-x_n, \frac{Bx}{\|Bx_n\|} - \frac{Bx_n}{\|Bx_n\|}\rangle \geq o, \ \forall x\in X,\ n=1,2,...$$

yields $\langle x-x_o, y\rangle \leq o$, $\forall\, x\in X$, i.e. $y=o$ which contradicts $\|y\|=1$. Now let $y_n \to y_o$ and let By_{n_k} be convergent subsequence of By_n. Say $By_{n_k} \to z$ as $k\to\infty$. From $\langle x-y_n, Bx-By_n\rangle \geq o$ we get $\langle x-y_o, Bx-z\rangle \geq o$, $\forall\, x\in X$. Arguing as for (1.20) (1.22), it follows (with $x=y_o+t(u-y_o)$) that

$$\langle u-y_o, By_o-z\rangle \geq o, \ \forall\, u\in X$$

which yields $z=By_o$. Thus $By_n \to By_o$, q.e.d.

Remark 1.3. If $D(B)$ is not the whole X, then the conclusion of Proposition 1.9 fails. Indeed, let $X=R^2$ and $B:R^2\to R^2$ defined by

$$B(x,y) = \begin{cases} \left(\frac{4x^3}{y}, \frac{x^4}{y^2}\right), & \text{if } y>o,\ x\in R \\ (0,0) & \text{if } x=y=o \end{cases} \tag{1.23}$$

This operator is monotone and hemicontinuous on $D(B) = \{(x,y)\in R^2,\ y>o\} \cup \{(0,0)\}$ but it is not continuous at $u=(0,0)$ (This example is due to Ursescu). The hemicontinuity of B is obvious, while the monotonicity is not. To show the latter property, one observes that $B(x,y)$ is the gradient (f_x,f_y) of the convex function $f(x,y)= \frac{x^4}{y}$, $y>o$. Or, the gradient (grad $f(x)=(f_x,f_y) = B(x,y)$) of a convex function on R^2 is monotone. Finally, the convexity of f follows from the fact that its Hessian matrix

$$\begin{pmatrix} f_{xx} & f_{xy} \\ f_{yx} & f_{yy} \end{pmatrix}$$

is positively defined. Indeed,

$$f_{xx} + f_{yy} \geq o,\ f_{xx}f_{yy} - f_{xy}^2 = \frac{8x^6}{y^4} \geq 0$$

on $D(f) = \{(x,y)\in R^2,\ y>o\}$

An important characterization of maximal monotone sets in reflexive spaces is given by

Theorem 1.4. Suppose that both X and X* are reflexive and strictly convex. Then a (nonempty) set $A \subset X \times X^*$ is maximal monotone iff

$$R(A+\lambda F)= X^* \ , \ \forall \lambda > 0 \tag{1.24}$$

(or equivalently, for some $\lambda > 0$, where F is the duality mapping of X).

Making use of Theorem 1.4 one can prove

Theorem 1.5. Let X and X* be reflexive. (1) iff $A \subset X \times X^*$ is maximal monotone and $B: X \to X^*$ is monotone, hemicontinuous and bounded on bounded sets, then A+B is maximal monotone.

(2) If in addition, either A or B is coercive, then A+B is surjective.

(3) Every maximal monotone and coercive set $C \subset X \times X^*$ is surjective.

Theorem 1.4 is due to Browder.

A good reference for Theorem 1.5 as well as for surjectivity of monotone operators is Browder, who established main results in this direction. A detailed and simplified proof of Theorem 1.4 and 1.5 can be found in author's lecture notes [12, Ch. 1].

Another useful notion is that of cyclically monotone sets

Definition 1.4. (1) $B \subset X \times X^*$ is said to be cyclically monotone if for every positive integer n and $[x_j, x_j^*] \in X$, $j=0,1,\ldots,n$,

$$\langle x_0 - x_1, x_0^* \rangle + \ldots + \langle x_{n-1} - x_n, x_{n-1}^* \rangle + \langle x_n - x_0, x_n^* \rangle \geq 0 \tag{1.25}$$

(2) B is said to be maximal cyclically monotone (m.c.m.) if it has no cyclically monotone extensions (i.e. if B is not properly contained in any other cyclically monotone set).

Remark 1.4. Clearly, a cyclically monotone set is monotone (this follows from (1.25) for n=1). The converse statement is not true (unless X=R). Indeed, for $x = (a,b) \in R^2$, set $x^* = Bx = (-b,a) \equiv x^{\perp}$ a rotation $Ro(\frac{\pi}{2})$ of $\pi/2$ in R^2). Obviously we have $\langle Bx, x \rangle = 0$, hence is monotone. However, it is easy to check that B is not cyclically monotone (e.g. for n=2, the operator B will not satisfy (1.25)). The subdifferential ∂f of a proper functional $f: X \to]-\infty, +\infty]$ is cyclically monotone.

Obviously, a cyclically monotone set which is maximal monotone, is maximal cyclically monotone. Therefore, the subdifferential ∂f of a l.s.c. proper function is (m.c.m.).

It turns out that the only (m.c.m.) sets are subdifferentials of proper l.s.c. functions. Precisely, we have

Theorem 1.6. A set $B \subset X \times X^*$ is (m.c.m.) if and only if there exists a proper (l.c.s.) function $f: X \to]-\infty,+\infty]$ such that $B = \partial f$, (such a function f is uniquely up to an additive constant)

Proof. Let $B \subset X \times X^*$ be (m.c.m.). Define

$$f(x) = \sup_{\substack{[x_i, x_i^*] \in B \\ i=\overline{1,n}, n \in N}} \{ \langle x_1 - x_o, x_o^* \rangle + \dots + \langle x_n - x_{n-1}, x_n^* \rangle + \langle x - x_n, x_n^* \rangle \}$$

Obviously f is l.s.c. Moreover, $f(x_o) \leq 0$ so f is proper. We have $B = \partial f$ To prove this it suffices to observe that $B \subset \partial f$. Indeed, take $[x,x^*] \in B$. We must verify

$$f(y) - f(x) \geq \langle x^*, y-x \rangle, \; \forall \, y \in X \tag{1.26}$$

Or, by the definition of f

$$f(y) \geq \langle x_1 - x_o, x_o^* \rangle + \dots + \langle x - x_n, x_n^* \rangle + \langle y - x, x^* \rangle$$

which implies (1.26) q.e.d.

Corollary 1.2. (1) Every monotone set of $R \times R$is cyclically monotone. (2) Every maximal monotone set $B \subset R \times R$ is the subdifferential of a l.s.c. function $f: R \to]-\infty,+\infty]$ (i.e. $B = \partial f$)

Proof. (1) In this case we may assume that $x_j \leq x_{j+1}$ fo $j=0,\dots,n-1$. Then $y_i \in Bx_i$ satisfy also $y_j \leq y_{j+1}$. Thus

$$(x_o - x_1) y_o + \dots + (x_n - x_o) y_n = (x_o - x_1)(y_o - y_n) + \dots +$$

$$+ (x_{n-1} - x_n)(y_{n-1} - y_n) \geq 0$$

Part (2) follows from Theorem 1.5.

Most of maximal monotone sets arising in applications to PDE are subdifferentials of proper l.s.c. function (The duality mapping F equals ∂f with $f(x) = \frac{1}{2}\|x\|^2$, Laplace operator Δ in Ch.3, § 3, and so on)

A simple example of a maximal monotone operator which is not a subdifferential is the rotation $Ro(\frac{\pi}{2})$ in R^2 defined above (This is because $Ro(\frac{\pi}{2})$ is not cylcically monotone - Remark 1.4). In this direction there is an important result, namely.

Theorem 1.7. Let $A: D(A) \subset H \to H$ be a linear (unbounded) maximal monotone operator. Then A is densely defined (i.e. $\overline{D(A)} = H$) and the following conditions are equivalent

(1) A is selfadjoint (i.e. $A=A^*$)

(2) There exists a proper (l.s.c.) function 1 $f:H \to]-\infty,+\infty]$ such that $A=\partial f$

Proof. (1)⇒(2). Conditions $\langle Ax,x\rangle \geq 0$, $\forall\, x\in D(A)$ and $A=A^*$ imply the existence of the square root $A^{1/2}$ of A. Set

$$f(x) = \tfrac{1}{2}\|A^{1/2}x\|^2 \quad \text{if } x\in D(A^{1/2}), \text{ and } f(x) = +\infty \quad , \text{otherwise} \qquad (1.27)$$

Clearly, f is proper and convex. Moreover since $A^{1/2}$ is closed, it follows that f is l.s.c. On the other hand, $\langle Ax,\, x_o - x\rangle = \langle A^{1/2}x, A^{1/2}x_o - A^{1/2}x\rangle$ for $x\in D(A)$ and $x_o\in D(A^{1/2})$. Consequently

$$\frac{1}{2}\|A^{1/2}x_o\|^1 - \frac{1}{2}\|A^{1/2}x\|^2 \geq \langle Ax, x_o - x\rangle \quad , \forall x_o\in D(A^{1/2})$$

which implies $A\subset\partial f$ (i.e. $A=\partial f$, since A is maximal monotone). Conversely (2)⇒(1). Indeed, if $A=\partial f$, the $o\in D(A)=D(\partial f)\subset D_e(f)$ so $f(0)$ is a finite number. We may assume that $f(o)=o$ (otherwise we can choose $f_1(x)=f(x)-f(o)$, $x\in H$). Let f_λ be the regularizant of f (see (2.32), next section). Then $A_\lambda = (\partial f)_\lambda = \dot{f}_\lambda$ and the Frechet derivative $\dot{f}_\lambda$ of f is a linear bounded operator (i.e. $x\to \dot{f}_\lambda(x)$ is in $L(H)$, since $x\to A_\lambda x$ is so). Therefore

$$\frac{d}{dt} f_\lambda(tx) = \langle \dot{f}_\lambda(tx), x\rangle = \langle A_\lambda(tx), x\rangle = t\langle A_\lambda x, x\rangle \;, \; x\in H$$

Integrating over $[o,1]$ and using $f_\lambda(o)=o$ we get $f_\lambda(x) = \tfrac{1}{2}\langle A_\lambda x, x_\lambda\rangle$. We now have

$$\langle \dot{f}_\lambda(x), h\rangle = \tfrac{1}{2}(\langle A_\lambda h, x\rangle + \langle A_\lambda x, h\rangle) = \tfrac{1}{2}\langle A_\lambda x + A_\lambda^* x, h\rangle \;, \; \forall h\in H$$

that is

$$A_\lambda = \dot{f}_\lambda = \frac{1}{2}(A_\lambda + A_\lambda^*), \text{ so } A_\lambda = A_\lambda^*$$

(which implies $A=A^*$) q.e.d.

<u>Corollary 1.3.</u> If $A:D(A)\subset H\to H$ is maximal monotone and nonselfadjoint, then A is not a subdifferential ∂f (of any proper, l.s.c. function $f:H\to]-\infty, +\infty]$).

<u>Remark 1.5.</u> Part "(2)⇒(1)" of Theorem 1.6 can be restated as: If the subdifferential ∂f of a proper l.s.c. function $f:H\to]-\infty,+\infty]$ is linear (i.e; $x\to\partial f(x)$ from $D(\partial f)$ into H is linear), then $\partial f=(\partial f)^*$, i.e. $x\to\partial f(x)$ is selfadjoint.

Moreover, from

$$f_\lambda(x) = \tfrac{1}{2} < (\partial f)_\lambda(x), x > \quad , \forall \lambda > o \tag{1.28}$$

and $(\partial f)_\lambda(x) \to f(x)$ as $\lambda \downarrow o$, (for $x \in D(\partial f)$) we derive

$$f(x) = \tfrac{1}{2} < \partial f(x), x > , \quad \forall x \in D(\partial f) \tag{1.29}$$

hence $f(x) \geq o$ for all $x \in D(\partial f)$. In particular $f(x) = ||x||^2$ satisfies (1.29) ($\partial f = 2F = 2I$)

Remark 1.6. The duality mapping F of some concrete spaces may be found in the author's book [15]. Here we mention in addition that the duality mapping of $L^1(\Omega)$ is given by

$$(Fu)(x) = ||u||_{L^1(\Omega)} \text{ sign } u(x), \; x \in \Omega \; , \; u \in L^1(\Omega) \tag{1.30}$$

where

$$\text{sign } r = \begin{cases} -1 & \text{if } r < o \\ [-1,+1], & \text{if } r = o \\ +1 & \text{if } r > o \end{cases} \tag{1.31}$$

Note that $\text{sign}_o \; r = \text{sign } r$ for $r \neq o$ and $\text{sign}_o \; o = o$.

By virtue of Proposition 2.1 in Chapter 1 it follows

Corollary 1.4. The norm of Banach space X is Gâteaux differentiable at $x \neq o$, i.e. $\lim\limits_{t \to o} \frac{||x+ty|| - ||x||}{t}$ exists for all $y \in X$ iff its duality mapping F is single values and we have (in this case)

$$\lim_{t \to o} \frac{||x+ty|| - ||x||}{t} = < \frac{1}{||x||} F(x), y > = \frac{d}{dt} ||x+ty|| \Big|_{t=o}, \; y \in X \tag{1.32}$$

(i.e. the Gâteaux derivative of the norm at x is just $\frac{1}{||x||} F(x)$).
Furthermore, Proposition 1.5 along with Corollary 1.4 show that if X* is strictly convex, then Gâteuax derivative of the norm at $x \neq o$ equals the subdifferential of the norm at x which in turn equals $\frac{1}{||x||} F(x)$. It follows that if X* is uniformly convex then $\frac{1}{||x||} F(x)$ is just the Fréchet derivative of the norm at $x \neq o$. Recall also the followinf result of Kato [1].

Lemma 1.1. Let $u: I \to X$ be a function from the interval $I \subset R$, into the real Banach space X. If u is weakly differentiable at $t_o \in I$ and $t \to ||u(t)||$ is differentiable at t_o, then

$$\frac{1}{2}\frac{d}{dt}\|u(t)\|^2\Big|_{t=t_0} = \|u(t_0\| \frac{d}{dt}\|u(t)\|\Big|_{t=t_o} = \langle u'(t_o),u(t_o)\rangle_s =$$

$$\langle u'(t_o),u(t_o)\rangle_i = \langle u'(t_o),x^*\rangle, \ \forall x^* \in F(u(t_0)) \tag{1.33}$$

where $u'(t_0)$ denotes the weak derivative of u at t_0.

Proof. Let $x^* \in F(u(t_0))$. Then we have

$$\langle u(t_o+h)-u(t_o),x^*\rangle = \langle u(t_o+h),x^*\rangle - \|u(t_o\|^2 \leq \tag{1.34}$$
$$(\|u(t_0+h)\| - \|u(t_0)\|)\|u(t_0)\|,$$

for all $h \in R$ with $t_0+h \in I$, and the result follows (For the notation $\langle y,x\rangle_i$, see § 2 in Ch.1). See also author's book [15], pp. 17-18 for other remarks on this lemma.

§ 2. Dissipative operators

Given a multivalued operator $A:D(A)\subset X \to 2^X$ denote by G(A) its graph, i.e.

$$G(A) = \bigcup_{x\in D(A)}(x,Ax) \tag{2.1}$$

Of course $(x,Ax) \equiv \bigcup_{y\in Ax}[x,y]$ and therefore $G(A)\subset X\times X$. For simplicity of writing, the multivalued operators A are not distinguished from their graphs G(A) (i.e. A is identified with G(A)).In other words $A:D(A)\subset X\to 2^X$ is regarded as a set of $X\times X$. It follows that given $A\subset X\times X$ we have to define

$$Ax=\{\, y\in X;\ [x,y]\in A\},\ D(A)=\{x\in X; Ax\neq\emptyset\}, R(A)=\bigcup_{x\in D(A)}Ax \tag{2.2}$$

Thus $[x,y]\in A$ means $x\in D(A)$ and $y\in Ax$.

If $A,B\subset X\times X$, and $a\in R$, then

$$aA = \bigcup_{x\in D(A)}(x,aAx),\ aAx=\{\,ay, y\in Ax\,\} \tag{2.3}$$

$$A^{-1}=\{\,[y,x];\ [x,y]\in A\,\} \tag{2.4}$$

$$A+B=\{\,[x,y+z];\ x\in D(A)\cap D(B), y\in Ax, z\in Bx\,\} \tag{2.5}$$

With these conventions, one says "the set $A\subset X\times X$" in place of "the operator $A:D(A)\subset X\to 2^X$". Before we proceed to define the notion of "dissipative operator (set)" let us observe that Proposition 2.1 in Chapter 1 yields(with $x^*(y) \equiv \langle y,x^*\rangle \equiv \langle x^*,y\rangle$, X a real Banach space)

Lemma 2.1. Let $x,y \in X$. The following properties are equivalent

(1) there exists $x^* \in F(x)$ such that $\langle y,x^*\rangle \leq 0$

(2) $\langle y,x\rangle_i = \|x\| \langle y,x\rangle_- = \|x\| \lim\limits_{t\uparrow 0} \dfrac{\|x+ty\| - \|x\|}{t} \leq 0$

(3) $\|x\| \leq \|x-\lambda y\|, \quad \forall \lambda > 0$

Definition 2.1. The set $A \subset X \times X$ is said to be dissipative if for every $x_1,x_2 \in D(A)$ there is $x^* \in F(x_1-x_2)$ such that

$$\langle y_1-y_2,x^*\rangle \leq 0, \text{ for all } y_j \in Ax_j, \ j = 1,2 \tag{2.6}$$

A is said to be accretive, if $-A$ is dissipative (If A is continuous, then the existence of x^* satisfying (2.6)implies (2.6) for all $x^* \in F(x_1-x_2)$ See Remark 5.3 in Ch. 2 and (2.58) in this section).

According to Lemma 2.1 we have

Proposition 2.1. The set $A \subset X \times X$ is dissipative if and only if

$$\|x_1-x_2\| \leq \|x_1-x_2 - \lambda(y_1-y_2)\|, \ \forall \lambda > 0, \ [x_j,y_j] \in A, j=1,2 \tag{2.7}$$

or

$$\langle y_1-y_2,x_1-x_2\rangle_i \leq 0, \ \forall \ [x_j,y_j] \in A, \ j=1,2 \tag{2.8}$$

or still

$$\lim_{t\uparrow 0} \frac{\|x_1-x_2+ t(y_1-y_2)\| - \|x_1-x_2\|}{t} \equiv \langle y_1-y_2,x_1-x_2\rangle_- \leq 0 \tag{2.9}$$

A set $B \subset X \times X$ is said to be ω-dissipative ($\omega \in R$) if $B-\omega I$ is dissipative. Thus, o-dissipativity is just dissipativity.

Proposition 2.2. $B \subset X \times X$ is ω-dissipative iff

$$\langle y_1-y_2,x_1-x_2\rangle_i \leq \omega\|x_1-x_2\|^2, \ \forall \ [x_j,y_j] \in B, \ j=1,2 \tag{2.10}$$

or

$$(1-\lambda\omega)\|x_1-x_2\| \leq \|x_1-x_2- \lambda(y_1-y_2)\|, \ \forall \ [x_j,y_j] \in B,$$

$$j=1,2; \lambda > 0 \tag{2.11}$$

Proof. $B-\omega I$ dissipative means the existence of $x^* \in F(x_1-x_2)$ such that

$$\langle y_1- \omega x_1-(y_2- \omega x_2), \ x^*\rangle \leq 0, \text{ for } y_j \in Ax_j, \ j=1,2 \tag{2.12}$$

i.e.

$$\langle y_1-y_2, x^*\rangle \leq \omega\langle x_1-x_2, x^*\rangle = \omega\|x_1-x_2\|^2 \qquad (2.12)'$$

which implies (2.10). We now prove that (2.10) $\Rightarrow$ (2.11). Indeed

$$\|x_1-x_2\|^2 = x^*(x_1-x_2) = x^*(x_1-x_2-\lambda(y_1-y_2)) + \lambda x^*(y_1-y_2) \leq$$

$$\|x_1-x_2\|(\|x_1-x_2-\lambda(y_1-y_2)\| + \lambda\omega\|x_1-x_2\|)$$

which is just (2.11). Conversely, (2.11)$\Rightarrow$(2.10). Indeed, (2.11) can be written as

$$(-\lambda)^{-1}\|x_1-x_2-\lambda(y_1-y_2)\| - \|x_1-x_2\| \leq \omega\|x_1-x_2\|$$

Letting $\lambda \downarrow o$ one obtains

$$\langle y_1-y_2, x_1-x_2\rangle_- \leq \omega\|x_1-x_2\| \qquad (2.13)$$

Multiplying by $\|x_1-x_2\|$, (2.13) leads to (2.10).
Clearly (2.10) implies (2.12) with $x^* \in F(x_1-x_2)$ such that

$$\langle y_1-y_2, x_1-x_2\rangle_i = \langle y_1-y_2, x^*\rangle \qquad (2.13)'$$

This completes the proof. Note that a L-Lipschitz continuous operator $A:D(A) \subset X \to X$, i.e.

$$\|As-Ay\| \leq L\|x-y\|, \ \forall\ x,y \in D(A) \qquad (2.14)$$

with $L > o$, is L-dissipative. This is because by (2.8) in Ch. 1, we have

$$\langle Ax-Ay, x-y\rangle_i \leq \|As-Ay\| \ \|x-y\| \leq L\|x-y\|^2 \qquad (2.15)$$

The converse implication fails. Indeed, $Ax = -\sqrt{x}+Lx$, $x \geq o$ is L-dissipative with $L > o$ but not L-Lipschitz continuous on $R_+ = [o,+\infty[$ for any $L > o$. Note that for $\omega < o$, the property (2.10) is said to be a "strong dissipativity" of B. The strong accretivity of B is defined by (2.54) (resp. (2.54)') with $\omega > o$. A notion which is strictly more general than "dissipativity" is that of "quasi-dissipativity" in the sense of Kobayashi [1] , namely

<u>Definition 2.2.</u> $A \subset X \times X$ is said to be ω-quasi-dissipative if for every $[x_j, y_j] \in A$, $j=1,2$

$$\langle y_1, x_1-x_2\rangle_i + \langle y_2, x_2-x_1\rangle_i \leq \omega\|x_1-x_2\|^2 \qquad (2.16)$$

o-quasi-dissipativity is simply called quasi-dissipativity. Obviously, if the duality mapping F is single-valued then (in view of Proposition 2.1 in Ch. 1), "dissipativity" is equivalent to "quasi-dissipativity" since in this case

$$<y_1,x_1-x_2>_i + <y_2,x_2-x_1>_i = <y_1-y_2, F(x_1-x_2)> \qquad (2.17)$$

According to the notions and properties (2.1)-(2.10) in Chapter 1, it is also clear that

Proposition 2.3. The duality mapping F is single-valued at $x\neq o$, iff for all $y \in X$

$$<y,x>_- \equiv \lim_{t\uparrow o} \frac{\|x+ty\|-\|x\|}{t} = \lim_{t\downarrow o} \frac{\|x-ty\|-\|x\|}{t} \equiv <y,x>_+ \qquad (2.18)$$

(i.e. iff the norm of X is Gâteaux differentiable at $x\neq o$) or

$$<y,x>_i = <y,x>_s, \text{ for all } y\in X \qquad (2.18)'$$

In this case, if $x\neq o$

$$\lim_{t\to o} \frac{\|x+ty\|-\|x\|}{t} = <y, \frac{1}{\|x\|}F(x)> = \frac{d}{dt}\|x+ty\|\Big|_{t=o} \qquad (2.19)$$

In words, (2.19) says that if the norm of X is Gâteaux differentiable at $x\neq o$, then its Gâteaux derivative is just $\frac{1}{\|x\|}F(x)$. Clearly, (2.19) allosw the determination of duality mapping F of some concrete spaces (see e.g. Pavel [15]).

Now let us turn our attention to Definition 2.2. It is easy to see that (2.10) (i.e. ω-dissipativity implies (2.17). Indeed, let x* as in (2.13)'. Then

$$\omega\|x_1-x_2\|^2 \geq <y_1-y_2, x_1-x_2>_i = <y_1,x^*> + <-y_2,x^*> \geq <y_1,x_1-x_2>_i$$

This inequality and $<-y_2,x_1-x_2>_i = <y_2,x_2-x_1>_i$ prove that (2.10) $\Rightarrow$(2.17). The converse implication fails. In this direction, we give the following example of Miyadera

Example 1.1. Consider $X=R^2$ endowed with the norm

$$\|(x_1,x_2)\| = \max\{|x_1|, |x_2|\}, \quad (x_1,x_2)\in R^2$$

let $u=(1,1)$ and $v=(o,o)\in R^2$. Define $A:D(A)\subset R^2 \to 2^{R^2}$ as follows

$$D(A)=\{u,v\},\ Au=\{(a,b);\ a\leq o \text{ or } b\leq o\}$$
$$Av=\{(c,d);\ c\geq o \text{ or } d\geq o\} \qquad (2.20)$$

One can prove that A is quasi-dissipative but not ω-dissipative (for any $\omega \in R$). Moreover

$$R(I-\lambda A) \supset D(A) = \overline{D(A)}, \ \forall \lambda > 0 \tag{2.21}$$

Indeed, in our case here quasi-dissipativity of A is equivalent to

$$2 \leq \|(1-\lambda a, 1-\lambda b)\| + \|(1+\lambda c, 1+\lambda d)\| \tag{2.22}$$

where either $a \leq 0$ or $b \leq 0$ and either $c \geq 0$ or $d \geq 0$ (see Proposition 2.4 below with $\omega = 0$). Therefore (2.22) holds, i.e. A is o-quasi-dissipative. However (for every $\omega \in R$) A is not ω-dissipative. This is because ω-dissipativity means (by (2.11) (with x_1 and x_2 replaced by u and v respectively)

$$(1-\lambda\omega) \leq \|(1+\lambda(c-a); 1+\lambda(d-b))\|, \ \lambda > 0 \tag{2.23}$$

which is not true.

Range Condition (2.21) is trivially satisifed since $z=(0,0) \in Au \cap Av$, i.e. $x-\lambda z = x$ for both $x=u$ and $x=v$. q.e.d.

<u>Proposition 2.4.</u> Let A be a (nonempty) subset of $X \times X$ and $\omega \in R$. (I) The following three conditions are equivalent

(1) A is ω-quasi-dissipative

(2) $(\lambda+\mu-\omega\lambda\mu)\|x_1-x_2\| \leq \lambda\|x_1-x_2-\mu y_1\| + \mu\|x_2-x_1-\lambda y_2\|$ for all $[x_j,y_j] \in A$, $j=1,2$ and $\lambda > 0$, $\mu > 0$

(3) $(2-\lambda\omega)\|x_1-x_2\| \leq \|x_1-x_2-\lambda y_1\| + \|x_2-x_1-\lambda y_2\|$ for all $[x_j,y_j] \in A$, $j=1,2$ and $\lambda > 0$

(II) ω-quasi-dissipativity of A implies

$$(1-\lambda\omega)\|x-u\| \leq \|x-u-\lambda y\| + \lambda|Au|, \ \forall [x,y] \in A, \ u \in D(A), \ \lambda > 0 \tag{2.24}$$

where

$$|Au| = \inf\{\|v\|, v \in Au\} \tag{2.25}$$

Proof. (1) $\Rightarrow$ (2). Let $x_1^* \in F(x_1-x_2)$ and $x_2^* \in F(x_2-x_1)$ be such that

$$\langle y_1, x_1^* \rangle + \langle y_2, x_2^* \rangle \leq \omega\|x_1-x_2\|^2 \tag{2.25'}$$

Then we have

$$(\lambda+\mu)\;\|x_1-x_2\|^2 = \lambda x_1^*(x_1-x_2) + \mu\, x_2^*(x_2-x_1) \leq$$

$$\lambda x_1^*(x_1-x_2-\mu y_1) + \mu x_2^*(x_2-x_1-\lambda y_2) + \lambda\mu\omega\|x_1-x_2\|^2$$

$$\leq (\lambda\|x_1-x_2-\mu y_1\| + \mu\|x_2-x_1-\lambda y_2\| + \lambda\mu\omega\|x_1-x_2\|)\|x_1-x_2\|$$

i.e. (2) holds. Clearly, for $\lambda=\mu$, (2) $\Rightarrow$ (3). We now prove that (3) $\Rightarrow$ (1). Indeed, (3) can be written in the form

$$\lambda^{-1}(\|x_1-x_2\|-\|x_1-x_2-\lambda y_1\|)+\lambda^{-1}(\|x_1-x_2\|-\|x_2-x_1-\lambda y_2\|)\leq\omega\|x_1-x_2\|$$

Letting $\lambda\downarrow o$ one obtains obviously

$$\langle y_1,x_1-x_2\rangle_- + \langle y_2,x_2-x_1\rangle_- \leq \omega\|x_1-x_2\| \tag{2.26}$$

which is equivalent to (2.16) (in view of $\langle y,x\rangle_i = \|x\|\,\langle y,x\rangle_-$). It remains to prove (2.24). Let $|x,y|\in A$, $u\in D(A)$, $\lambda>o$ and $v\in Au$. Then on the basis of (3)

$$(2-\lambda\omega)\|x-u\| \leq \|x-u-\lambda y\|+\|x-u-\lambda v\| \leq \|x-u-\lambda y\|+\|u-x\|+\lambda\|v\|$$

for all $v\in Au$, which yields (2.24), q.e.d.

Remark 2.1. In applicationsto PDE we do not know (at elast so far) quasi-dissipative operators which are not dissipative. However, in the proof od the fundamental estimates(i.e. (2.40) in Ch. 1 and its consequence (1.16) in Ch.2) we have naturally used only quasi-dissipativity condition (see Remark 2.3 in Ch. 1) In otehr words, in the proof of the generation of evolution operators and nonlinear semigroups, the generality from quasi-dissipative sets to dissipative sets costs nothing in terms of analytical difficulty. See also Remark 1.2 in Ch. 2.

Remark 2.2. Except for minor modifications, the proof of (2.14)$\Leftrightarrow$(2.14)' in Ch. 1 is the same as the proof of (1)$\Leftrightarrow$(2) in proposition 2.4 above. The equivalence of (1.9) with (1.9)' in CHapter 2 is just (1)$\Leftrightarrow$(2) above.

The operators J_λ and A_λ. Let $A\subset X\times X$ be ω-dissipative and $\lambda>o$ such that $\lambda\omega<1$. On the bais of (2.11), for every $u\in(RI-\lambda A)$, there exist a unique $x_1\in D(A)$ and a unique $y_1\in Ax_1$ such that $u=x_1-\lambda y_1$. Set

$$x_1 = J_\lambda u \equiv (1-\lambda A)^{-1}u,\quad y_1=A_\lambda u = \frac{(I-J_\lambda)u}{\lambda} \tag{2.27}$$

for $u\in R(I-\lambda A)$.

Therefore $u=J_\lambda u-\lambda A_\lambda u$ and

$$J_\lambda : R(I-\lambda A) \to D(A),\ J_\lambda(x-\lambda y)=x,\ A_\lambda(x-\lambda y)=y \qquad (2.28)$$

for every $\lambda > 0$, $|x,y| \in A$. Obviously, $y_1 \in Ax_1$ means

$$A_\lambda u \in AJ_\lambda u,\ u \in R(I-\lambda A),\ \lambda > 0 \qquad (2.29)$$

If B is accretive (i.e. A=-B is dissipative) then

$$u=x_1+\lambda y_1 \text{ and } x_1=J_\lambda u=(1+\lambda B)^{-1}u,\ y_1=B_\lambda u=\frac{u-J_\lambda u}{\lambda} \qquad (2.30)$$

$$u=J_\lambda u+\lambda B_\lambda u,\ u \in R(I+\lambda B),\ \lambda > 0 \qquad (2.31)$$

Combining the definition of J_λ and A_λ with (2.11) we can easily prove

<u>Proposition 2.5.</u> Let $A \subset X \times X$ be ω-dissipative (with $\omega \in R$). Then for every $\lambda > 0$ such that $\lambda\omega < 1$ the following properties hold

(1) $\|J_\lambda u-J_\lambda v\| \le (1-\lambda\omega)^{-1}\|u-v\|$, $u,v \in R(I-\lambda A)$

(2) $\|J_\lambda u-u\| \le \lambda(1-\lambda\omega)^{-1}|Au|$, $u \in D(A) \cap R(I-\lambda A)$

where $|Au|$ is defined by (2.25)

(3) $(1-\lambda\ )\|A_\lambda u\| \le (1-\lambda\mu)\|A_\mu u\|$, $\lambda\omega < 1$, $\mu\ \omega\ 1$, $0 < \mu \le \lambda$
$u \in R(I-\lambda A) \cap R(I-\mu A)$

(4) $\|A_\lambda u-A_\lambda v\| \le \lambda^{-1}|1+(1-\lambda\omega)^{-1}|\,\|u-v\|$, $u,v \in R(I-\lambda A)$

(5) if $u \in R(I-\lambda A) \cap D(A)$ then $A_\lambda u \in AJ_\lambda u$ and $\|A_\lambda u\|\ \ (1-\lambda\omega)^{-1}|Au|$

(6) $\langle A_\lambda u-A_\lambda v, u-v\rangle_s \le \frac{\omega}{1-\lambda\omega}\|u-v\|^2$, $u,v \in R(I-\lambda A)$

(7) If $u \in R(I-\lambda A)$, then $\frac{\mu}{\lambda}u+\frac{\lambda-\mu}{\lambda}J_\lambda u \in R(I-\mu A)$ and
$J_\lambda u=J_\mu\left(\frac{\mu}{\lambda}u+\frac{\lambda-\mu}{\lambda}J_\lambda u\right)$, $\mu > 0$
(the "resolvent formula")

The proof of these properties (as well as other properties) of J_λ and A_λ can be found in author's books [12], [15]. Let us comment some of them.

Property (1) is a direct consequence of (2.11). For Property (2), one uses (1) and $u=J_\lambda(u-\lambda v), v \in Au$. Obviously (2) implies (5).

<u>Definition 2.3.</u> (1) The dissipative set $A \subset X \times X$ is said to be maximal dissipative if it is not properly contained in any other dissipative set of $X \times X$ (2) $A \subset X \times X$ is said to be m-dissipative if it is dissipative and $R(I-\lambda A)=X$, $\forall \lambda > 0$ (or equivalently, if $R(I-A)=X$).
(3) A is said to be m-accretive if -A is m-dissipative.

Remark 2.3. It is easy to prove that m-dissipativity implies maximal dissipativity. The converse implication fails. The frist counterexample xas given by Calvert [1] in 1^p. Later, Cernes [1]1 has proved that if both X and X* are even uniformly convex (but X is not a Hilbert space), there are maximal dissipative sets $A \subset X \times X$ which are not m-dissipative.

However, one of the main results of Minty [1] asserts that in Hilbert spaces , "maximal dissipativity" is equivalent to "m-dissipativity" The following theorem collects some properties of maximal dissipative sets.

Theorem 2.1. If $A \subset X \times X$ is maximal dissipative then

(1) A is closed (hence, Ax is a closed subset of X, $\forall \ x \in D(A)$)

(2) If $x_\lambda \in D(A) \cap R(I-\lambda A)$, $x_\lambda \rightharpoonup x, A_\lambda x_\lambda \to y$ as $\lambda \downarrow 0$, then $x \in D(A)$ and $y \in Ax$.

(3) If X* is strictly convex then Ax is convex (ans closed) for each $x \in D(A)$.

(4) If X* is uniformly convex then
A is demiclosed (i.e. $y_n \in Ax_n$, with $x_n \to x$ and $y_n \rightharpoonup y$, imply $y \in Ax$)

(5) Let A be m-dissipative. Then $\lim_{\lambda \downarrow 0} J_\lambda x = x$ for every $x \in \overline{D(A)}$.
$x_{\lambda_p} \to x$ and $A_{\lambda_p} x_{\lambda_p} \rightharpoonup y$, yield $[x,y] \in A$ (with $\lambda_p \downarrow 0$ as $p \to \infty$)

(6) Let $A \subset X \times X$ be m-dissipative. Then

(6.1) If X* is uniformly convex, then for every $x \in D(A)$ we have $\|A_\lambda x\| \uparrow \|Ax\|$ (see (2.25)). If $x \notin D(A)$, then $\|A_\lambda x\| \to +\infty$ as $\lambda \downarrow 0$.

(6.2) If both X and X* are uniformly convex then $A_\lambda x$ is (strongly) convergent to the element $A_0 x$ of the least norm of Ax (i.e. $A_0 x \in Ax$ and $\|A_0 x\| = |Ax|$) for every $x \in D(A)$.

(7) If A is m-dissipative, X and X* uniformly convex, then $\overline{D(A)}$ is a convex subset of X (see also Remark 2.3)

The complete proof of Theorem 2.1. can be found in some existing books on this subject (e.g. Barbu [2], Pavel [15]). We shall prove here only Part (6). According to Proposition 2.5 with $\omega = 0$, we have $\|A_\lambda x\| \le \|A_\mu x\| \le |Ax|$ for $0 < \mu < \lambda$ and $x \in D(A)$. Therefore $\lim \|A_\lambda x\| \equiv L$ exists and $L \le |Ax|$

Let $A_{\lambda_k} x \to u$, $\lambda_k \downarrow 0$ as $k \to \infty$. Since $A_{\lambda_k} x \in AJ_{\lambda_k} x$ it follows $y \in Ax$, hence $|Ax| \le \|y\| \le \lim_{k\to\infty} \|A_{\lambda_k} x\| \le |Ax|$ and (6.1) follows. Clearly, for the conclusion $y \in Ax$ we have used uniform convexity of X^*, which implies the uniform continuity of the duality mapping on bounded subsets of X.

In order to prove (6.2) we observe that $\|y\| = |Ax|$ and $y \in Ax$ yield $y = A_o x$. In other words every weakly convergent subsequence of $A_\lambda x$ converges to $A_o x$. This means that even $A_\lambda x$ is weakly convergent to $A_o x$. But $A_\lambda x \rightharpoonup A_o x$, $\|A_\lambda x\| \le \|A_o\|$ and X uniformly convex, imply $A_\lambda x \to A_o x$ (strongly, as $\lambda \downarrow o$). Clearly $J_\lambda x \to x$ as $\lambda \downarrow o$, for all $x \in D(A)$ (Proposition 2.5 (2)). Since $x \to J_\lambda x$ is non expansive, it follows $\lim_{\lambda\downarrow o} J_\lambda x = x$, $\forall x \in \overline{D(A)}$. See Remark 2.3.

We are now in a position to recall and prove the following useful theorem (in the real Hilbert space H)

<u>Theorem 2.2.</u> Let $f: H \to]-\infty, +\infty]$ be a proper lower semicontinuous convex function (l.s.c.). Set

$$f_\lambda(y) = \min_{x\in H} \left\{ f(x) + \frac{1}{2\lambda} \|y-x\|^2 \right\}, \; y \in H, \; \lambda > o \tag{2.32}$$

Then we have

$$f_\lambda(y) = f(J_\lambda y) + \frac{\lambda}{2} \|A_\lambda y\|^2, \; y \in H, \; \lambda > o \tag{2.33}$$

where $A = \partial f$ and A_λ, J_λ are defined by (2.30) and (2.31) with $B = A = \partial f$. Moreover, f_λ is convex, Frechet differentiable and (with $\dot{f}$=Frechet derivative)

$$\dot{f}_\lambda = \partial f_\lambda = (\partial f)_\lambda \equiv A_\lambda, \; f_\lambda(y) \uparrow f(y) \text{ as } \lambda \downarrow o, \; f(J_\lambda y) \le f_\lambda(y) \le f(y), \; \forall y \in H \tag{2.34}$$

$$D(\partial f) \subset D_e(f) \subset \overline{D_e(f)} = \overline{D(\partial f)} \tag{2.35}$$

(f_λ is said to be the regularizant of f).

Proof. In view of Proposition (1.17) with $a = \frac{1}{\lambda}$ we conclude that (2.32) implies (2.33). The convexity of f_λ is obvious. Inasmuch as $A_\lambda y \in AJ_\lambda y$ $\partial f(J_\lambda y)$, we have

$$f(J_\lambda x) - f(J_\lambda y) \geq \langle A_\lambda y, J_\lambda x - J_\lambda y \rangle, \; \forall x \in H \tag{2.36}$$

Combining (2.33), (2.36) and $J_\lambda x = x - \lambda A_\lambda x$ it follows

$$f_\lambda(x) - f_\lambda(y) = f(J_\lambda x) - f(J_\lambda y) + \frac{\lambda}{2}(\|A_\lambda x\|^2 - \|A_\lambda y\|^2) \geq$$
$$\frac{\lambda}{2}(\|A_\lambda x\|^2 - \|A_\lambda y\|^2) + \langle A_\lambda y, x-y\rangle + \lambda \langle A_\lambda y, A_\lambda y - A_\lambda x\rangle$$

Therefore

$$f_\lambda(x)-f_\lambda(y)-\langle A_\lambda y,\ x-y\rangle \geq \frac{\lambda}{2}\|A_\lambda x-A_\lambda y\|^2 \geq o,\ \forall x\in H \tag{2.37}$$

That is $A_\lambda y\in\partial f_\lambda(y)$. On the other hand, by (2.37) we can also derive

$$o\leq f_\lambda(x)-f_\lambda(y)-\langle A_\lambda y,\ x-y\rangle\leq\langle A_\lambda x-A_\lambda y, x-y\rangle\leq\frac{2}{\lambda}\|x-y\|^2 \tag{2.38}$$

where the Lipschitz continuity of $x\to A_\lambda x$ with $L=\frac{2}{\lambda}$ (Proposition 2.5, (4) with $\omega = o$) has been used.

By (2.38) we see that f_λ is Frechet differentiable and Frechet derivative $\dot{f}_\lambda$ is just A_λ. It follows that f_λ is also subdifferentiable and $\partial f_\lambda=\dot{f}_\lambda=A_\lambda$ (Proposition 1.6). According to (2.33) and (2.32) we see that $f(J_\lambda y)\leq f_\lambda(y)\leq f(y)$. If $y\in\overline{D(\partial f)}$, then $J_\lambda y\to y$ as $\lambda\downarrow o$ and therefore $f(y)\leq\liminf_{\lambda\downarrow o} f(J_\lambda y)\leq\liminf_{\lambda\downarrow o} f_\lambda(y)\leq\limsup_{\lambda\downarrow o} f_\lambda(y)\leq f(y)$ that is $f_\lambda(y)\uparrow f(y)$ as $\lambda\downarrow o$ (and $y\in\overline{D(\partial f)}$). If $y\bar{\in}\overline{D(\partial f)}$, then $\lambda\|A_\lambda y\|^2\to +\infty$ as $\lambda\downarrow o$. Since $J_\lambda y$ is bounded (see Proposition 2.7 below) and f is bounded from below by an affine function, (2.33) shows that $f_\lambda(y)\to +\infty$ as $\lambda\downarrow o$ (one observes that)

$$\lambda\|A_\lambda y\| = \|x-J_\lambda x\|\geq d|y,D(A)|>o,\ \forall\ \lambda>o,\ y\bar{\in}\overline{D(A)} \tag{2.38'}$$

It follows that $f(y)= +\infty$ for $y\bar{\in}\overline{D(\partial f)}$ ($A=\partial f$) which yields

$$D_e(f)\subset\overline{D(\partial f)} \tag{2.39}$$

Now the obvious inclusion $D(\partial f)\subset D_e(f)$ along with (2.39) imply (2.35) q.e.d.

<u>Corollary 2.1.</u> Let $f,\ g:H\to]-\infty,+\infty]$ be l.s.c. proper functions. If $\partial f=\partial g$ then there exists a constant C such that $f=g+C$.

Proof. It follows by Theorem 2.2 that $\dot{f}_\lambda=(\partial f)_\lambda=(\partial g)_\lambda=\dot{g}_\lambda$ for all $\lambda>o$. Therefore there exists C_λ such that $C_\lambda=f_\lambda-g_\lambda$. Let $x\in D(\partial g)= D(\partial f)$. Then $C_\lambda=f_\lambda(x)-g_\lambda(x)\to C=f(x)-g(x)$ as $\lambda\downarrow o$. Since C_λ is indipendent of $x\in D(\partial f)$, so is C. Thus $C_\lambda\to C\in R$ as $\lambda\to\infty$ and since $C_\lambda=f_\lambda(x)-g_\lambda(x)$, $\forall\ x\in H$, we have $C=f(x)-g(x)$, for all $x\in H$, q.e.d.

<u>Corollary 2.2.</u> If $f,g:H\to]-\infty,+\infty]$ are l.s.c. proper functions with $D_e(f)\cap\mathrm{Int}\,D_e(g)\neq\emptyset$, then $\partial(f+g)=\partial f+\partial g$.

This corollary is a consequence of Theorem 8.9 in Ch. 2 in conjunction with

<u>Proposition 2.6.</u> Let $f:\ X\to]-\infty,+\infty]$ be a proper l.s.c. function.

Then

(1) f is continuous at $x \in D_e f$ iff $x \in \text{Int } D_e(f)$

(2) $\text{Int } D_e f = \text{Int } D(\partial f)$

The proof can be found in the book of Brezis |3| as well as in author's lecture notes |12,p.492|. Other results which have been used in this work are given by

<u>Theorem 2.3.</u> Let X be uniformly convex, $x_n, x \in X$, $n=1,\dots$

(1) If $x_n \rightharpoonup x$ (weakly) as $n \to \infty$ and $\lim\sup \|x_n\| \leq \|x\|$ (i.e. $\|x_n\| \to \|x\|$), then $x_n \to x$ (strongly in X)

(2) If C is a (nonempty) closed convex subset of X, then for each $y \in X$ there is a unique element (call it $P_C y \equiv \text{Proj}_C y$) of C such that

$$d[y;C] = \|y-P_C y\| \equiv \text{ the distance from } y \text{ to } C \tag{2.40}$$

(3) If X is a Hilbert space H (of inner product $\langle .,. \rangle$ and norm $\|.\|$) then for $y \in H$ and $x_o \in C$, (3.1) $x_o = P_C y$ iff

$$\text{Re} \langle y-x_o, z-x_o \rangle \leq 0, \ \forall z \in C \tag{2.41}$$

(where $\text{Re} \langle y,z \rangle \equiv$ the real part of $\langle y,z \rangle$, $y,z \in H$) (3.2). If $A \subset H \times H$ is maximal monotone then

$$\lim_{\lambda \downarrow o} J_\lambda x = \text{Proj}_{\overline{D(A)}} x, \ \forall x \in H \tag{2.42}$$

Proof. Part (1) can be found in many books (see e.g. Brezis [3] or author's book [15]). For the proof of (2) we refer e.g. to author's book |15|. The proof of (3.1) is an elementary fact. Indeed, let $z \in C$. Then for all $t \in |o,1|$, $tz+(1-t)x_o \in C$ and (2.41) is equivalent to

$$\|y-x_o\|^2 \leq \|y-(tz+(1-t)x_o\|^2, \ \forall t \in [o,1], \ \forall z \in C \tag{2.42'}$$

Clearly, $\text{Proj}_C y = x_o$ implies (2.42)' (i.e. (2.42)). COnversely (2.42)' with t=1 yields

$$\|y-x_o\|^2 \leq \|y-z\|^2, \ \forall z \in C, \text{ i.e. } x_o = \text{Proj}_C y..$$

Part (3.2). Let $z \in D(A)$ and $y \in Az$. We know that $A_\lambda x \in AJ_\lambda x$, i.e. $\frac{1}{\lambda}(x-J_\lambda x) \in AJ x$. This and the monotonicity of A imply (H-a real space)

$$\langle \frac{1}{\lambda}(x-J_\lambda x)-y, \ J_\lambda x-z \rangle \geq o, \ \forall \lambda > o$$

or equivalently

$$||J_\lambda x||^2 \le \langle x, J_\lambda x - z\rangle + \langle J_\lambda x, z\rangle - \lambda \langle y, J_\lambda x - z\rangle \tag{2.43}$$

Obviously (2.43) shows that $J_\lambda x$ is bounded as $\lambda \downarrow o$ (for every $x \in H$). Suppose (relabeling if necessary) that $J_\lambda x \rightharpoonup x_o$ as $\lambda \downarrow o$. Inasmuch as $||x_o|| \le \liminf_{\lambda \downarrow o} ||J_\lambda x||$, letting $\lambda \downarrow o$ in (2.43) one obtains

$$||x_o||^2 \le \langle x, x_o - z\rangle + \langle x_o, z\rangle, \ \forall z \in D(A) \tag{2.43'}$$

i.e.

$$\langle x - x_o, \ z - x_o\rangle \le o, \ \forall z \in \overline{D(A)} \tag{2.44}$$

which is equivalent to (2.42). Taking into account that $\overline{D(A)}$ is convex (Th. 2.1, (7)), it is also weakly closed, so $x_o \in \overline{D(A)}$. Moreover, (2.43) yields

$$\limsup_{\lambda \downarrow o} ||J_\lambda x||^2 \le \langle x, x_o - z\rangle + \langle x_o, z\rangle, \forall z \in \overline{D(A)} \tag{2.45}$$

For $z = x_o$ it follows from (2.45) that $||J_\lambda x|| \to ||x_o||$. This property, in conjunction with Part (1) and $J_\lambda x \rightharpoonup x_o$ imply $J_\lambda x \to x_o$ (i.e. (2.42)).

<u>Remark 2.4.</u> (1) If X is a real Banach space with X* uniformly convex and $A \subset X \times X$ is m-dissipative then

$$||A_\lambda x|| \text{ is bounded (as } \lambda \downarrow o) \text{ iff } x \in D(A) \tag{2.46}$$

(see also Remark 4.2 in Ch. 2)

(2) If in addition $\overline{D(A)}$ is convex (this is the case if e.g. both X and X* are uniformly convex, cf. Th. 2.1, (7)), then

$$\lambda ||A_\lambda x|| = ||J_\lambda x - x|| \ge d[x, \overline{D(A)}] > o, \ \forall x \bar{\in} \overline{D(A)} \tag{2.47}$$

which yields

$$\lambda ||A \ x||^2 \to +\infty \text{ as } \lambda \downarrow o, \ \forall \ x \bar{\in} \overline{D(A)} \tag{2.48}$$

proof of (2.46). Since $||A_\lambda x|| \le |Ax|$, $\forall x \in D(A)$ it remains to prove that $||A_\lambda x||$ bounded implies $x \in D(A)$. Indeed, if $A_\lambda x$ is bounded then $J_\lambda x - x = \lambda A_\lambda x$ gives $J_\lambda x \to x$ and we may assume that $A_\lambda x \rightharpoonup y$ as $\lambda \downarrow o$. The dissipativity of A means

$$\langle A_\lambda x - w, \ F(J_\lambda x - v)\rangle \le o, \ \forall \lambda > o, \ v \in D(A) \text{ and } w \in Av \tag{2.49}$$

Letting $\lambda \downarrow o$ and taking into account that $F(J_\lambda x - v) \to F(x - v)$ as $\lambda \downarrow o$, we get

$$\langle y - w, \ F(x - v)\rangle \le o, \ \forall \ [w, v] \in A \tag{2.50}$$

Since A is maximal dissipative, it follows $x \in D(A)$ (and $y \in Ax$) q.e.d.
If $\overline{D(A)}$ is convex and X is reflexive then

$$d[x;\overline{D(A)}] > 0, \forall x \bar{\in} \overline{D(A)} \tag{2.51}$$

and (2.47) follows.

According to Definition 2.1, A is said to be accretive, if there exists $x^* \in F(x_1-x_2)$ such that

$$< y_1-y_2, x^* > \geq 0, \forall y_j \in Ax_j, j=1,2 \tag{2.52}$$

By virtue of Propositions 2.1 and 2.2 it follows

<u>Proposition 2.7.</u> (1) $A \subset X \times X$ is accretive iff

$$\|x_1-x_2\| \leq \|x_1-x_2+\lambda(y_1-y_2)\|, \forall \lambda > 0, [x_j,y_j] \in A, j=1,2 \tag{2.53}$$

or

$$< y_1-y_2, x_1-x_2 >_s \geq 0, \forall [x_j,y_j] \in A, j=1,2 \tag{2.53)'$$

or still

$$< y_1-y_2, x_1-x_2 >_+ \geq 0 \tag{2.53)''$$

(2) For $\omega \in R$, the following two conditions are equivalent

$$< y_1-y_2, x_1-x_2 >_s \geq \omega\|x_1-x_2\|^2 \tag{2.54}$$

$$(1+\lambda\omega)\|x_1-x_2\| \leq \|x_1-x_2+\lambda(y_1-y_2)\| \tag{2.54)'$$

for all $\lambda > 0$ and $[x_j,y_j] \in A$, $j=1,2$.
Clearly, (2.54)' is a more general form of (2.53). The function $<.,.>_s$ is defined in Ch. 2, § 2. Condition (2.54) means that $A-\omega I$ is accretive (we also say that A is ω-accretive). In other words, A is ω-accretive if $-A$ is $(-\omega)$-dissipative. In some recent papers (see e.g. Morales [1]), k-accretivity of A (i.e. (2.54)') is defined by the following inequality

$$(\mu-k)\|x_1-x_2\| \leq \|(\mu-1)(x_1-x_2)+y_1-y_2\| \tag{2.55}$$

for all $\mu > k$, $y_j \in Ax_j$, $j=1,2$.
For k=1 it is obvious that (2.55) means just (2.53) (i.e. the accretivity of A).
For $k < 1$ (2.55) is called "strong accretivity" (cf. Morales [1]).This is actually more restrictive than the standard strong accretivity (i.e.

(2,54) with $\omega > 0$). Indeed, for $k < 1$ (2.55) is equivalent to (2.54)' with $\omega > 0$, plus

$$(\omega - \lambda)\|x_1 - x_2\| \leq \|\lambda(x_1 - x_2) - (y_1 - y_2)\|, \tag{2.56}$$

with $\omega = 1-k$, $\lambda = 1-\mu$, $k < \mu \leq 1$, $y_j \in Ax_j$, $j=1,2$

In particular for $\lambda = 0$ (2.56) shows that if $x_1 \neq x_2$, then $Ax_1 \cap Ax_2 = \emptyset$ i.e. A is invertible. Therefore (2.55) implies (2.54)' with $\omega = 1-k > 0$ and the expansivity of A, i.e.

$$\|y_1 - y_2\| \geq (1-k)\|x_1 - x_2\|, \quad k < 1, \ y_j \in Ax_j, j=1,2 \tag{2.57}$$

Note that according to Definition 2.1 and Remark 1.6 it follows that $A \subset L^1(\Omega) \times L^1(\Omega)$ is accretive iff for all $[x_j, y_j] \in A$, $j=1,2$, there exists $f \in L^\infty(\Omega)$, with $f(x) \in \text{sign}\,(x_1(x) - x_2(x))$, a.e. on Ω such that

$$\int_\Omega (y_1 - y_2) f \, dx \geq 0 \tag{2.58}$$

In Chapter 2 (§ 7 , § 10) the following result has been used

Lemma 2.2. Let $f: H \to]-\infty, +\infty]$ be a lower-semicontinuous convex proper function. Set $A = \partial f$. Suppose that

(1) There is $u \in W^{1,2}(o,T;H)$ such that $u(t) \in D(A)$ a.e. on $]o,T[$

(2) There is $g \in L^2(o,T;H)$ such that $g(t) \in Au(t)$ a.e. on $]o,T[$

Then $t \to f(u(t))$ is absolutely continuous on $[o,T]$ and

$$\frac{d}{dt} f(u(t)) = \langle v, \frac{du}{dt}(t)\rangle, \text{ for all } v \in Au(t), \text{ and } t \in K \tag{2.59}$$

where K is the set of all $t \in]o,T[$ with the properties that $u(t) \in D(A)$ and $t \to f(u(t))$ is differentiable at t.

Proof. We know that $\|A_\lambda u(t)\| \leq \|A_o u(t)\| \leq \|g(t)\|$ and $A_\lambda u(t) \to A_o u(t)$ as $\lambda \downarrow o$ a.e. on $]o,T[$. On the basis of (2.34)

$$\frac{d}{dt}(f_\lambda(u(t)) = \langle \dot{f}_\lambda(u(t)), \frac{du}{dt}(t)\rangle = \langle A_\lambda u(t), \frac{du}{dt}(t)\rangle \quad \text{a.e. on }]o,T[$$

which gives

$$f_\lambda(u(t)) - f_\lambda(u(s)) = \int_s^t \langle A_\lambda u(\tau), u'(\tau)\rangle \, d\tau \quad (\text{with } u' = \frac{du}{dt}, \ t,s \in [o,T])$$

Letting $\lambda \downarrow o$ one obtains

$$f(u(t)) - f(u(s)) = \int_s^t \langle A_o u(\tau), u'(\tau)\rangle \quad d\tau$$

hence $t \to f(u(t))$ is absolutely continuous on $[o,T]$. Let $t_o \in K$. According to the definition of $Au(t_o) \equiv \partial f(u(t_o))$, for every $v \in Au(t_o)$

we have

$$f(y) - f(u(t_o)) \geq \langle v, y-u(t_o)\rangle \quad , \forall\, y \in H \tag{2.60}$$

Replacing $y=u(t_o \pm h)$, $h > o$ in (2.60), one derives (2.59) q.e.d.

§ 3. The regularization of a function

Let X be a Banach space, T o and R the real axis. Denote by D(R,R) the set of all real-valued infinitely differentiable functions on R, with compact support in R. Let n be an arbitrary natural number. Define $g_n : [o,T] \to R$ by

$$g_n(t) = \begin{cases} e^{\frac{1}{n^2t^2-1}} & , \text{ for } |t| < \frac{1}{n} \\ 0 & , \text{ for } |t| \geq \frac{1}{n} \end{cases} \tag{1}$$

Set

$$f_n(t) = a_n g_n(t), \quad t \in R, \tag{2}$$

where a_n is defined by the condition

$$a_n \int_R g_n(t)dt = 1. \tag{3}$$

Obviously, f_n is a sequence of real-valued functions with the properties

$$f_n(t) \geq o, \ \forall\, t \in R, \ \text{supp } f_n \subset]-\frac{1}{n}, \frac{1}{n}[, \ \int_R f_n(t)dt = 1 \tag{4}$$

$$f_n(t) = f_n(-t), \ \forall\, t \in R, \ f_n(t_1) > f_n(t_2) \text{ if } o \leq t_1 < t_2 \tag{5}$$

A sequence $\{f_n\} \subset D(R,R)$ with the properties (4) and (5) is said to be a regularizant sequence. Now let $v \in L^P(o,T;X)$, $1 \leq p < +\infty$ and $\{f_n\}$ a regularizant sequence. The function

$$v_n(t) = \int_o^T f_n(t-s)v(s)ds, \quad t \in [o,T] \tag{6}$$

is called the regularization of v. Some fundamental properties of v_n are given by

<u>Proposition 1.</u> (i) The regularization v_n of v is infinitely differentiable on $[o,T]$.

(ii) $\|v_n\|_p \leq \|v\|_p$, $v_n \to v$ (in $L^p(o,T;X)$ as $n \to \infty$.

(iii) If $v:[o,T] \to X$ is continuous, then $v_n(t) \to v(t)$ (in X) as $n \to \infty$ uniformly on every compact $[a,b] \subset]o,T[$, $a < b$.

Proof. The property (i) is obvious. Moreover

$$\frac{d^k}{dt^k}(v(t)) = \int_o^T \frac{d^k}{dt^k}(f_n(t-s))v(s)ds, \; t \in [o,T].$$

(ii) Clearly, we have denoted by $\|.\|_p$ the norm of $L^p(o,T;X)$ (and by $\|.\|$ the norm of X). Obviously if p and q satisfy $\frac{1}{p} + \frac{1}{q} = 1$ (with $p \geq 1$), then according to Schwartz-Cauchy's inequality, we have

$$\|v_n(t)\| \leq \int_o^T (f_n(s-t))^{1/q} \left[(f_n(s-t))^{1/p} \|v(s)\|\right] ds \leq$$

$$\left\{\int_o^T f_n(s-t)\|v(s)\|^p ds\right\}^{1/p} \quad , \quad o \leq t \leq T$$

since $\int_o^T f_n(s-t)ds \leq 1$, $\forall t \in [o,T]$. Therefore

$$\int_o^T \|v_n(t)\|^p dt \leq \int_o^T \left(\int_o^T f_n(s-t)dt\right)\|v(s)\|^p ds \leq \int_o^T \|v(s)\|^p ds$$

that is $\|v_n\|_p \leq \|v\|_p$. Now let us observe that

$$\int_o^T f_n(t-s)ds = 1, \quad \text{for } \frac{1}{n} \leq t \leq T < -\frac{1}{n}$$

Indeed, in this case (denoting s-t=u) it follows

$$-t \leq -\frac{1}{n} \leq u \leq \frac{1}{n} \leq T - t,$$

hence

$$\int_o^T f_n(s-t)ds = \int_{-t}^{T-t} f_n(u)dt = \int_R f_n(u)du=1$$

We can write (in view of (7))

$$v_n(t)-v(t) = \int_o^T f_n(s-t)v(s)ds - \int_o^T f_n(s-t)v(t)ds$$

for each $\frac{1}{n} \leq t \leq T - \frac{1}{n}$, therefore

$$\|v_n(t)-v(t)\| \le \int_0^T f_n(s-t)\|v(s)-v(t)\|ds, \quad t \in \left[\frac{1}{n}, T - \frac{1}{n}\right] \tag{8}$$

Using once again the identity

$$f_n(s-t) = (f_n(s-t))^{1/q}(f_n(s-t))^{1/p} \quad (\frac{1}{p} + \frac{1}{q} = 1)$$

(8) yields

$$\|v_n(t)-v(t)\|^p \le \int_{|u| \le 1/n} f_n(u)\|v(t+u)-v(t)\|^p du, \quad t \in \left[\frac{1}{n}, T-\frac{1}{n}\right] \tag{9}$$

Integrating (9) over $\left[\frac{1}{n}, T - \frac{1}{n}\right]$ and using Fubini's theorem one obtains

$$\int_{1/n}^{T-1/n} \|v_n(t)-v(t)\|^p dt \le \int_{|u| \le 1/n} (\int_{1/n}^{T-1/n} \|v(t+u)-v(t)\|^p dt) f_n(u))du \tag{10}$$

which obviously implies $v_n \to v$(in $L^p(o,T;X)$) as $n \to \infty$. The property

$$\int_0^T \|v(t+u)-v(t)\|^p dt \to o \text{ as } u \to o$$

has to be also used. (iii) Let $\varepsilon > o$. Since v is assumed to be (uniformly) continuous on $[o,T]$, there is $\delta = \delta(\varepsilon) > o$, such that

$$\|v(t)-v(s)\| < \varepsilon \quad , \text{ for all } t,s \in [o,T], \ |t-s| < \delta \tag{11}$$

Choose an arbitrary compact $[a,b] \subset]o,T[$, $a < b$.
Let $n_o = n\,(a,b,\varepsilon)$ be a natural number such that

$$\frac{1}{n} < \delta \quad , \ \frac{1}{n} \le a < b \le T - \frac{1}{n}, \ \forall\, n \ge n_o \tag{12}$$

Comparing (7) and (11) it follows

$$v(t) = \int_0^T f_n(t-s)v(t)ds, \ n \ge n_o, t \in [a,b]$$

therefore

$$\|v_n(t)-v(t)\| \le \int_0^T f_n(t-s)\|v(s)-v(t)\|ds, \ n \ge n_o, \ t \in [a,b] \tag{13}$$

For each $t \in [a,b]$, denote

$$M(t) = \{ s \in [o,T]; \ \|v(s) - v(t)\| < \varepsilon \} .$$

Then for each $s \in [o,T]-M(t)$ we have (by (11) $|s-t| \ge \delta > \frac{1}{n}$ if $n > n_o$, hence

$$f_n(s-t)=o,\ \forall s \in [o,T]-M(t),\ n \geq n_o \tag{14}$$

Combining (13) and (14) pne obtains

$$\|v_n(t)-v(t)\| \leq \int_{M(t)} f_n(t-s)\|v(s)-v(t)\|ds < \varepsilon\int_{M(t)} f_n(t-s)ds \leq \varepsilon \tag{15}$$

for all $n > n_o$ and $t \in [a,b]$, since

$$\int_{M(t)} f_n(t-s)ds \leq \int_o^T f_n(t-s)ds = 1 \text{ for } n > n_o,\ t \in [a,b].$$

Remark 1. In general $v_n(o)$ (or $v_n(T)$ does not converge to $v(o)$ (resp. $v(T)$). This phenomena can be easily illustrated by the following simple example:

Take $x \in X$, $x \neq 0$ and $v(t) = x$, $\forall t \in [o,T]$.

Inasmuch as

$$\int_o^{1/n} f_n(s)ds = \frac{1}{2} \int_{-1/n}^{1/n} f_n(s)ds = \frac{1}{2}$$

it follows

$$v_n(o) = \int_o^T f_n(s)v(s)ds = x \int_o^T f_n(s)ds = x\int_o^{1/n} f_n(s)ds = \frac{x}{2}$$

for all $n \geq N$, therefore $\lim_{n\to\infty} v_n(o) = \frac{x}{2} \neq v(o)=x$. The same fact holds in T since $v_n(T) = \frac{x}{2}$, too.

If $v \in C_o(R,X)$ (i.e. if v continuous from the real axis R into X, having compact support) then

$$\tilde{v}_n(t) = \int_R f_n(t-s)v(s)ds$$

tends to $v(t)$ as $n \to \infty$, uniformly on supp v.

The proof of this fact is similar to that of (iii) observing that

$$\int_R f_n(t-s)ds = 1,\ \forall\ t \in R$$

See also Yosida [1].

The regularization of a continuous finition $v : [o,T] \times [o,T] \to X$ is defined by

$$v_n(t,s) = \int_o^T\int_o^T f_n(t-p)f_n(s-q)v(p,q)dpdq,\ t,s \in [o,T]. \tag{16}$$

Clearly v_n is infinitely differentiable on $[o,T] \times [o,T]$.

Using the property

$$\int_0^T \int_0^T f_n(t-p)f_n(s-q)dp\,dq = 1, \tag{17}$$

for $\qquad t \in \left[\frac{1}{n}, T - \frac{1}{n}\right], \; s \in \left[\frac{1}{n}, T - \frac{1}{n}\right]$

one can prove (arguing as in the proof of the part (iii) of Proposition 1) that $v_n(t,s) \to v(t,s)$ as $n\to\infty$, uniformly on every compact $[a,b]\times[c,d] \subset]o,T[\times]o,T[$.

Proposition 2. Let $v,w \in L^1(o,T;R)$ be such that

$$\int_a^b v(s)ds \le \int_c^d w(s)ds, \; \forall\, a,b,c,d \in [o,T], \;\; a \le b, \; c \le d. \tag{18}$$

Then

$$\int_a^b v_n(s)ds \le \int_c^d w_n(s)ds \tag{19}$$

for each a,b,c,d satisfying

$$\frac{1}{n} \le a \le b \le T - \frac{1}{n}, \quad \frac{1}{n} \le c \le d \le T - \frac{1}{n}, \quad n \ge 1 \tag{20}$$

where v_n and w_n are the regularizations of v and w respectively.

Proof. Let a,b,c, and d as in (20). Then for $|u| \le \frac{1}{n}$, a+u,b+u,c+u, d+u $\in [o,T]$, therefore by (18)

$$\int_{a+u}^{b+u} v(s)ds \le \int_{c+u}^{d+u} w(s)ds$$

or

$$\int_a^b v(\tau+u)d\tau \le \int_c^d w(\tau+u)d\tau \tag{21}$$

Multiplying (21) by $f_n(u)$ and integrating over $|u| \le \frac{1}{n}$ we get

$$\int_{-1/n}^{1/n} f_n(u)du \int_a^b v(\tau+u)d\tau \le \int_{-1/n}^{1/n} f_n(u)du \int_c^d w(\tau+u)d\tau \tag{22}$$

Since $-\tau \le -\frac{1}{n} < \frac{1}{n} \le T-\tau$ (22) yields

$$\int_a^b d\tau \int_{-\tau}^{T-\tau} f_n(u)v(\tau+u)du \leq \int_c^d d\tau. \int_{-\tau}^{T-\tau} f_n(u)w(\tau+u)du \qquad (23)$$

Inasmuch as

$$g_n(\tau) = \int_o^T f_n(s-\tau)g(s)ds = \int_{-\tau}^{T-\tau} f_n(u)g(\tau+u)du$$

for g = v or g = w, (23) is just (19).

A result similar to Proposition 2 holds for v=v(t,s) and w=w(t,s).

Notes and References

The historical comments on most of the results have been inserted at the right places inside the book. Here we will present some of them separately, along with additional information regarding the sources of the material.

CHAPTER 1. Hypotheses H(2.1) and H(2.2) in Section 2 are taken from the author's paper [9]. These are included in more general conditions of Kobayasi, Kobayashi and Oharu [1]. Actually, the fundamental estimate in Lemma 2.3 is taken from their paper. Definition 3.2 of the integral solution to the problem (1.1) + (1.1)' is also taken from the author's paper [9]. This definition is an extention to the nonautonomous case of Benilan's [2] notion of integral solution from the autonomous case.

The characterization of the compactness of an evolution operator in Section 5 is given by the author. It extends the result of Brezis [4] from the autonomous case (see Ch. 2). The material in Section 3.3. on the general properties of an evolution operator is based on the paper by Iwamiya, Oharu and Takahaski [1]. Other aspects on evolution equations can be found in some papers mentioned in the references, e.g., Altman [1], Attouch and Danlamian [1], Browder [1-2], Haraux [1], Kato [1], Krein [1], Evans [1], Evans and Massey [1], Martin [2], Pierre [1], Schechter [1] and so on.

It is interesting to discuss the paper by Evans and Massey [1] in which a standard trick of converting a nonautonomous into an autonomous differential equation is employed. Precisely, given $(n.a)$ $u' + A(t)u = 0$ set

$$Bw = \begin{pmatrix} A(t)x \\ -1 \end{pmatrix} \quad \text{for} \quad w = \begin{pmatrix} x \\ t \end{pmatrix}, \quad w \in D(B) \subset X \times R \text{ with } \| w \| = | t | + \| x \| \quad (1)$$

where

$$D(B) = \{w;\, w = \begin{pmatrix} x \\ t \end{pmatrix},\ 0 \le t < \infty,\ x \in D(A(t)) \subset X\}.$$

Then nonautonomous equation $(n.a)$ in the Banach space X is equivalent to the autonomous equation $(a) w' + Bw = 0$ in $X \times R$. Suppose that $A(t)$ is m-accretive (and for simplicity - single valued) and that (6.1) holds with L independent of x i.e.,

$$\| J_\lambda(t)x - J_\lambda(s)x \| \le \lambda \| f(t) - f(s) \| L,\ x \in X,\ t, s \ge 0,\ \lambda > 0. \quad (2)$$

It is easy to check that

$$R(I + \lambda B) \supset X \times [0, +\infty[,\ \forall\lambda > 0 \quad (3)$$

Indeed, given $y = \binom{x}{r}$, then $\overline{y} = \begin{pmatrix} J_\lambda(r+\lambda)x \\ r+\lambda \end{pmatrix} \in D(B)$ and we have $\overline{y} + B\overline{y} = y$. It follows that $\overline{J}_\lambda y = \begin{pmatrix} J_\lambda(r+\lambda)x \\ r+\lambda \end{pmatrix}$ with $y = \binom{x}{r}$, $x \in D(A(r))$ satisfies for $f(t) = tI$

$$\| \overline{J}_\lambda y_1 - \overline{J}_\lambda y_2 \|_{X \times R} \le (1 - L)^{-1} \| y_1 - y_2 \|_{X \times R}$$

hence $B + LI$ is accretive in $X \times R$. In that manner, by the above trick, the hypotheses on $(n.a)$ imply that Theorem 5.1 in Chapter 2 can be applied to (a), that is $(n.a)$ reduces to (a). However, let us observe that *Condition*2 is very restrictive. For example, in linear case replacing x by kx, dividing by $k > 0$ and then letting $k \to \infty$ it follows obviously $J_\lambda(t)x = J_\lambda(s)x$ for all $x \in X$ and $t, s \geq 0$, i.e., $J_\lambda(t)$ is independent of t. Consequently, in accretivity conditions $(n.a)$ can be reduced to (a) but only in very strong conditions of the form (2) which do not cover the usual cases.

CHAPTER 2. Most of the results in this chapter (devoted to the autonomous differential equation-inclusion $x' \in Ax$) are derived from nonautonomous case in Chapter 1. In the theory of generation of nonlinear semigroups, the fundamental result is the exponential formula of Crandall-Liggett [1], that is

$$S(t)x_0 = \lim_{n\to\infty}(I - \frac{t}{n}A)^{-n}x_0, \quad t \geq 0, \ x_0 \in \overline{D(A)} \tag{4}$$

(see Formula (3.10)). Its simplest proof is that based on the sharp estimate (1.16) in Lemma 1.1 due to Kobayashi [1] as shown in Section 3.2. Let us give some details in this direction. As in (2.14), suppose that A is m-dissipative, and set $u_n(t) = (I - \frac{t}{n}A)^{-n}x_0^n$. As indicated by (2.12)-(2.14), this u_n corresponds to the DS-approximate solution u_n given by (1.2) for $p_K^n = 0 \in X$, $t_0^n = 0$, $t_K^n = K\frac{t}{n}$, $K = 0, 1, \ldots, N_n$ (with N_n as in (2.15)) for a fixed $t > 0$. Therefore, $d_n = \frac{t}{n}$, $t_n^n = n\frac{t}{n} = t$ so $u_n(t) = u_n(t_n^n) = x_n^n = (I - \frac{t}{n}A) :^{-n} x_0^n$. Accordingly, for $\omega = 0$, $\hat{p}_j^m = 0$, $\hat{t}_j^m = j\frac{t}{m}$ (so $\hat{d}_m = \frac{t}{m}$ and $\hat{t}_m^m = t$) the estimate (1.16) of Kobayashi yields

$$\begin{aligned} \| x_n^n - \hat{x}_m^m \| &= \| u_n(t) - u_m(t) \| \\ &\geq \| x_0^n - u \| + \| \hat{x}_0^m - u \| + t \mid Au \mid (\frac{1}{n} + \frac{1}{m})^{1/2} \\ &\forall u \in D(A) \end{aligned} \tag{5}$$

for all positive integers n and m (with $t = t_n^n = t_m^m$) which shows that the limit in (4) exists uniformly with respect to t in bounded intervals of $[0, +\infty[$.

Recall that if A is an arbitrary linear bounded operator from X into X, then the series

$$\sum_{n=0}^{\infty} \frac{(tA)^n}{n!} \equiv e^{At}, \qquad t \in R \tag{6}$$

gives the unique (strong) solution $u(t) = e^{At}x_0$ to the Cauchy Problem

$$(CP) \qquad u'(t) = Au(t), \qquad t \in R, \ u(0) = x_0 \in X.$$

If, in addition, A is dissipative (i.e. $\langle Ax, x\rangle_i \leq 0, \ \forall x \in X$) then it follows

$$\sum_{n=0}^{\infty} \frac{(tA)^n}{n!} = \lim_{n\to\infty}(I - \frac{t}{n}A)^{-n} = e^{At}, \quad t \geq 0 \tag{7}$$

in the uniform operator topology.

This is because A is m-dissipative, and from (5) we derive

$$\| (I - \frac{t}{n}A)^{-n} - (I - \frac{t}{m}A)^{-m} \| \leq t \| A \| (\frac{1}{n} + \frac{1}{m})^{1/2} \tag{8}$$

for $t \geq 0$, $m, n = 1, 2, \ldots$ Tangential condition (2.8) guarantees the existence of DS-approximate solution u_n, but it has nothing to do with the convergence of u_n (which is guaranteed by the dissipativity of A). The relationship of the semigroup $S_A(t)$ (generated by A via the exponential formula) and $(I - tA)^{-1}$ was established by Brezis [4]. By using this important relationship (i.e., Theorem 3.1) Brezis has given the characterization of compactness of $S_A(t)$ (Theorem 6.1) extending the result of Pazy [1] from the linear case. The characterization of compactness of $S_A(t)$ in linear case, solely in terms of the resolvent of A (a problem raised by Pazy), that is Theorem 6.4, is due to Vrabie [2]. Its proof here is given by the author and it is similar to that of Vrabie. Theorems 8.11 and 8.12 are given by the author [16].

Another general result is that of Bruck [1], on the asymptotic behaviour of $S_A(t)$ in the case $A = \partial f$ (see Theorem 9.3). In the proof one uses an interesting lemma of Opial (Lemma 9.2). In this direction we refer the reader to the books by Haraux [1] and Moroşanu [1].

The notion of integral solution has been introduced by Benilan [1]. He proved that $u(t; t_0, x_0) = S(t - t_0)x_0$ with $x_0 \in \overline{D(A)}$ and $t \geq t_0$ is the unique integral solution to the problem $u' \in Au$, $u(t_0) = x_0$. These arguments are now called "Benilan's method" or "Benilan's uniqueness theorem" (cf. Crandall and Evans [1] which contains a general and simple method to prove the exponential formula). As we have already mentioned, most of the results in Sections 6, 7 and 10 are due to Brezis.

Now a few comments on maximal monotone (dissipative) operators. It is clear that the notion of "monotone operator" is an extension of that of nondecreasing function from R into R. The theory of nonlinear monotone operators has started with the papers of Vainberg [1], Kachurovski [1] and Zarantonello [1]. Fundamental results on monotone operators have been otbained by G. Minty [1,2]. One of the main results of Minty, asserts that in Hilbert spaces, maximal accretivity (monotonicity) is equivalent to m-accretivity. This is not the case in general Banach spaces. A first counterexample has been constructed by Calvert [1] in 1^p. Moreover, Cernes [1] pointed out that if both X and X^* are uniformly convex (but X is not a Hilbert space), then there always exists a maximal accretive set A which is not m-accretive.

An impressive development of the Theory of monotone operators and its applications to partial differential equations has been given by Felix E. Browder in a great deal of papers. This theory is now called "the monotonicity method of Minty-Browder" (cf. Leray and Lions [1]). An important feature of this theory is that we do not have to restrict our attention to singlevalued operators. It allows A to be a multivalued operator, i.e., for $x \in D(A)$, Ax may contain more than one element. This generality (from singlevalued to multivalued) is essential in many applications to PDE and control theory and in most cases, it costs nothing in terms of analytical difficulty.

CHAPTER 3. The results in Section 1 are established by the author. Section 2 follows Oharu and Takahashi [1] except for some simplifications of the proofs and some remarks given by the author. Note that the (model) equation (2.7) i.e., Kortweg-deVries equation (KdV) is unanimously accepted, while (BBM) equation (2.6) - not. For Section 3, we have used Vasques [1], Hirsh [1] and Benilan [2]. For most of the results in Sections 4, 5, and 6 we have used some papers and books by Brezis [1-6]. Theorem 5.14 is proved by the author and extends a result of Ball [2].

APPENDIX It contains some of the basic results of nonlinear analysis on which this work is based. These facilitate the reading of the Chapters 1-3.

REFERENCES

ADAMS, R.A.

1. Sobolev spaces, Academic Press, New York 1975.

AGMON, S., DOUGLIS, A. and NIRENBERG, L.

1. Estimates near the boundary for solutions of elliptic partial differential equations satisfying general boundary conditions, Comm. Pure Appl. Math. 12 (1959), 623-727.

AIZAWA, S.

1. A semigroup treatment of the Hamilton-Jacobi equation in one space variable, Hiroshima Math. J., 3 (1973), 367-386.

ALTMAN, M.

1. Quasi-linear evolution equations in non-reflexive Banach spaces. I. Nonlinear Anal. 5 (1981), 411-421.

AMES, W.F.

1. Nonlinear partial differential equations in engineering, Academic Press, New York, 1965.

ARONSON, G. and MELLANDER, I.

1. A deterministic model in biomathematics: asymptotic behaviour and threshold conditions, Math. Biosci. 49 (1980), 207-222.

ATTOUCH , H.

1. On the maximality of the sum of two maximal monotone operators. Nonlinear Anal. 5(1981), 143-147.

ATTOUCH, H. and DAMLAMIAN, A.

1. On multivalued evolution equations in Hilbert spaces, Israel J. Math., 12 (1972), 373-390.

AUBIN, J.P. and CELLINA, A.

1. Differential Inclusions, Springer Verlag (1984).

BAILLON, J.B.

1. Générateurs et semi-groupes dans les espaces de Banach uniformément lisses. J. Functional Anal. 29(1978), 199-213.

BAILLON, J.B. and HARAUX, A.

1. Comportément à l'infini pour les équations d'évolution avec forcing périodique. Arch. Rational Mech. Anal. 67(1977), 101-109.

BALL, J.M.

1. Strongly continuous semigroups, weak solutions and the variation of constants formula, Proc. Amer. math. Soc., 63(1977), 370-373.
2. Finite time blow-up in nonlinear problems, Proc. Symp. of Nonlin. evolution equations held at Univ. of Wisconsin-Madison, Acad. Press (1978), 189-205.

BARAS, P.

1. Compacité de l'opérateur f→u solution d'une équation non-linéaire u'(t)+Au∋f, C.R. Acad. Sci. Paris, Série A, 286(1978), 1113-1116.

BARAS, P., HASSAN, J.C. and VERON, L.

1. Compacité de l'opérateur définissant la solution d'une équation d'évolution non homogène. C.R. Acad. Sci. Paris, 284(1977),799-802.

BARBU, V.

1. Continuous perturbations of nonlinear m-accretive operators in Banach spaces, Boll. Un. Mat. Ital. 6(1972) 270-278.
2. Nonlinear semigroups and differential equations in Banach spaces Noordhoff and Leyden, (1976).

BARBU,V. and PRECUPANU, Th.

1. Convexity and optimization in Banach spaces, Sÿthoff and Hoordhoff (1978).

BARDOS, C., PENEL, P., FRISCH, U. and SULEM, P.L.

1. Modified dissipativity for a non-linear evolution equation arising in turbulence, Arch. Rational Mech. Anal. 7(1979) 237-256.

BARONE, E.

1. The initial value problem for the neutron transport equation by the semi-group method, Atti Acad. Naz. Lincei Rend. Cl. Sci. Fis. Mat. Natur., LV(1973), 429-436.

BENILAN, Ph.

1. Solutions intégrales d'équations d'évolution dans un espace de Banach. C.R. Acad. Sci. Paris 274(1972), 47-50.
2. Equations d'évolution dans un espace de Banach quelconque et applications. Thèse, Orsay 1972.

BENILAN, Ph., BREZIS, H. and CRANDALL, M.G.

1. A semilinear equation in $L^1(R^N)$. Ann. Scuola Norm. Sup. Pisa Cl. Sci. (4), Vol. III (1975).

BENJAMIN, T.B., BONA, J.L. and MAHONY, J.J.

1. Model equation for long waves in nonlinear dispersive systems. Philos. Trans. Roy. Soc. London, Ser. A 272(1972), 47-78.

BERRYMAN, J.G. and HOLLAND, C.I.

1. Stability of the separable solution for fast diffusion. Arch. Rat. Mech. Anal. 74(1980), 279-288.

BONA, J.L. and BRYANT, P.L.

1. A mathematical model for long waves generated by wavemakers in non-linear dispersive systems. Proc. Camb. Phil. Soc. 73(1973), 391-405.

BONA, J.L. and SMITH, R.

1. The initial-value problem for Korteweg-de-Vries equation. Philos. Trans. Royal Soc. London 278 A(1975), 555-601.

BREZIS, H.

1. Propriétés régularisantes de certains semi-groupes nonlinéaires, Israel J. Math., 9(1971), 513-534.
2. Problèmes unilatéraux, J. Math. Pures Appl. 51(1972), 1-164.
3. Opérateurs maximaux monotones et semigroupes de contractions dans les espaces de Hilbert, Math. Studies, 5, North Holland, 1973.
4. New results concerning monotone operators and nonlinear semi-groups. proceedings of RIMS (on "Analysis of nonlinear problems"), Kyoto University, 1975.
5. Analyse Fonctionnelle. Théorie et Applications. Masson, Paris 1983.
6. Courses delivered at: Autumn course on Semigroups, Theory and Applications, ICTP, Trieste (Italy), 12 Nov.-14 Dec., 1984.

BREZIS, H. and CRANDALL, M.G.

1. Uniqueness of solutions of the initial-value problem for $u_t-\Delta\varphi(u)=0$. J. Math. Pures Appl. 58(1979), 153-163.

BREZIS, H., CRANDALL, M.G. and PAZY, A.

1. Perturbations of nonlinear maximal monotone sets in Banach space. Comm. Pure Appl. Math. 23(1970), 123-144.

BREZIS, H. and GALLOUET, T.

1. Nonlinear Schrödinger evolution equations, Nonlinear Anal., 4 (1988),677-681.

BREZIS, H. and STRAUSS, W.A.

1. Semilinear second-order elliptic equations in L^1, J. Math. Soc. Japan, 25 (1973), 565-592.

BROWDER, F.E.

1. Nonlinear equations of evolution, Annal. Math. 80(1964), 485-523.
2. Nonlinear equations of evolution and nonlinear accretive operators in Banach spaces. Bull. Amer. Math. Soc. 73(1967), 867-874.

BRUCK, R.

1. Asymptotic convergence of nonlinear contraction semigroups in Hilbert space. J. Funct. Anal. 18(1975), 15-26.

BURCH, B.C.

1. A semigroup treatment of the Hamilton-Jacobi equations in several space variables. J. Differential Equations, 23(1977), 102-124.

BUTZER, P.L. and BERENS, H.

1. Semigroups of operators and approximation, Springer Verlag (1967).

CAFFARELLI, A.L. and FRIEDMAN, A.

1. Continuity of the density of a gas flow in a porous medium, Trans Amer. Math. Soc. 252(1979), 99-113.

CALVARET, B.

1. Maximal accretive is not m-accretive, Boll. Un. Mat. Ital. 6 (1970), 1042-1044.
2. Nonlinear equations of evolution. Pacific J. Math. 39 (1971), 243-350.

CAPASSO, V. and MADDALENA, L.

1. Convergence to equilibrium states for a reaction-diffusion system modelling the spatial spread of a class of bacterial and viral diseases, J. Math. Biology 13(1981), 173-184.

CAZENAVE, T. and HARAUX, A.

1. Equations d'évolution avec non linéarité logaritmique. Ann. Fac. Sci. Toulouse Vol. II (1980), 21-51.

CERNES, A.

1. Ensembles maximaux accrétifs et m-accrétifs, Israel J. Math. 19 (1974), 335-348.

CHERNOFF, P.

1. Note on product formulas for semigroups. J. Funct. Anal.2(1968), 238-242.

CLARKE, F.H.

1. Generalized gradients and applications. Trans. Amer. Mat.Soc. 205(1975), 247-262.

CRANDALL, M.G.

1. The semigroup approach to the first order quasilinear equations in several space variables, Israel J. Math. 12(1972), 108-132.
2. A generalized domain for semigroups generators. Proc. Amer. Soc. 37(1973), 434-440.
3. Semigroups of nonlinear transformations and evolution equations. Proc. Int. Congress Math, Vancouver, 1974, pp. 257-262.

CRANDALL, M.G. and EVANS, L.C.

1. On the relation of the operator $\partial/\partial s + \partial/\partial\tau$ to evolution governed by accretive operators. Israel J. Math. 21(1975), 261-178.
2. A singular semilinear equation in $L^1(R)$. Trans. Amer. Math. Soc. 225(1977), 145-153.

CRANDALL, M.G. and LIGGETT, T.M.

1. Generation of semigroups of nonlinear transformations on general Banach spaces. Amer. J. Math. 93(1971), 265-298.

CRANDALL, M.G. and PAZY, A.

1. Nonlinear evolution equations in Banach spaces. Israel J. Math. 11(1972).
2. Semi-groups of nonlinear operators and dissipative sets. J. Funct. Anal. 3(1969), 376-418.

DAFERMOS, C. and SLEMROD, M.

1. Asymptotic behaviour of nonlinear contraction semigroups. J. Funct. Anal. 12(1973), 96-106.

DA PRATO , G.

1. Applications croissantes et équations d'évolutions dans les espaces de Banach. Ist. Nazionale di Alta Matematica, Vol. II, 1976.

DA PRATO, G. and GRISVARD, P.

1. Equations d'évolution abstraites non linéaires de type parabolique. Ann. Mat. Pura Appl. CXX(1979) 329-396.

DEBRUNNER, H. and FLOR, P.

1. Ein Erweiterungssatz für monotone Mogen, Archiv. Math., 15 (1964), 445-447.

DEIMLING, K.

1. Ordinary differential equation in Banach spaces, Lecture Notes

in Mathematics, 596, Springer Verlag (1977)

DESCH, W. , LASIECKA, I. and SCHAPPACHER, W.

1. Feedback boundary control problems for linear semigroups. Israel J. Math. 51(1985), 177-207.

DIAZ, J.I.

1. Nonlinear Partial Differential Equations and Free Boundaries. Research Notes in Math., Vol. 106, Pitman Adv. Publ. Program

DI BLASIO, G.

1. Nonlinear perturbations and abstract evolution equation in general Banach spaces. Ann. Math. Pura Appl. 4(1976), 317-355.

DINCA, G.

1. Operatori monotoni în teoria plasticităţii. Ed. Academici, Bucureşti, 1972.

DORROH, J.R.

1. A nonlinear Hille-Yosida-Phillips theorem. J. Funct. Anal. 3 (1969), 345-353.

DUNFORD, N. and SCHWARTZ, J.

1. Linear operators, part I, Interscience, New York 1964

ENGLER, H.

1. Invariant sets for functional differential equations in Banach spaces and appl. Nonlinear Anal., 5(1981), 1225-1243.

EVANS, L.C.

1. Nonlinear evolution equations in an arbitrary Banach space. Israel J. Math., 26(1977), 1-42.

EVANS, L.C. and MASSEY, F.J.

1. A remark on the construction of nonlinear evolution operators, Huston J. Math. 4(1978), 35-40.

FILIPPOV, A.F.

1. Righ-hand side discontinuous differential equations. Trans. Amer. Math. Soc. 42(1964), 199-227.

FOIAŞ, C., GUSSI, G. and POENARU, V.

1. Sur les solution généralisées de certaines équations linéaires et quasi-linéaires dans l'espace de Banach. Rev. Roumaine Math. Pures Appl. 3(1959), 283-304.

FRIEDMAN, A.

1. Partial Differential Equations of Parabolic Type. Prentice-Hall, Englewood Cliffs, N.Y. 1964.

GLASSEY, R.T.

1. On the blowing up of solutions to the Cauchy problem for the nonlinear Schrödinger equation, J. Math. Phys. 18(1977).

GOLDSTEIN, J.A.

1. Semigroups of Linear Operators and Applications, Oxford University Press, 1985.

GOSSEZ, J.P.

1. Opérateurs monotones non-linéaires dans les espaces de Banach nonréflexifs, J. Math. Anal. Appl. 34(1971), 371-395.

GUTMAN, S.

1. Evolution governed by m-accretive plus compact operators, Nonlin. Anal., 7(1983), 707-717.

HALE, J.K.

1. Functional Differential Equations, Appl. Math. Sci. Vol. 3, Springer Verlag, New York, 1971.

HALLAM, Th. G.

1. On the nonexistence of L^p-solutions of certain nonlinear differential equations. Glasgow Math. J., 8(1967), 133-138.

HARAUX, A.

1. Nonlinear Evolution Equations. Global Behaviour of Solutions, Lecture Notes in Math. Vol. 841, Springer-verlag 1981.

HARAUX, A. and KIRANE, M.

1. Estimations C^1 pour des problèmes paraboliques semi-linéaires. Ann. Fac. Sci. Toulouse, Vol. V(1983), 265-280.

HENRY, D.

1. Geometric theory of semilinear parabolic equations. Lecture Notes in Math. Vol. 840; Springer Verlag, 1981.

HILLE, E. and PHILIPS, R.S.

1. Functional Anlaysis and Semigroups. Amer. Math. Soc., Providence, R.I. 1957.

HIRSCH, M.W.

1. The dynamical system approach to differential equations, Bull.

AMS, 11(1984), 1-64.

IANELLI, M.

1. Note on some nonlinear semigroups. Boll. Un. Mat. Ital. 6(1970), 1015-1025.
2. On a certain class of semilinear evolution systems, J. Math. Anal. Appl. 56(1976), 351-367.

ICHIKAWA, A. and PRITCHARD, A.J.

1. Existence, uniqueness and stability of nonlinear evolution equations. J. Math. Anal. Appl. 68(1979), 454-475.

ITAYA, N.

1. A note on the blow-up probelms in nonlinear parabolic equations, Proc. Japan Acad., Ser. A, Math. Sci. 55(1979), 241-244.

IWAMIYA, T., OHARU, A. and TAKAHASHI, T.

1. On the class of nonlinear evolution operators in Banach spaces, Nonl. Anal. (to appear).

JOHN, F.

1. Blow-up for quasilinear wave equations in three space dimensions Comm. Pure Appl. Math., 34(1981), 29-51.

KACHUROVSKII, R.I.

1. On monotone operators and convex functional (Russian), Uspechi Math. Nauk 15(1960), 213-215.

KARTSATOS, A.G.

1. Perturbed evolution equations and Galerkin's method. Math. Nachr. 91(1979), 337.

KARTSATOS, A.G. and PARROTT, M.E.

1. Convergence of the Kato approximation for evolution equation involving functional perturbations. J. Diff. Equations, 47 (1983), 358-377.
2. A method of lines for a nonlinear abstract functional evolution equation. Trans. Amer. Mat. Soc. 286(1984), 73-89.

KATO, T.

1. Nonlinear semigroups and evolution equations. J. Math. Soc. Japan, 19(1967), 508-520.
2. On Korteweg-de Vries equation. Manuscripta Math., 28(1979), 89-99.

KENMOCHI, N. and TAKAHASHI, T.

1. Nonautonomous differential equations in Banach spaces. Nonl. Anal. 4(1980), 1109-1121.

KIRK, W.A.

1. A fixed point theorem for mappings which do not increase distances. Amer. Math. Monthly, 72(1965)., 1004-1006.

KOBAYASI , K., KOBAYASHI, Y. and OHARU, S.

1. Nonlinear evolution operators in Banach spaces. Osaka J. Math. 21(1984), 281-310.

KOBAYASHI, Y.

1. Difference approximation of Cauchy problems for quasi-dissipative operators and generation of nonlinear semigroups. J. Math.Soc. Japan, 27(1975), 641-663.
2. A remark on convergence of nonlinear semigroups. Proc. Japan Acad. Ser. A, Math. Sci. 55(1979), 45-48.

KOMURA, Y.

1. Nonlinear semi-groups in Hilbert space. J. Math. Soc. Japan, 19(1967), 493-507.

KONISHI, Y.

1. On the nonlinear semigroups associated with $u_t = \Delta\beta(u)$ and $\phi(u_t) = \Delta u$., J. Math. Soc. Japan 25(1973), 622-628.
2. Compacité des résolvantes des opérateurs cycliquement monotones. Proc. Japan. Acad. Ser. A, Math. Sci. 49(1973), 303-305.

KREIN, S.G.

1. Linear Differential Equations in Banach Spaces, Translations of Math. Monographs, Vol. 29(1971), AMS.

KRUZKOV, S.N.

1. First order quasilinear equations in several independent variables. Math. USSR, Sb. 10(1970), 217-243.

KURTZ, T.

1. Convergence of sequences of semigroups of nonlinear operators with an application to gas kinetics. Trans. Amer. Math. Soc. 186(1973), 259-272.

LADAS, G. and LAKSHMIKANTHAM, V.

1. Differential equations in abstract spaces. Academic Press, New York, 1972.

LAKSHMIKANTHAM, V. and LEELA, S.

1. Nonlinear differential equations in abstract spaces. Pergamon Press, Oxford, New York, 1981.

LA PELETIER,

1. Large time behaviour of solutions of the porous media equation, in Recent contribution to nonlinear partial differential equations (Vol. 50, pp.157-163), Research Notes in Math., Pitman, Beresticki and Brézis (Editors)

LERAY, J. and LIONS, J.L.

1. Quelques résultats de Visik sur les problèmes elliptiques quasi linéaires par la méthode de Minty-Browder, Séminaires de Collège de France, 1964.

LIGHTBOURNE, J.H. and MARTIN, R.H. Jr.

1. Relatively continuous nonlinear perturbations of analytic semigroups. Nonlinear Analysis 1(1977), 277-292.

LIONS, J.L.

1. Quelques méthodes de résolution de problèmes aux limites nonlinéaires. Dunnod, Paris 1969.
2. Quelqeus problèmes de la théorie des équations non-linéaires d'évolution. Problems in non-linear analysis. CIME, Varenna 1970, pp. 242-271.

LIONS, J.L. and MAGENES, E.

1. Problèmes aux limites nonhomogènes et applications.Vol. I. Dunod, Paris 1968.

MARTIN, R.H. Jr.

1. A global existence theorem for autonomous differential equations in a Banach space. Proc. Amer. Math. Soc. 26(1970), 307-314.
2. Nonlinear Operators and Differential Equations in Banach Spaces. John Wiley and Sons. New York, 1976.

MINTY, G.J.

1. Monotone nonlinear operators in Hilbert spaces. Duke Math. J., 29(1962), 341-346.
2. On the monotonicity method for the solution of nonlinear equations in Banach spaces. Proc. Nat. Acad. Sci. USA 50(1963), 1034-1014.

MIYADERA , I. and KOBAYASHI, Y.

1. Convergence and approximation of nonlinear semigroups. Japan-France Seminar, Tokyo and Kyoto 1976; H. Fujita (Ed.): Japan Society for the Promotion of Science, 1978, 277-295.

MORALES, C.

1. Existence theorems for semicontinuous accretive operators in Banach spaces, Huston J. Math. Vol. 10(1984), 535-543.

MOROȘANU, G.

1. Nonlinear evolution equation and applications, Reidel Publishing Co., Dordrecht/Boston/London (to appear)

MOTREANU, D. and PAVEL, N.H.

1. Quasi-tangent vectors in flow-invariance and optimization problems, J. Math. Anal. Appl. 88(1982), 116-132.

MUSKAT,

1. The flow of homogeneous fluids through porous media, McGraw-Hill New York, 1937.

OHARU, S.

1. On the generation of semigroups of nonlinear contractions. J. Math. Soc. Japan, 22(1970), 526-550.

OHARU, S. and TAKAHASHI, T.

1. On nonlinear evolution operators associated with some nonlinear dispersive equations (to appear)

OKOCHI, H.

1. A Note on Asymptotic Strong Convergence of Nonlinear Contraction Semigroups, Proc. Japan, Acad. 56. Ser. A(1980) 83-84.

OPIAL, Z.

1. Lecture notes on nonexpansive and monotone mappings in Banach spaces, Brown University 1967

ÔTANI, M.

1. Nonmonotone perturbations for linear parabolic equations associated with subdifferential operators. J. Diff. Equations 48 (1982), 268-279.

PASCALI, D. and SBURLAN, S.

1. Nonlinear mappings of monotone type. Sijthoff and Noordhoff 1978.

PAVEL, N.H.

1. Sur certaines équations différentielles abstraites. Boll. Un. Mat. Ital. 6 (1972), 397-409.
2. Equation non-linéaires d'évolution. Mathematica Cluj.14(1972), 289-300.
3. Approximate solutions of Cauchy Problem for some differential equations on Banach spaces. Funkcial. Ekvak. 17(1974), 85-94.
4. Nonlinear Evolution Equations Associated with Accretive Operators on General Banach Spaces. Mimeographed Lecture Notes, Univ. Iasi, 1974.
5. Nonlinear boundary value problems for second order differential equations. J. Math. Anal. Appl. (1975), 373-383.

6. Mixed boundary value problem for second order differential equations with monotone operators. Bull. Math. Soc. Sci. Math. de RSR 19(1975), 127-145.

7. Integral solutions for nonlinear evolution equations on Banach spaces. Studia Mathematica 55(1976), 141-149.

8. Ecuaţii Diferenţiale Asociate unor Operatori Neliniari pe Spaţii Banach. Editura Acad. RSR, Bucureşti 1977.

8'. Differential equations associated with continuous and dissipative time-dependent domain operators. Proc. Helsinki Symp. on Volterra Equations. Lecture Notes in Math, Springer-Verlag 727.

9. Nonlinear evolution equations governed by f-quasi-dissipative operators. Nonlinear Analysis, TMA 5(1981), 449-468.

10. Addendum-Nonlinear evolution equations governed by f-quasi dissipative operators. Nonlinear Anal., TMA 5(1981), 1389.

11. Toward the unification of the theory of nonlinear semigroups. Bul. Inst. Polit. Iasi XXVII (XXXI) 1981, 35-40.

12. Analysis of some Nonlinear Problems in Banach Spaces and Applications. Mimeographed Lecture Notes, Univ. Iasi, 1982.

13. Semilinear equations with dissipative time-dependent domain pertrubations. Israel J. Math. 46 (1983), 103-122.

(See REFERENCES - ADDENDUM)

PAVEL, N.H. and URSESCU, C.

1. Existence and uniqueness for some nonlinear functional equations in a Banach space, An. Stiint Univ. Iasi, Sect. I-a Mat. 20 (1974).

2. Flow-invariant sets for autonomous second order differential equations and applications in Mechanics. Nonlinear Anal. 6(1982), 35-74.

PAVEL, N.H. and VRABIE, I.I.

1. Semilinear evolution equations with multivalued right-hand side in Banach spaces, An. Stiint Univ. Iasi Sect. I-a Mat XXV (1979) 137-157.

PAZY, A.

1. On the differentiability and compactness of semigroups of linear operators, J. Math. Mech? 11 (1968), 1131-1142.

2. A class of semilinear equations of evolution, Israel J. Math. 20(1975), 26-36

3. The Lyapunov method for semigroups of nonlinear contractions in Banach spaces, J. Analyse Math., 40(1981), 239-262.

4. Semigroups of linear operators and applications to partial differential equations, Springer-Verlag, New York, Berlin, Heidelberg, Tokyo, Appl. Math. Sci. Vol. 44(1983).

PIERRE, M.

1. Enveloppe d'une famille de semigroupes nonlinéaires et équatio[n] d'évolution, séminaire d'Analyse non-linéaire, Univ. de Besanç[on] (25), (1976-1977).

PLANT, A.T.

1. Nonlinear semigroups of translations in Banach space generated by functional differential equations. J. Math. Anal. Appl., 60(1977) 67-74.

QUINN, B.K.

1. Solutions with shocks: An example of L_1-contractive semigroups. Comm. Pure Appl. Math. 24(1971), 125-132.

REICH, S.

1. Strong convergence theorems for resolvents of accretive operators in Banach spaces, J. Math. Anal. Appl., 75(1980), 287-292.

ROCKAFELLAR, R.T.

1. Local boundedness of nonlinear monotone operators, The Michigan Math. J. 16(1969), 397-407.

SCHECHTER, E.

1. Evolution generated by semilinear dissipative plus compact operators, Trans. Amer. Math. Soc., 275 (1983), 297-308.

SEGAL, I.

1. Nonlinear semigroups, Ann. Math. 78(1963), 339-364.

SINESTRARI, E.

1. Accretive differential operators. Boll. Un. Mat. Ital. 13 (1976), 19-31.

TROTTER, H.

1. Approximation of semigroups. Pacific J. Math. 8 (1958), 887-919.

URSESCU, C.

1. Tangent sets and differentiable functions. Banach Center Publications. 1(1974), 151-155.

VAINBERG, M.

1. On the convergence of the process of steepest descent for nonlinear equations, Sibirsk Math. J. 2(1961), 201-220.

VASQUES, J.L.

1. Topics on nonlinear diffusion. The porous medium equation (autumn course on semigroups, theory and applications, 12 November - 14 December 1984, ICTP Trieste, Italy).

VRABIE, I.I.

1. The nonlinear version of Pazy's local existence theorem, Israel J. Math. 32(1979), 221-235.
2. A characterization of Compactness of Linear Semigroups, Personal Communication. (See also REFERENCES-ADDENDUM).

WATANABE, J.

1. On certain nonlinear evolution equations. J. Math. Soc. Japan, 25(1973), 446-463.

WEBB, G.F.

1. Continuous nonlinear perturbations of linear accretive operators in Banach spaces, J. Funct. Anal. 10(1972), 191-203.
2. Theory of Nonlinear Age-dependent Population Dynamics, New York, Dekker 1985.

WEISSLER, F.B.

1. L^p-Energy and blow-up for a semilinear heat equation, Proc. Symposia in Pure Math., Vol. 45, Part 2, pp. 545-551, 1986.

WESTPHAL, U.

1. Sur la saturation pour des semi-groupes non-linéaires, C.R. Acad. Sci. Paris 274(1972), 1351-1353.

YAMADA, Y.

1. On evolution equations governed by subdifferential operators, J. Fac. Sci. Univ. Tokyo 23(1976), 491-515.

YOSIDA, K.

1. Functional Analysis (6th edition), Springer-Verlag 1980.

ZARANTONELLO, E.

1. Contribution to Nonlinear Analysis, Zarantonello, ed. Academic Press 1971, 237-424.

ZEIDLER, E.

1. Nonlinear Functional Analysis and its Applications, Vol. 1-5, Springer-Verlag (1986).

ZELDOVICH, Y.B. and RAIZER, Y.P.

1. Physics of shock waves and high temperature. Hydrodynamic Phenomena, Vol. II. Academic Press, New York (1969).

REFERENCES(ADDENDUM)

PAVEL , N.H.

14. Some problems on nonlinear semigroups anf the blow-up of integral solutions. Proc. Conf. on Operators, Semigroups and Appl., Retzhof-Graz 1983, Lecture Notes in Math. 1076(1984), Springer-Verlag (Kappel & Schappacher editors). (Internal Report IC/70 (1983), ICTP Trieste (Italy)).
15. Differential Equations, Flow-invariance and Applications, PITMAN Research Notes in Math. 113(1984), London-Boston.
16. Compact semigroups, invariance and positivity in differential equations. Proc. Symposium: Aspects of positivity in functional analysis, Tübingen, Germany, 24-28 June 1985.

VRABIE , I.I.

3. Compactness Methods for Nonlinear Evolutions. Pitman Advanced Publishing Program (to appear).

SCHAPPACHER,, W.

1. Translation semigroups and functional differential equations. PITMAN Res.Notes in Math. ,Vol.48 (Kappel and Schappacher edit.), 175-190 .

Index

LECTURE NOTES IN MATHEMATICS
Edited by A. Dold and B. Eckmann

Some general remarks on the publication of monographs and seminars

In what follows all references to monographs, are applicable also to multiauthorship volumes such as seminar notes.

1. Lecture Notes aim to report new developments - quickly, informally, and at a high level. Monograph manuscripts should be reasonably self-contained and rounded off. Thus they may, and often will, present not only results of the author but also related work by other people. Furthermore, the manuscripts should provide sufficient motivation, examples and applications. This clearly distinguishes Lecture Notes manuscripts from journal articles which normally are very concise. Articles intended for a journal but too long to be accepted by most journals, usually do not have this "lecture notes" character. For similar reasons it is unusual for Ph.D. theses to be accepted for the Lecture Notes series.

   Experience has shown that English language manuscripts achieve a much wider distribution.

2. Manuscripts or plans for Lecture Notes volumes should be submitted either to one of the series editors or to Springer-Verlag, Heidelberg. These proposals are then refereed. A final decision concerning publication can only be made on the basis of the complete manuscripts, but a preliminary decision can usually be based on partial information: a fairly detailed outline describing the planned contents of each chapter, and an indication of the estimated length, a bibliography, and one or two sample chapters - or a first draft of the manuscript. The editors will try to make the preliminary decision as definite as they can on the basis of the available information.

3. Lecture Notes are printed by photo-offset from typed copy delivered in camera-ready form by the authors. Springer-Verlag provides technical instructions for the preparation of manuscripts, and will also, on request, supply special staionery on which the prescribed typing area is outlined. Careful preparation of the manuscripts will help keep production time short and ensure satisfactory appearance of the finished book. Running titles are not required; if however they are considered necessary, they should be uniform in appearance. We generally advise authors not to start having their final manuscripts specially tpyed beforehand. For professionally typed manuscripts, prepared on the special stationery according to our instructions, Springer-Verlag will, if necessary, contribute towards the typing costs at a fixed rate.

   The actual production of a Lecture Notes volume takes 6-8 weeks.

.../...

4. Final manuscripts should contain at least 100 pages of mathematical text and should include

   - a table of contents
   - an informative introduction, perhaps with some historical remarks. It should be accessible to a reader not particularly familiar with the topic treated.
   - subject index; this is almost always genuinely helpful for the reader.

5. Authors receive a total of 50 free copies of their volume, but no royalties. They are entitled to purchase further copies of their book for their personal use at a discount of 33 1/3 %, other Springer mathematics books at a discount of 20 % directly from Springer-Verlag.

   Commitment to publish is made by letter of intent rather than by signing a formal contract. Springer-Verlag secures the copyright for each volume.

Vol. 1117: D.J. Aldous, J.A. Ibragimov, J. Jacod, Ecole d'Été de Probabilités de Saint-Flour XIII – 1983. Édité par P.L. Hennequin. IX, 409 pages. 1985.

Vol. 1118: Grossissements de filtrations: exemples et applications. Seminaire, 1982/83. Edité par Th. Jeulin et M. Yor. V, 315 pages. 1985.

Vol. 1119: Recent Mathematical Methods in Dynamic Programming. Proceedings, 1984. Edited by I. Capuzzo Dolcetta, W.H. Fleming and T. Zolezzi. VI, 202 pages. 1985.

Vol. 1120: K. Jarosz, Perturbations of Banach Algebras. V, 118 pages. 1985.

Vol. 1121: Singularities and Constructive Methods for Their Treatment. Proceedings, 1983. Edited by P. Grisvard, W. Wendland and J.R. Whiteman. IX, 346 pages. 1985.

Vol. 1122: Number Theory. Proceedings, 1984. Edited by K. Alladi. VII, 217 pages. 1985.

Vol. 1123: Séminaire de Probabilités XIX 1983/84. Proceedings. Edité par J. Azéma et M. Yor. IV, 504 pages. 1985.

Vol. 1124: Algebraic Geometry, Sitges (Barcelona) 1983. Proceedings. Edited by E. Casas-Alvero, G.E. Welters and S. Xambó-Descamps. XI, 416 pages. 1985.

Vol. 1125: Dynamical Systems and Bifurcations. Proceedings, 1984. Edited by B.L.J. Braaksma, H.W. Broer and F. Takens. V, 129 pages. 1985.

Vol. 1126: Algebraic and Geometric Topology. Proceedings, 1983. Edited by A. Ranicki, N. Levitt and F. Quinn. V, 423 pages. 1985.

Vol. 1127: Numerical Methods in Fluid Dynamics. Seminar. Edited by F. Brezzi, VII, 333 pages. 1985.

Vol. 1128: J. Elschner, Singular Ordinary Differential Operators and Pseudodifferential Equations. 200 pages. 1985.

Vol. 1129: Numerical Analysis, Lancaster 1984. Proceedings. Edited by P.R. Turner. XIV, 179 pages. 1985.

Vol. 1130: Methods in Mathematical Logic. Proceedings, 1983. Edited by C.A. Di Prisco. VII, 407 pages. 1985.

Vol. 1131: K. Sundaresan, S. Swaminathan, Geometry and Nonlinear Analysis in Banach Spaces. III, 116 pages. 1985.

Vol. 1132: Operator Algebras and their Connections with Topology and Ergodic Theory. Proceedings, 1983. Edited by H. Araki, C.C. Moore, Ş. Strătilă and C. Voiculescu. VI, 594 pages. 1985.

Vol. 1133: K.C. Kiwiel, Methods of Descent for Nondifferentiable Optimization. VI, 362 pages. 1985.

Vol. 1134: G.P. Galdi, S. Rionero, Weighted Energy Methods in Fluid Dynamics and Elasticity. VII, 126 pages. 1985.

Vol. 1135: Number Theory, New York 1983–84. Seminar. Edited by D.V. Chudnovsky, G.V. Chudnovsky, H. Cohn and M.B. Nathanson. V, 283 pages. 1985.

Vol. 1136: Quantum Probability and Applications II. Proceedings, 1984. Edited by L. Accardi and W. von Waldenfels. VI, 534 pages. 1985.

Vol. 1137: Xiao G., Surfaces fibrées en courbes de genre deux. IX, 103 pages. 1985.

Vol. 1138: A. Ocneanu, Actions of Discrete Amenable Groups on von Neumann Algebras. V, 115 pages. 1985.

Vol. 1139: Differential Geometric Methods in Mathematical Physics. Proceedings, 1983. Edited by H. D. Doebner and J. D. Hennig. VI, 337 pages. 1985.

Vol. 1140: S. Donkin, Rational Representations of Algebraic Groups. VII, 254 pages. 1985.

Vol. 1141: Recursion Theory Week. Proceedings, 1984. Edited by H.-D. Ebbinghaus, G.H. Müller and G.E. Sacks. IX, 418 pages. 1985.

Vol. 1142: Orders and their Applications. Proceedings, 1984. Edited by I. Reiner and K. W. Roggenkamp. X, 306 pages. 1985.

Vol. 1143: A. Krieg, Modular Forms on Half-Spaces of Quaternions. XIII, 203 pages. 1985.

Vol. 1144: Knot Theory and Manifolds. Proceedings, 1983. Edited by D. Rolfsen. V, 163 pages. 1985.

Vol. 1145: G. Winkler, Choquet Order and Simplices. VI, 143 pages. 1985.

Vol. 1146: Séminaire d'Algèbre Paul Dubreil et Marie-Paule Malliavin. Proceedings, 1983–1984. Edité par M.-P. Malliavin. IV, 420 pages. 1985.

Vol. 1147: M. Wschebor, Surfaces Aléatoires. VII, 111 pages. 1985.

Vol. 1148: Mark A. Kon, Probability Distributions in Quantum Statistical Mechanics. V, 121 pages. 1985.

Vol. 1149: Universal Algebra and Lattice Theory. Proceedings, 1984. Edited by S. D. Comer. VI, 282 pages. 1985.

Vol. 1150: B. Kawohl, Rearrangements and Convexity of Level Sets in PDE. V, 136 pages. 1985.

Vol 1151: Ordinary and Partial Differential Equations. Proceedings, 1984. Edited by B.D. Sleeman and R.J. Jarvis. XIV, 357 pages. 1985.

Vol. 1152: H. Widom, Asymptotic Expansions for Pseudodifferential Operators on Bounded Domains. V, 150 pages. 1985.

Vol. 1153: Probability in Banach Spaces V. Proceedings, 1984. Edited by A. Beck, R. Dudley, M. Hahn, J. Kuelbs and M. Marcus. VI, 457 pages. 1985.

Vol. 1154: D.S. Naidu, A.K. Rao, Singular Pertubation Analysis of Discrete Control Systems. IX, 195 pages. 1985.

Vol. 1155: Stability Problems for Stochastic Models. Proceedings, 1984. Edited by V.V. Kalashnikov and V.M. Zolotarev. VI, 447 pages. 1985.

Vol. 1156: Global Differential Geometry and Global Analysis 1984. Proceedings, 1984. Edited by D. Ferus, R.B. Gardner, S. Helgason and U. Simon. V, 339 pages. 1985.

Vol. 1157: H. Levine, Classifying Immersions into $\mathbb{R}^4$ over Stable Maps of 3-Manifolds into $\mathbb{R}^2$. V, 163 pages. 1985.

Vol. 1158: Stochastic Processes – Mathematics and Physics. Proceedings, 1984. Edited by S. Albeverio, Ph. Blanchard and L. Streit. VI, 230 pages. 1986.

Vol. 1159: Schrödinger Operators, Como 1984. Seminar. Edited by S. Graffi. VIII, 272 pages. 1986.

Vol. 1160: J.-C. van der Meer, The Hamiltonian Hopf Bifurcation. VI, 115 pages. 1985.

Vol. 1161: Harmonic Mappings and Minimal Immersions, Montecatini 1984. Seminar. Edited by E. Giusti. VII, 285 pages. 1985.

Vol. 1162: S.J.L. van Eijndhoven, J. de Graaf, Trajectory Spaces, Generalized Functions and Unbounded Operators. IV, 272 pages. 1985.

Vol. 1163: Iteration Theory and its Functional Equations. Proceedings, 1984. Edited by R. Liedl, L. Reich and Gy. Targonski. VIII, 231 pages. 1985.

Vol. 1164: M. Meschiari, J.H. Rawnsley, S. Salamon, Geometry Seminar "Luigi Bianchi" II – 1984. Edited by E. Vesentini. VI, 224 pages. 1985.

Vol. 1165: Seminar on Deformations. Proceedings, 1982/84. Edited by J. Ławrynowicz. IX, 331 pages. 1985.

Vol. 1166: Banach Spaces. Proceedings, 1984. Edited by N. Kalton and E. Saab. VI, 199 pages. 1985.

Vol. 1167: Geometry and Topology. Proceedings, 1983–84. Edited by J. Alexander and J. Harer. VI, 292 pages. 1985.

Vol. 1168: S.S. Agaian, Hadamard Matrices and their Applications. III, 227 pages. 1985.

Vol. 1169: W.A. Light, E.W. Cheney, Approximation Theory in Tensor Product Spaces. VII, 157 pages. 1985.

Vol. 1170: B.S. Thomson, Real Functions. VII, 229 pages. 1985.

Vol. 1171: Polynômes Orthogonaux et Applications. Proceedings, 1984. Edité par C. Brezinski, A. Draux, A.P. Magnus, P. Maroni et A. Ronveaux. XXXVII, 584 pages. 1985.

Vol. 1172: Algebraic Topology, Göttingen 1984. Proceedings. Edited by L. Smith. VI, 209 pages. 1985.

Vol. 1173: H. Delfs, M. Knebusch, Locally Semialgebraic Spaces. XVI, 329 pages. 1985.

Vol. 1174: Categories in Continuum Physics, Buffalo 1982. Seminar. Edited by F.W. Lawvere and S.H. Schanuel. V, 126 pages. 1986.

Vol. 1175: K. Mathiak, Valuations of Skew Fields and Projective Hjelmslev Spaces. VII, 116pages. 1986.

Vol. 1176: R.R. Bruner, J.P. May, J.E. McClure, M. Steinberger, H_∞ Ring Spectra and their Applications. VII, 388 pages. 1986.

Vol. 1177: Representation Theory I. Finite Dimensional Algebras. Proceedings, 1984. Edited by V. Dlab, P. Gabriel and G. Michler. XV, 340 pages. 1986.

Vol. 1178: Representation Theory II. Groups and Orders. Proceedings, 1984. Edited by V. Dlab, P. Gabriel and G. Michler. XV, 370 pages. 1986.

Vol. 1179: Shi J.-Y. The Kazhdan-Lusztig Cells in Certain Affine Weyl Groups. X, 307 pages. 1986.

Vol. 1180: R. Carmona, H. Kesten, J.B. Walsh, École d'Été de Probabilités de Saint-Flour XIV – 1984. Édité par P.L. Hennequin. X, 438 pages. 1986.

Vol. 1181: Buildings and the Geometry of Diagrams, Como 1984. Seminar. Edited by L. Rosati. VII, 277 pages. 1986.

Vol. 1182: S. Shelah, Around Classification Theory of Models. VII, 279 pages. 1986.

Vol. 1183: Algebra, Algebraic Topology and their Interactions. Proceedings, 1983. Edited by J.-E. Roos. XI, 396 pages. 1986.

Vol. 1184: W. Arendt, A. Grabosch, G. Greiner, U. Groh, H.P. Lotz, U. Moustakas, R. Nagel, F. Neubrander, U. Schlotterbeck, One-parameter Semigroups of Positive Operators. Edited by R. Nagel. X, 460 pages. 1986.

Vol. 1185: Group Theory, Beijing 1984. Proceedings. Edited by Tuan H.F. V, 403 pages. 1986.

Vol. 1186: Lyapunov Exponents. Proceedings, 1984. Edited by L. Arnold and V. Wihstutz. VI, 374 pages. 1986.

Vol. 1187: Y. Diers, Categories of Boolean Sheaves of Simple Algebras. VI, 168 pages. 1986.

Vol. 1188: Fonctions de Plusieurs Variables Complexes V. Séminaire, 1979–85. Edité par François Norguet. VI, 306 pages. 1986.

Vol. 1189: J. Lukeš, J. Malý, L. Zajíček, Fine Topology Methods in Real Analysis and Potential Theory. X, 472 pages. 1986.

Vol. 1190: Optimization and Related Fields. Proceedings, 1984. Edited by R. Conti, E. De Giorgi and F. Giannessi. VIII, 419 pages. 1986.

Vol. 1191: A.R. Its, V.Yu. Novokshenov, The Isomonodromic Deformation Method in the Theory of Painlevé Equations. IV, 313 pages. 1986.

Vol. 1192: Equadiff 6. Proceedings, 1985. Edited by J. Vosmansky and M. Zlámal. XXIII, 404 pages. 1986.

Vol. 1193: Geometrical and Statistical Aspects of Probability in Banach Spaces. Proceedings, 1985. Edited by X. Femique, B. Heinkel, M.B. Marcus and P.A. Meyer. IV, 128 pages. 1986.

Vol. 1194: Complex Analysis and Algebraic Geometry. Proceedings, 1985. Edited by H. Grauert. VI, 235 pages. 1986.

Vol.1195: J.M. Barbosa, A.G. Colares, Minimal Surfaces in $\mathbb{R}^3$. X, 124 pages. 1986.

Vol. 1196: E. Casas-Alvero, S. Xambó-Descamps, The Enumerative Theory of Conics after Halphen. IX, 130 pages. 1986.

Vol. 1197: Ring Theory. Proceedings, 1985. Edited by F.M.J. van Oystaeyen. V, 231 pages. 1986.

Vol. 1198: Séminaire d'Analyse, P. Lelong – P. Dolbeault – H. Skoda. Seminar 1983/84. X, 260 pages. 1986.

Vol. 1199: Analytic Theory of Continued Fractions II. Proceedings, 1985. Edited by W.J. Thron. VI, 299 pages. 1986.

Vol. 1200: V.D. Milman, G. Schechtman, Asymptotic Theory of Finite Dimensional Normed Spaces. With an Appendix by M. Gromov. VIII, 156 pages. 1986.

Vol. 1201: Curvature and Topology of Riemannian Manifolds. Proceedings, 1985. Edited by K. Shiohama, T. Sakai and T. Sunada. VII, 336 pages. 1986.

Vol. 1202: A. Dür, Möbius Functions, Incidence Algebras and Power Series Representations. XI, 134 pages. 1986.

Vol. 1203: Stochastic Processes and Their Applications. Proceedings, 1985. Edited by K. Itô and T. Hida. VI, 222 pages. 1986.

Vol. 1204: Séminaire de Probabilités XX, 1984/85. Proceedings. Edité par J. Azéma et M. Yor. V, 639 pages. 1986.

Vol. 1205: B.Z. Moroz, Analytic Arithmetic in Algebraic Number Fields. VII, 177 pages. 1986.

Vol. 1206: Probability and Analysis, Varenna (Como) 1985. Seminar. Edited by G. Letta and M. Pratelli. VIII, 280 pages. 1986.

Vol. 1207: P.H. Bérard, Spectral Geometry: Direct and Inverse Problems. With an Appendix by G. Besson. XIII, 272 pages. 1986.

Vol. 1208: S. Kaijser, J.W. Pelletier, Interpolation Functors and Duality. IV, 167 pages. 1986.

Vol. 1209: Differential Geometry, Peñíscola 1985. Proceedings. Edited by A.M. Naveira, A. Ferrández and F. Mascaró. VIII, 306 pages. 1986.

Vol. 1210: Probability Measures on Groups VIII. Proceedings, 1985. Edited by H. Heyer. X, 386 pages. 1986.

Vol. 1211: M.B. Sevryuk, Reversible Systems. V, 319 pages. 1986.

Vol. 1212: Stochastic Spatial Processes. Proceedings, 1984. Edited by P. Tautu. VIII, 311 pages. 1986.

Vol. 1213: L.G. Lewis, Jr., J.P. May, M. Steinberger, Equivariant Stable Homotopy Theory. IX, 538 pages. 1986.

Vol. 1214: Global Analysis – Studies and Applications II. Edited by Yu.G. Borisovich and Yu.E. Gliklikh. V, 275 pages. 1986.

Vol. 1215: Lectures in Probability and Statistics. Edited by G. del Pino and R. Rebolledo. V, 491 pages. 1986.

Vol. 1216: J. Kogan, Bifurcation of Extremals in Optimal Control. VIII, 106 pages. 1986.

Vol. 1217: Transformation Groups. Proceedings, 1985. Edited by S. Jackowski and K. Pawalowski. X, 396 pages. 1986.

Vol. 1218: Schrödinger Operators, Aarhus 1985. Seminar. Edited by E. Balslev. V, 222 pages. 1986.

Vol. 1219: R. Weissauer, Stabile Modulformen und Eisensteinreihen. III, 147 Seiten. 1986.

Vol. 1220: Séminaire d'Algèbre Paul Dubreil et Marie-Paule Malliavin. Proceedings, 1985. Edité par M.-P. Malliavin. IV, 200 pages. 1986.

Vol. 1221: Probability and Banach Spaces. Proceedings, 1985. Edited by J. Bastero and M. San Miguel. XI, 222 pages. 1986.

Vol. 1222: A. Katok, J.-M. Strelcyn, with the collaboration of F. Ledrappier and F. Przytycki, Invariant Manifolds, Entropy and Billiards; Smooth Maps with Singularities. VIII, 283 pages. 1986.

Vol. 1223: Differential Equations in Banach Spaces. Proceedings, 1985. Edited by A. Favini and E. Obrecht. VIII, 299 pages. 1986.

Vol. 1224: Nonlinear Diffusion Problems, Montecatini Terme 1985. Seminar. Edited by A. Fasano and M. Primicerio. VIII, 188 pages. 1986.

Vol. 1225: Inverse Problems, Montecatini Terme 1986. Seminar. Edited by G. Talenti. VIII, 204 pages. 1986.

Vol. 1226: A. Buium, Differential Function Fields and Moduli of Algebraic Varieties. IX, 146 pages. 1986.

Vol. 1227: H. Helson, The Spectral Theorem. VI, 104 pages. 1986.

Vol. 1228: Multigrid Methods II. Proceedings, 1985. Edited by W. Hackbusch and U. Trottenberg. VI, 336 pages. 1986.

Vol. 1229: O. Bratteli, Derivations, Dissipations and Group Actions on C*-algebras. IV, 277 pages. 1986.

Vol. 1230: Numerical Analysis. Proceedings, 1984. Edited by J.-P. Hennart. X, 234 pages. 1986.

Vol. 1231: E.-U. Gekeler, Drinfeld Modular Curves. XIV, 107 pages. 1986.

Vol. 1232: P.C. Schuur, Asymptotic Analysis of Soliton Problems. VIII, 180 pages. 1986.

Vol. 1233: Stability Problems for Stochastic Models. Proceedings, 1985. Edited by V.V. Kalashnikov, B. Penkov and V.M. Zolotarev. VI, 223 pages. 1986.

Vol. 1234: Combinatoire énumérative. Proceedings, 1985. Edité par G. Labelle et P. Leroux. XIV, 387 pages. 1986.

Vol. 1235: Séminaire de Théorie du Potentiel, Paris, No. 8. Directeurs: M. Brelot, G. Choquet et J. Deny. Rédacteurs: F. Hirsch et G. Mokobodzki. III, 209 pages. 1987.

Vol. 1236: Stochastic Partial Differential Equations and Applications. Proceedings, 1985. Edited by G. Da Prato and L. Tubaro. V, 257 pages. 1987.

Vol. 1237: Rational Approximation and its Applications in Mathematics and Physics. Proceedings, 1985. Edited by J. Gilewicz, M. Pindor and W. Siemaszko. XII, 350 pages. 1987.

Vol. 1238: M. Holz, K.-P. Podewski and K. Steffens, Injective Choice Functions. VI, 183 pages. 1987.

Vol. 1239: P. Vojta, Diophantine Approximations and Value Distribution Theory. X, 132 pages. 1987.

Vol. 1240: Number Theory, New York 1984–85. Seminar. Edited by D.V. Chudnovsky, G.V. Chudnovsky, H. Cohn and M.B. Nathanson. V, 324 pages. 1987.

Vol. 1241: L. Gårding, Singularities in Linear Wave Propagation. III, 125 pages. 1987.

Vol. 1242: Functional Analysis II with contributions by J. Hoffmann-Jørgensen et al. Edited by S. Kurepa, H. Kraljević and D. Butković. VII, 432 pages. 1987.

Vol. 1243: Non Commutative Harmonic Analysis and Lie Groups. Proceedings, 1985. Edited by J. Carmona, P. Delorme and M. Vergne. V, 309 pages. 1987.

Vol. 1244: W. Müller, Manifolds with Cusps of Rank One. XI, 158 pages. 1987.

Vol. 1245: S. Rallis, L-Functions and the Oscillator Representation. XVI, 239 pages. 1987.

Vol. 1246: Hodge Theory. Proceedings, 1985. Edited by E. Cattani, F. Guillén, A. Kaplan and F. Puerta. VII, 175 pages. 1987.

Vol. 1247: Séminaire de Probabilités XXI. Proceedings. Edité par J. Azéma, P.A. Meyer et M. Yor. IV, 579 pages. 1987.

Vol. 1248: Nonlinear Semigroups, Partial Differential Equations and Attractors. Proceedings, 1985. Edited by T.L. Gill and W.W. Zachary. IX, 185 pages. 1987.

Vol. 1249: I. van den Berg, Nonstandard Asymptotic Analysis. IX, 187 pages. 1987.

Vol. 1250: Stochastic Processes – Mathematics and Physics II. Proceedings 1985. Edited by S. Albeverio, Ph. Blanchard and L. Streit. VI, 359 pages. 1987.

Vol. 1251: Differential Geometric Methods in Mathematical Physics. Proceedings, 1985. Edited by P.L. García and A. Pérez-Rendón. VII, 300 pages. 1987.

Vol. 1252: T. Kaise, Représentations de Weil et GL_2 Algèbres de division et GL_n. VII, 203 pages. 1987.

Vol. 1253: J. Fischer, An Approach to the Selberg Trace Formula via the Selberg Zeta-Function. III, 184 pages. 1987.

Vol. 1254: S. Gelbart, I. Piatetski-Shapiro, S. Rallis. Explicit Constructions of Automorphic L-Functions. VI, 152 pages. 1987.

Vol. 1255: Differential Geometry and Differential Equations. Proceedings, 1985. Edited by C. Gu, M. Berger and R.L. Bryant. XII, 243 pages. 1987.

Vol. 1256: Pseudo-Differential Operators. Proceedings, 1986. Edited by H.O. Cordes, B. Gramsch and H. Widom. X, 479 pages. 1987.

Vol. 1257: X. Wang, On the C*-Algebras of Foliations in the Plane. V, 165 pages. 1987.

Vol. 1258: J. Weidmann, Spectral Theory of Ordinary Differential Operators. VI, 303 pages. 1987.

Vol. 1259: F. Cano Torres, Desingularization Strategies for Three-Dimensional Vector Fields. IX, 189 pages. 1987.

Vol. 1260: N.H. Pavel, Nonlinear Evolution Operators and Semigroups. VI, 285 pages. 1987.

This series reports new developments in mathematical research and teaching – quickly, informally and at a high level. The type of material considered for publication includes.

1. Research monographs
2. Lectures on a new field or presentations of a new angle in a classical field
3. Seminar work-outs
4. Reports of meetings, provided they are
   a) of exceptional interest and
   b) devoted to a single topic.

Texts which are out of print but still in demand may also be considered if they fall within these categories.

The timeliness of a manuscript is more important than its form, which may be unfinished or tentative. Thus, in some instances, proofs may be merely outlined and results presented which have been or will later be published elsewhere. If possible, a subject index should be included. Publication of Lecture Notes is intended as a service to the international mathematical community, in that a commercial publisher, Springer-Verlag, can offer a wide distribution of documents which would otherwise have a restricted readership. Once published and copyrighted, they can be documented in the scientific literature.

Manuscripts

Manuscripts should be no less than 100 and preferably no more than 500 pages in length.
They are reproduced by a photographic process and therefore must be typed with extreme care. Symbols not on the typewriter should be inserted by hand in indelible black ink. Corrections to the typescript should be made by pasting in the new text or painting out errors with white correction fluid. The typescript is reduced slightly in size during reproduction; best results will not be obtained unless on each page a typing area of 18 x 26.5 cm (7 x 10 ½ inches) is respected. On request, the publisher can supply paper with the typing area outlined. More detailed typing instructions are also available on request.

Manuscripts generated by a word-processor or computerized typesetting are in principle acceptable. However if the quality of this output differs significantly from that of a standard typewriter, then authors should contact Springer-Verlag at an early stage.

Authors of monographs and editors of proceedings receive 50 free copies.

Manuscripts should be sent to Prof. A. Dold, Mathematisches Institut der Universität Heidelberg, Im Neuenheimer Feld 288, 6900 Heidelberg, Germany; Prof. B. Eckmann, Eidgenössische Technische Hochschule, CH-8092 Zürich, Switzerland; or directly to Springer-Verlag Heidelberg.

Springer-Verlag, Heidelberger Platz 3, D-1000 Berlin 33
Springer-Verlag, Tiergartenstraße 17, D-6900 Heidelberg 1
Springer-Verlag, 175 Fifth Avenue, New York, NY 10010/USA
Springer-Verlag, 37-3, Hongo 3-chome, Bunkyo-ku, Tokyo 113, Japan

---

ISBN 3-540-17974-7
ISBN 0-387-17974-7